Springer Optimization and Its Applications

VOLUME 55

Aims and Scope
Optimization has been expanding in all directions at an astonishing rate during the last few decades. New algorithmic and theoretical techniques have been developed, the diffusion into other disciplines has proceeded at a rapid pace, and our knowledge of all aspects of the field has grown even more profound. At the same time, one of the most striking trends in optimization is the constantly increasing emphasis on the interdisciplinary nature of the field. Optimization has been a basic tool in all areas of applied mathematics, engineering, medicine, economics, and other sciences.

The series *Springer Optimization and Its Applications* publishes undergraduate and graduate textbooks, monographs and state-of-the-art expository work that focus on algorithms for solving optimization problems and also study applications involving such problems. Some of the topics covered include nonlinear optimization (convex and nonconvex), network flow problems, stochastic optimization, optimal control, discrete optimization, multi-objective programming, description of software packages, approximation techniques and heuristic approaches.

For further volumes:
http://www.springer.com/series/7393

Springer Optimization and Its Applications

VOLUME 55

Managing Editor
Panos M. Pardalos (University of Florida)

Editor–Combinatorial Optimization
Ding-Zhu Du (University of Texas at Dallas)

Advisory Board
J. Birge (University of Chicago)
C.A. Floudas (Princeton University)
F. Giannessi (University of Pisa)
H.D. Sherali (Virginia Polytechnic and State University)
T. Terlaky (McMaster University)
Y. Ye (Stanford University)

Aims and Scope
Optimization has been expanding in all directions at an astonishing rate
during the last few decades. New algorithmic and theoretical techniques
have been developed, the diffusion into other disciplines has proceeded at
a rapid pace, and our knowledge of all aspects of the field has grown even
more profound. At the same time, one of the most striking trends in
optimization is the constantly increasing emphasis on the interdisciplinary
nature of the field. Optimization has been a basic tool in all areas of applied
mathematics, engineering, medicine, economics, and other sciences.

The series Springer Optimization and Its Applications publishes undergraduate and graduate textbooks, monographs and state-of-the-art
expository work that focus on algorithms for solving optimization problems and
also study applications involving such problems. Some of the topics covered
include nonlinear optimization (convex and nonconvex), network flow problems, stochastic optimization, optimal control, discrete optimization, multi-
objective programming, description of software packages, approximation
techniques and heuristic approaches.

For further volumes:
http://www.springer.com/series/7393

D.A. Klyushin · S.I. Lyashko · D.A. Nomirovskii
Yu.I. Petunin · V.V. Semenov

Generalized Solutions of Operator Equations and Extreme Elements

Springer

D.A. Klyushin
Department of Cybernetics
Kyiv National Taras Shevchenko University
01601 Kyiv
Ukraine
dokmed5@gmail.com

S.I. Lyashko
Department of Cybernetics
Kyiv National Taras Shevchenko University
01601 Kyiv
Ukraine
silsil1@yandex.ru

D.A. Nomirovskii
Department of Cybernetics
Kyiv National Taras Shevchenko University
01601 Kyiv
Ukraine
kashpir@mail.ru

Yu.I. Petunin
Department of Cybernetics
Kyiv National Taras Shevchenko University
01601 Kyiv
Ukraine
vm214@dcp.kiev.ua

V.V. Semenov
Department of Cybernetics
Kyiv National Taras Shevchenko University
01601 Kyiv
Ukraine
semenov.volodya@gmail.com

ISSN 1931-6828
ISBN 978-1-4614-2964-7   ISBN 978-1-4614-0619-8 (eBook)
DOI 10.1007/978-1-4614-0619-8
Springer New York Dordrecht Heidelberg London

Printed on acid-free paper

Springer is part of Springer Science+Business Media (www.springer.com)

In memory of my parents, Nina Andreevna and Anatoliy Arkhipovich.

Dmitry Klyushin

Dedicated in memory of my father, Ivan Ivanovich and to my dear family: mother, Vera Stepanovna; wife, Natalie; and children, Lena, Viktor, and Vera.

Sergey Lyashko

To my daughter, Alina.

Dmitry Nomirovskii

In memory of my parents, Zoya Ivanovna and Ivan Petrovich.

Yuriy Petunin

To my son, Andrey.

Vladimir Semenov

Preface

"F-friends," said Fyodor Simeonovich ...
"But this is the Ben B-Betzalel's p-problem.
C-Cagliostro has proved that it does not have a s-solution indeed."
"We do know that it does not have a solution," said Junta...
"We wish to know how to solve it."
"You are somehow arguing oddly, C-Christo. . .
H-how to s-search for a s-solution, when it does not exist?
It's a nonsense.
"I am sorry, Fyodor, but it's you who are arguing very strangely.
The nonsense is to search for a solution when it exists anyway.
The question is how to deal with a problem that does not have a solution.
This is a profoundly principled question ..."

A. Strugatsky and B. Strugatsky, "Monday Begins on Saturday"

At the International Mathematical Congress in Paris (1900), D. Hilbert put forth his famous 23 problems. In Hilbert's opinion, these problems had to predefine the mainstream of mathematics in the twentieth century. By now, most of Hilbert's problems have been solved successfully. However, despite the fact that many mathematical disciplines have arisen and new important problems were put forth in the twentieth century, Hilbert's problems remain fundamental [3].

Among Hilbert's problems, the 20th problem – "the general problem of boundary values" – takes its deserved place. This problem is formulated in the following way: "has not every regular variation problem a solution, provided certain assumptions regarding the given boundary conditions are satisfied (say that the functions concerned in these boundary conditions are continuous and have in sections one or more derivatives) and provided also if need be that the notion of a solution shall be suitably extended?" (see [25]).

The 20th problem is outstanding because D. Hilbert put it on extending the classical solution when there was neither the concept of completion of metric space nor the concept of normed space that serves as a basis of such a notion as "generalized solution of operator equations". The idea of the generalized solution is quite simple: consider an operator equation $A(x) = y$, where A is a continuous operator (linear or nonlinear) from metric or Banach space E into F. Operator equations cover wide classes of differential equations (including boundary value problems), integral equations, integro-differential equations and more. In many situations, the operator equation $A(x) = y$ does not have a classical solution, since the right-hand side y does not belong to the range $R(A) \subset F$ of the operator A, but we can introduce a weaker topology in E, so that the completion $\widetilde{E}$ of E in this topology is a wider space: $E \subset \widetilde{E}$ and the operator A can be extended by continuity to $\widetilde{E}$, so that the right-hand side y belongs to the range $R(\widetilde{A})$ of the extended operator $\widetilde{A}$. Thus, the operator equation $\widetilde{A}(x) = y$ ($x \in \widetilde{E}$, $y \in F$, $\widetilde{A} : \widetilde{E} \to F$) has a classical solution $\widetilde{x} \in \widetilde{E}$ called a generalized solution of the original equation $A(x) = y$. This is exactly such an extension of the concept of solution about which D. Hilbert wrote.

The concept of generalized solution is closely related to the concept of a near-solution x_ε of the operator equation $A(x) = y$; this is such an element in E that $A(x_\varepsilon) = y_\varepsilon$ differs less than ε from y: $\rho(y, y_\varepsilon) < \varepsilon$. In some cases, x_ε may be considered as an approximate solution of the equation $A(x) = y$. If we put $\varepsilon = \varepsilon_n \to 0$ as $n \to \infty$ and consider a sequence of the near-solutions x_{ε_n}, then in $\widetilde{E}$ (but not in E!) the sequence x_{ε_n} converges to a generalized solution $\widetilde{x}$. In the case of linear operator A, the computation of the near-solution is reduced to the problem of computation of the approximate (or precise) solution of a system of linear algebraic equations. This is why we give so much attention to these issues and propose various methods for solving this problem.

Along with the investigation of generalized solutions, we study the so-called generalized extreme elements which are closely related to this concept. Let D be a region in a Banach or metric space E and a continuous functional $f(x)$ is defined on D. As a rule, the region D is non-compact in an infinite dimensional space, therefore the extreme element x^* from D, at which $f(x)$ attains its minimum or maximum value may not exist. Determination of a "generalized" extreme element resembles the construction of generalized solution. We introduce a weaker topology $\mathscr{T}_D$ on the set D, such that the completion $\widetilde{D}$ of D with respect to the topology $\mathscr{T}_D$ is a compact topological space, and the functional f may be extended on $\widetilde{D}$ by continuity, such that there is a classical extreme element x^* in $\widetilde{D}$. This element is considered as a generalized extreme element, since $x^* \notin D$. Note that the concept of a generalized extreme element may be defined in other ways. These ways are considered in the book as well.

By an *operator equation* we will always mean an equation where some known operator $\mathscr{L}$ from E into F acts on an unknown element u (a vector, sequence or function), where F may differ from E. The spaces E and F may be finite or infinite dimensional spaces, normed spaces (in particular, Banach), metric spaces, topological vector spaces, topological or differentiable manifold, and so on. In a general way, an operator equation has the following form

$$\mathscr{L}u = f,$$

where u is an unknown element in E, f is the known element in F, and $\mathscr{L}$ is the known operator which acts from E into F. The most important problems related to operator equations are the existence and uniqueness of a solution. The uniqueness of a solution is ensured by the condition of invertibility of the operator $\mathscr{L}$, that may be satisfied by the corresponding factorization of the space E (at least theoretically). It is clear that a solution of the equation $\mathscr{L}u = f$ exists iff the right-hand side f belong in the range $R(\mathscr{L})$ of the operator $\mathscr{L}$. Thus, if $f \in R(\mathscr{L})$ then the issue of the existence of a solution of the equation $\mathscr{L}u = f$ has, in principle, a positive answer. However, in many cases the right-hand side f does not belong to the set $R(\mathscr{L})$, so this equation does not have a solution in a classical sense. Nevertheless, from the practical point of view such equations may have "intuitive solutions", that must be defined correctly. The problem of construction of a generalized solution of the operator equation is closely related with the problem of introducing the "natural" notion of a generalized solution of the equation $\mathscr{L}u = f$ for all $f \in F$; in particular, when $f \in F \setminus R(\mathscr{L})$, and with the investigation of the properties of such generalized solutions. The point is that the description of a function set of $R(\mathscr{L})$ is extremely difficult. Therefore it is impossible to establish the criteria for the solvability of the equation $\mathscr{L}u = f$. We could say that it is possible to formulate the criterion of the solvability of the equation $\mathscr{L}u = f$ only in exceptional cases. For example, even in the simplest case of the investigation of the classical solvability of an ordinary differential equation $u'(t) = f(t)$ when $1 > t > 0$ and $u(0) = 0$, it is necessary to test the convergence of an integral (possibly improper integral)

$$\int_0^1 f(t)\, dt.$$

However, as is well known, there are no general effective criteria for testing the convergence of improper integrals.

Consider one of the approaches to the formalization of such solutions. Suppose that in any ε-neighborhood f (in topological space F – in any neighborhood f) there exists such an element f_ε, that $\mathscr{L}u_\varepsilon = f_\varepsilon$ for some $u_\varepsilon \in E$. Then for small $\varepsilon > 0$ one could think that $f_\varepsilon \approx f$, since the distance $\rho(f_\varepsilon, f) < \varepsilon$, therefore the element u_ε can be accepted as a "generalized" solution of the operator equation $\mathscr{L}u = f$ (if topological space F is non-metrizable, then these reasonings must be slightly modified, but this is not a principal issue).

Consider the issue of the existence of classical and generalized solutions on concrete examples. Suppose that we want to obtain the best unbiased linear estimation x^* of an unknown mathematical expectation of a continuous random process $x(t)$ ($t \in [0, T]$) with a constant mathematical expectation and a correlation function $K(t, s)$. If we look for this estimation in the form

$$x^* = \int_0^T x(t)u(t)\, dt,$$

then the problem is reduced to looking for the solution $u(t)$ of the integral equation

$$\int_0^T K(t,s)u(t)\,dt = 1 \qquad\qquad (\text{P.1})$$

in the function class $L_2(0,T)$. In general case, the matter concerns the equation

$$\int_D K(t,s)u(t)\,dt = f(s), \quad t \in \bar{D}. \qquad\qquad (\text{P.2})$$

However, solutions of such equations have the square integrability property very seldom (see example [23]). For example, it is shown in [36] that (P.1) never has a classical solution if a correlation function $K(\tau)$ of the stationary random process $x(t)$ has a spectral density. Nevertheless, it has the generalized solution. In some cases, the fact that the integral equation (P.1) does not have a solution in the class of square-integrable functions can be proved directly. For example, if a correlation function has the form $K(t,s) = e^{-\beta|t-s|}$, which corresponds to a stationary Markov process when all probability distributions are normal, then it is impossible to construct a function $u(t)$ that sets the best unbiased estimation x^* of an unknown mathematical expectation. To prove this statement let us consider the integral equation

$$\int_0^T e^{-\beta|t-s|}\,dF(t) = \frac{2}{2+\beta T}.$$

It is easy to examine that this equation is satisfied by the following function of bounded variation

$$F(t) = \frac{\Theta(t)+\Theta(t-T)+\beta t}{2+\beta T},$$

where $\int_0^T dF(t) = 1$ and $\Theta(t)$ is the Heaviside function:

$$\Theta(t) = \begin{cases} 0, & \text{if } t < 0 \\ 1, & \text{if } t \geq 0. \end{cases}$$

Hence, the expression

$$x^* = \frac{x(0)+x(T)+\beta \int_0^T x(t)\,dt}{2+\beta T}$$

defines an unbiased estimation x^* having the least variance in the class of unbiased linear estimations (actually, this estimation is also the best in a much more wider estimation class [23]). Since the estimation x^* is unique and the formula for x^* contains Dirac delta-functions $\delta(t)$ and $\delta(t-T)$, that do not belong to $L_2(0,T)$, it is impossible to construct the function $u(t)$ from $L_2(0,T)$, that defines the estimation x^* and is a solution of (P.1). Therefore, (P.1) does not have the classical solution in $L_2(0,T)$. The issues related with the problem described above are listed in [96]: "The problems are: in which functional spaces should one look for the solution?

Is the solution unique? Is the solution of the equation also is the solution of the estimation problem? Does the solution depend continuously on the initial data, for example, on f and K? How can the solution be found analytically and numerically? What are the properties of the solutions? For example, what is the order of singularity? How can the properties of the integral operator be described, for example, in $L^2(D)$?"

Another possible example that requires the introduction of a generalized solution is the problem of optimal control of a system with a generalized external impact

$$\mathscr{L}u = f(h), \tag{P.3}$$

$$J(h) = \Phi(u(h), h) \to \min_{h}, \qquad h \in U, \tag{P.4}$$

where h is a control from an admissible set U, $\mathscr{L} : E \to F$ is some operator, J is a performance functional. To express this problem correctly it is necessary to ensure the solvability of (P.3) for all $h \in U$, i.e., it is necessary to ensure the inclusion $f(U) \subset R(\mathscr{L})$. However, generally it is very difficult to describe the range of f and $\mathscr{L}$; therefore, it is very hard to check the condition $f(U) \subset R(\mathscr{L})$. Moreover, often such an inclusion does not occur at all (in spite of the fact that a physical interpretation of the equation is natural and reasonable from the practical point of view). Thus, we must develop a theory of generalized solvability of (P.3) for an arbitrary right-hand side f from the set $f(U)$, or (much better) for all $f \in F$. In a general sense, (P.3) has a solution $u(h)$ for an arbitrary control $h \in U$. It is clear, that we must know peculiarities of these generalized solutions to prove some meaningful statements about the problem of the minimization of (P.4).

Now, problems of complex system control with singular impacts have a fundamental importance. For example, simulation of devices with laser and pulse impacts, correction of space vehicles movement, modelling of water transport in porous media with point sources and sinks are closely related with the equations with a singular right-hand side. The singularity of a control impact means that a control map f takes on a value in a space of generalized function. Traditionally, the natural range of the operator L does not contain generalized functions. So, lumped singularity in space and time bring us outside of the classical problem definitions. So, we face with the need to develop a theory of generalized solvability of (P.3).

The problem of construction of generalized solutions becomes the most important in the case of linear operator $\mathscr{L}$ (e.g., differential or integral) which acts between linear topological spaces E, F, in particular, between Banach or Hilbert spaces. Note that the "naturalness" of generalized solution means the conservation of the main properties of operator $\mathscr{L}$ (linearity, continuity, injectivity and so on) under extension on the class of generalized solutions. Thus, the offered problem fundamentally differs from various definitions of approximate solutions, pseudo-solutions, quasi-solution, and so on. [47, 107, 112].

The problems of construction of generalized solutions of equations with linear differential and integral operators are quite typical. They have been investigated successfully for a long time. For example, this problem for the classical operator of differentiation $\frac{d}{dt} : C^1([0,1]) \to L_2(0,1)$ may be solved by introducing of the

Sobolev generalized derivative and corresponding Sobolev spaces. In this sense, the theory of generalized functions may be considered as the first step in solving the posed problem.

The method of a priori inequations (e.g. [5, 39, 54]) is a very effective tool for the investigation of existence and uniqueness of solutions of various classical linear problems with generalized impacts. It was often used in the context of rigged Hilbert spaces. In [5], the theory of generalized solvability for equations with elliptic differential operators acting in the Sobolev discrete scale of Hilbert spaces was constructed. This theory is based on the concept of weak solutions (in the context of the theory of generalized functions). Berezanskii proved the theorems of a unique generalized solvability of elliptic operator equations for major problems of mathematical physics and investigated the smoothness of the generalized solutions. The theorems proved are the criteria of solvability (i.e., an operator determines a topological isomorphism). For example, the theorems of a unique solvability (in L_2 and other spaces) of the equations of mathematical physics of various types were proved in [44, 45]. Some criteria of solvability of parabolic equations are described in [2]. The issues of generalized solvability for pseudo-parabolic equations of order more than two were investigated in [58, 100], for pseudo-hyperbolic equations – in [73, 79, 85, 100], for Sobolev type system – in [59, 76, 100], for wave systems of fifth order – [60, 61, 64, 80, 82], and in many other papers (see also [62, 63]). Note that in these papers were used a priori inequalities in negative norms when a generalized solution belongs to Sobolev type spaces.

The generalized solvability of linear integral equations is closely related with Fredholm and Volterra; integral equations of the first kind [23] and [68, 87, 88, 92]. It must be stressed that in many above-mentioned papers the proofs of existence and uniqueness of a generalized solution are based on the classical idea of relations between direct and "adjoint" equations and the coercive inequality. Therefore, these theorems can be considered as the developing of classical results of S.G. Krein (e.g., see [39]).

There is one more important aspect of the theory of generalized solutions. It is related to the problem of optimal control (P.3), (P.4), rather than with only (P.3). As it is well known, there are problems of calculus of variations and optimal control which have no solutions in "traditional" sets of curves (in spaces of smooth functions). This problem was solved in classical papers on optimal control theory in the generalized statement. For example, the general plan of looking for generalized extreme curves is described in [116]. The plan involves the following activities: to densely embed the control space (and therefore an admissible set of controls) in a new topological space such that a functional in question is still sequentially continuous and an admissible set is sequentially compact. This idea naturally connects the optimal control problems with the Schwarz distributions spaces. We have to mention L. Young [118] among the authors who began to apply the ideas of the theory of generalized functions to the calculus of variations problems and the optimal control problems. From the Young's point of view, the spaces of curves with "traditional" topologies are poorly adaptable for the calculus of variations. More convenient are the topologies which induce so-called "generalized curves" (by Young) that are equivalent to the concepts of weak controls and gliding regimes. In the optimal control theory for ordinary and partial differential equations, Filippov and Gamkrelidze considered the weak solutions and analogous constructions [17, 22] ("gliding regimes"), Warga studied the "generalized curves" [115], McShane investigated the "generalized controls" [70, 71], Chouila-Houri considered "boundary controls" [10], and this list might be continued (see, e.g. books of A. Chikrii [9], M. Zgurovsky, V. Mel'nik [120], V. Kuntsevich [42], J.-L. Lions [51, 52], and B. Mordukhovich [72]).

These results naturally pose the general problem of looking for generalized extreme elements in various classes of functionals. These problems are interesting even in the simplest case when we look for an extremum of a continuous functional defined on a bounded set in a Banach space.

Thus, there are many papers, where existence and uniqueness of generalized solution of an operator equation or extremal problem solutions were investigated. Multiplicity and similarity of these papers suggest that there is a general approach to the construction of the concept of generalized solvability. The major elements of this approach are described in our book.

The book consists of the preface, eight chapters, divided into sections, and a bibliography. The numbering of definitions, lemmas, theorems, and so on, is continuous. Chapter 1 contains major definitions, concepts, and auxiliary facts used in the book. Chapter 2 is an introduction to the theory of generalized solutions of operator equations. It describes the simple schemes of generalized solutions for linear operator equations. In Chap. 3, we investigate the method of a priori estimates for generalized solutions. Chapter 4 describes some applications of the theory of generalized solvability of linear equations. Chapter 5 is devoted to numerical aspects of the theory. Chapter 6 describes the general topological method of construction of generalized solutions of linear operator equations. In Chap. 7, the issues of generalized solvability of nonlinear operator equations are considered. Chapter 8 is devoted to the generalized solvability of extreme problems.

Kiev, Ukraine

Dmitry Klyushin
Sergey Lyashko
Dmitry Nomirovskii
Yuriy Petunin
Vladimir Semenov

Yuriy Ivanovich Petunin

Our co-author, friend and master Yuriy Ivanovich Petunin suddenly died on June 1, 2011.

Yuriy Ivanovich Petunin was born on September 30, 1937, in Michurinsk (Russia). He graduated Tambov State Pedagogical Institute and passed Ph.D. defense under the supervision of S.G. Krein in 1962. He became a Doctor of Science on 1968. Since 1970 Yuriy Ivanovich has been the professor of the department of computational mathematics of the faculty of cybernetics of the Kiev National Taras Shevchenko University.

Yu.I. Petunin started his scientific activity in the area of functional analysis under the supervision of S.G. Krein. His main achievements in this field are the creation of the theory of scales of Banach spaces, development of the theory of interpolation of linear operators (with S.G. Krein and E.M. Semenov) and the theory of characteristics of linear manifolds in conjugate Banach spaces (with A.M. Plichko). In addition he was the first who rigorously justified the empirical three-sigma rule for unimodal distributions, proving the famous problem posed by K.F. Gauss more than 150 years ago (the classical Vysochanskii-Petunin inequality), developed the theory of confidence intervals for a bulk of general population and parameters using order statistics, developed the theory of linear estimations of unknown mathematical expectation, the theory of quadratic estimations of unknown variance, statistical tests which use the procedure of indecision and individual statistical tests. Yu.I. Petunin solved the Banach's problem of norming subspaces in conjugate Banach spaces and the Calderon-Lions problem of interpolation in factor spaces. He also developed the lattice approach to solving the sixth Hilbert's problem (with D.A. Klyushin). Yu.I. Petunin had significant achievements in the theory of pattern recognition, in particular, in its application to differential diagnostics of oncological diseases (with B.V. Rublev, D.A. Klyushin, K.P. Ganina, N.V. Boroday, and R.I. Andrushkiw). It should be stressed that one of the main ideas in this book – the concept of generalized solution of operator equations in Banach space – was developed by Yuriy Ivanovich.

Yuriy Ivanovich was notable for his exceptional honesty, nobility, devotion to science and religiosity. His sincerity and helpfulness have always attracted many people. Death of such a man is a great grief for the whole mathematical community.

Cherished memories of our colleague, friend and master will always stay in our hearts!

Dmitry Klyushin
Sergey Lyashko
Dmitry Nomirovskii
Vladimir Semenov

Acknowledgements

We are very grateful to our colleagues, especially to the correspondent member of the National Academy of Sciences of Ukraine, A.A. Chikrii and academicians I.N. Kovalenko, V.M. Kuntsevich, I.V. Sergienko, Yu.M. Yermolyev, M.Z. Zgurovsky. We want to mention also our teachers and colleagues who are not with us: Prof. V.P. Didenko, Prof. S.G. Krein, Academician I.I. Lyashko, Dr. D.L. Pikus, Academician B.N. Pshenichny, Academician N.Z. Shor, and correspondent member V.V. Skopetsky. We thank Yuriy Malitsky and Varvara Obolonchikova for the help in the preparation of this manuscript.

We would like to express our sincere thanks to Prof. Panos Pardalos for his kindly support of our book.

Special thanks go to Elizabeth Loew (Senior Editor, Mathematics, Springer), Nathan Brothers (Assistant Editor) and Jacob Gallay (Editorial Assistant) for all their help with the cover, production and manufacturing of our book.

Contents

Chapter 1
The Major Definitions, Concepts and Auxiliary Facts

Let E be **a linear (vector)** space on the field $\mathbb{R}$, i.e. on the set E an operation of addition of every two elements x, y from E (denoted as $x + y$) and an operation of multiplication of every element x from E on an arbitrary λ from $\mathbb{R}$ (denoted as λx) are defined as follows:

(1) $(x+y)+z = x+(y+z)$.
(2) $x+y = y+x$.
(3) there exists $\theta \in E$, such that $0x = \theta$ for $x \in E$.
(4) $(\lambda + \mu)x = \lambda x + \mu x$.
(5) $\lambda(x+y) = \lambda x + \lambda y$.
(6) $(\lambda \mu)x = \lambda(\mu x)$.
(7) $1x = x$.

In the linear space E the difference $x - y$ means $x + (-1)y$.

A map $f : E \to \mathbb{R}$ is called **a linear functional** on E (when E is a linear space) if the following conditions are satisfied:

1 (Additivity).   For all $x, y \in E$

$$f(x+y) = f(x) + f(y).$$

2 (Homogeneity).   For all $\lambda \in \mathbb{R}, x \in E$

$$f(\lambda x) = \lambda f(x).$$

The algebraically conjugate to a linear space E is a space E' of linear functionals f defined on E. The set E' is a linear space if $f + g$ and λf are the functionals defined as follows:

(1) $(f+g)(x) = f(x) + g(x)$ for all $f, g \in E', x \in E$.
(2) $(\lambda f)(x) = \lambda f(x)$ for all $f \in E', \lambda \in \mathbb{R}, x \in E$.

In a similar way, **the second algebraically conjugate** space $E'' = (E')'$ is defined and so on. Every $x \in E$ uniquely defines some element of E'' – a linear functional

D.A. Klyushin et al., *Generalized Solutions of Operator Equations and Extreme Elements*,
Springer Optimization and Its Applications 55, DOI 10.1007/978-1-4614-0619-8_1,
© Springer Science+Business Media, LLC 2012

$L_x \in E''$ – according to the following rule: $L_x(f) = f(x)$ for all $f \in E'$. Thus, it is meaningful to say about **embedding** $E \subset E''$. A functional $f : E \to \mathbb{R}$ is called **convex** if $\lambda f(x) + (1 - \lambda) f(y) \geq f(\lambda x + (1 - \lambda) y)$, for every $x, y \in E$ and an arbitrary real number $\lambda \in [0, 1]$.

Let us consider two linear spaces E, F and their Cartesian product in which **a bilinear form** $\langle x, y \rangle$, $x \in E, y \in F$ is defined. By a bilinear form we mean a two-variate function which is linear in each argument separately. The linear spaces E, F are said to be **dual** with respect to this bilinear form (in other words, (E, F) is a **dual pair**) if:

(1) For all $x \in E$ ($x \neq 0$) there exists $y \in F$ such that $\langle x, y \rangle \neq 0$.
(2) For all $y \in F$ ($y \neq 0$) there exists $x \in F$ such that $\langle x, y \rangle \neq 0$.

If E is a linear space and E' is an algebraic conjugate to E then the expression $f(x) = \langle f, x \rangle$, where $f \in E'$, $x \in E$ defines a bilinear form over E' and E. In other words, the spaces E' and E are dual each other.

If linear spaces E and F are dual to each other with respect to the bilinear form $\langle x, y \rangle$, where $x \in E, y \in F$, then every element $y \in F$ can be identified with some linear functional $f \in E'$ using the equality $f(x) = \langle x, y \rangle$ for all $x \in E$. Thus, $F \subset E'$ always.

Let linear spaces E and F be dual with respect to the bilinear form $\langle x, y \rangle$, $x \in E, y \in F$, and M be a subset of E. **A polar** of M (with respect to this duality) is a set

$$M^\circ = \{ y \in F \mid |\langle x, y \rangle| \leq 1, x \in M \}.$$

Let linear spaces E and F be dual with respect to the bilinear form $\langle x, y \rangle$, $x \in E, y \in F$, and let M be a subset of E. A set is said to be **total** in E if the linear spaces M and F are dual also (with respect to $\langle x, y \rangle$).

An operator $\mathscr{L}$ which acts from the linear space E into the linear space F is called *linear* if it satisfies the following two conditions:

1 (Additivity) $\mathscr{L}(x + y) = \mathscr{L}(x) + \mathscr{L}(y)$ for all $x, y \in E$,
2 (Homogeneity) $\mathscr{L}(\lambda x) = \lambda(\mathscr{L}x)$ for all $x \in E$, $\lambda \in \mathbb{R}$.

The algebraically adjoint to a linear operator $\mathscr{L} : E \to F$ is an operator $\mathscr{L}' : F' \to E'$ such that:

$$(\mathscr{L}'\varphi)(x) = \varphi(\mathscr{L}x) \quad \text{for all } x \in E, \varphi \in F'.$$

Thus, $\mathscr{L}'\varphi = \varphi \circ \mathscr{L}$.

Suppose that in a linear space E a topology $\mathscr{T}$ (a system of open subsets O of the set E) is introduced, such that E is a linear topological space (see [40]). A set $M_x \subset E$ is called **a neighborhood of a point** $x \in E$ if there exist such an open set $O \in \mathscr{T}$ that $x \in O \subset M_x$.

Suppose that two topologies $\mathscr{T}_1$ and $\mathscr{T}_2$ are given on a set E. The topology $\mathscr{T}_1$ is said to be **stronger than the topology** $\mathscr{T}_2$ (this fact is denoted as $\mathscr{T}_2 \subset \mathscr{T}_1$) if for any $O_2 \in \mathscr{T}_2$ there exists such $O_1 \in \mathscr{T}_1$ that $O_1 \subset O_2$. If one topology is stronger than the other, then these topologies are said to be **comparable**.

Let $\mathscr{L}$ be an operator which acts from the linear topological space E into the linear topological space F. The operator $\mathscr{L}$ is called **continuous** if for any open set $O_F \subset F$ there exists such an open set $O_E \subset E$ that $\mathscr{L}(O_E) \subset O_F$. In particular, if $F = \mathbb{R}$, then the operator $\mathscr{L}$ is **a continuous functional**. A set E^* of linear continuous functionals defined on E is called **conjugate to** E.

A subset $M \subset E$ is called **bounded** in the linear topological space E if for any sequence of elements $x_n \in M$ and for any numerical sequence $\lambda_n \in \mathbb{R}$ the sequence $\lambda_n x_n$ converges to zero.

Let E be a linear set and $F \subset E'$ be a total subset. An important example of a linear topological space is a space E with a weak topology $\sigma(E, F)$. Denote as $O_{\varepsilon, f_1, f_2, \ldots, f_n}$, where $f_i \in F$, $\varepsilon > 0$, a set of points $x \in E$ such that $|f_i(x)| < \varepsilon$ for all $i \in \{1, \ldots, n\}$. The topology $\sigma(E, F)$ is given by the sets of neighborhoods of zero in the following way. **A neighborhood of zero** is a set containing $O_{\varepsilon, f_1, f_2, \ldots, f_n}$. **Neighborhoods of an arbitrary point** $x \neq 0$ are defined as a shift of the neighborhood of zero on element x, in other words an arbitrary neighborhood V_x of a point x has the form $V_x = V_0 + x$, where V_0 is a neighborhood of zero. In such linear topological space, linear (and only linear) functionals $f \in F$ are continuous, i.e. $E^* = F$.

Let E and F be normed spaces and $\mathscr{L}$ be a linear continuous operator which acts from E into F. If $\varphi(f)$ is a linear continuous functional on F ($\varphi \in F^*$), then the functional $l(x) = \varphi(\mathscr{L}x)$ is a linear continuous functional on E ($l \in E^*$). Thus, we have constructed a map $\mathscr{L}^* : F^* \ni \varphi \mapsto l \in E^*$ which is called **an adjoint operator**. Adjoint operators exist also when E and F are linear topological spaces (see, for example, [97]). Remember that **a subdifferential** of a convex functional $f : E \to \mathbb{R}$ in a point $x_0 \in E$ is a set $\partial f(x_0) \subset E^*$ of linear continuous functionals $x_0^* \in E^*$ such that $f(x) - f(x_0) \geq x_0^*(x - x_0)$ for all $x \in E$. If E is a Banach space and a functional f is continuous in a point $x_0 \in E$ then $\partial f(x_0)$ is a non-empty convex and compact with respect to the topology $\sigma(E^*, E)$ set [15].

Under the investigations of generalized solutions of operator equations the theory of embedding and intermediate Banach spaces plays an important role (see [86]). Remember the major concepts of this theory. Let E and F be two sets. An operator $\mathscr{L} : E \to F$ is called **injective** if $\mathscr{L}x \neq \mathscr{L}y$ when $x \neq y$; **surjective** if $\mathscr{L}(E) = F$ and **bijective** if it is injective and surjective.

A linear normed space E is said to be **embedded** into a linear normed space F with the help of an embedding operator j_{EF} if j_{EF} is a linear bounded injective operator with a domain coinciding with the space E. Let the space E be embedded into a space G that in its turn is embedded into the space F using corresponding embedding operators. The space G is called **intermediate** between E and F if the following diagram is commutative:

$$
\begin{array}{ccc}
E & \xrightarrow{\ j_{EG}\ } & G \\
\| & & \downarrow{\scriptstyle j_{GF}} \\
E & \xrightarrow{\ j_{EF}\ } & F
\end{array}
$$

The concepts of embedded and intermediate spaces have another, more simple interpretation. Let the space E be embedded into F. Let us consider an image $j_{EF}(E) \subset F$ with a norm $\|y\|_0 = \|x\|_E$ where $y = j_{EF}(x)$. Obviously, the spaces $(E, \|\cdot\|)$ and $(j_{EF}(E), \|\cdot\|_0)$ are isometric. So, elements of E and $j_{EF}(E)$ can be identified and we may consider that E is a subspace of F. Since the operator j_{EF} is bounded, there exists some positive number C that does not depended on y such that $\|y\|_F \leq C\|y\|_0$ for all $y \in j_{EF}$. Therefore, the definition of embedded spaces can be reformulated in the following way: a linear normed space E is embedded into F if E is a subspace of F and there exists $C > 0$ for which $\|x\|_F \leq C\|x\|_E$, where x is an arbitrary element from E.

The concept of an intermediate space is interpreted in the following way: a space G is an intermediate space between E and F if $E \subset G \subset F$ and $\|x\|_G \leq C_1\|x\|_E$ for all $x \in E$, and $\|x\|_F \leq C_2\|x\|_G$ for all $x \in G$, where the constants C_1, C_2 do not depend on x.

A space E is said to be **embedded densely** into F if the set E considered as a subset of the space F is dense in F with respect to the norm $\|\cdot\|_F$. If E is embedded densely into F, then a restriction to the set E of any linear continuous functional $f \in F^*$ induces a continuous linear functional over E. Indeed, taking into consideration the inequality $\|x\|_F \leq C\|x\|_E$ we have

$$\|f\|_{E^*} = \sup_{\|x\|_E \leq 1} \frac{|f(x)|}{\|x\|_E} \leq C \sup_{\|x\|_E \leq 1} \frac{|f(x)|}{\|x\|_F} \leq C \sup_{\|x\|_F \leq C} \frac{|f(x)|}{\|x\|_F}$$

$$= C \sup_{\|x\|_F \leq 1} \frac{|f(x)|}{\|x\|_F} = C\|f\|_{F^*}.$$

The injectivity of the map $j_{F^*E^*}$ which maps a functional $f \in F^*$ to a functional from E^* follows from the density of the embedding E into F. Thus, F^* is embedded into E^* and $\|f\|_{E^*} \leq C\|f\|_{F^*}$. It is easy to see that the density of the embedding $E \subset F$ implies the density of the embedding F^* into E^* if the space E^* is endowed with topology $\sigma(E^*, E)$; an embedding $F^* \subset E^*$ which is dense with respect to the norm of E^* may not exist. Consider an example. Let $E = l_1$, $F = c_0$ then E is densely embedded into F (with $C = 1$). However, the set $F^* = l_1$ is not dense in the space $E^* = l_\infty$, since the element $e = (1, 1, \dots) \in l_\infty$ is away from l_1 over the distance 1.

Let us consider some examples of embedded and intermediate Banach spaces. A Banach space $E_1 = C^1(0, 1)$ is embedded into $E_0 = C(0, 1)$ using the operator of natural embedding: if $x(t) \in C^1(0, 1)$ then $j_{E_1 E_0}(x(t)) = x(t)$. Indeed,

$$\|x\|_{E_0} = \max_{0 \leq t \leq 1} |x(t)| \leq \max_{0 \leq t \leq 1} |x(t)| + \max_{0 \leq t \leq 1} |x'(t)| = \|x\|_{E_1}.$$

Note that $C^1(0, 1)$ is embedded densely into $C(0, 1)$, since by the Weierstrass theorem every continuous function can be approximated uniformly with arbitrary accuracy by algebraic polynomials which obviously are the elements of $C^1(0, 1)$. Moreover, $C^1(0, 1)$ is embedded compactly into $C(0, 1)$ (E is embedded compactly into F if the unit ball $S_1(E)$ in E is a relatively compact set in F, i.e. its closing in F with respect to the norm F is compact in F). Indeed, if $x(t) \in S(\theta, 1)$, where $S(\theta, 1)$ is the unit ball in $C^1(0,1)$ then $|x(t)| + |x'(t)| \leq 1$ for all $t \in [0,1]$. This implies the inequalities $|x(t)| \leq 1$ and $|x(t) - x(\tau)| \leq |t - \tau|$ for all $t, \tau \in [0,1]$. Thus, the set of functions $x(t)$ is uniformly bounded and equicontinuous in $C(0,1)$, hence according to the Arzela theorem this set if relatively compact in $C(0,1)$.

The other examples of an embedding of Banach spaces are $C(0,1)$ and $L_p(0,1)$ $(1 \leq p < \infty)$. Let us define the embedding operator j of $C(0,1)$ into $L_p(0,1)$ mapping a continuous function $x(t)$ to a class $\tilde{x}(t) \in L_p(0,1)$ consisting of functions, which differ from $x(t)$ on a zero Lebesgue measure set; then $C(0,1)$ in embedded into $L_p(0,1)$. Actually, j is an algebraic isomorphism, and moreover

$$\|j(x)\|_{L_p(0,1)} = \left(\int_0^1 |x(t)|^p dt \right)^{1/p} \leq \max_{0 \leq t \leq 1} |x(t)| = \|x\|_{C(0,1)}.$$

The density of $C(0,1) \subset L_p(0,1)$ with respect to the metric of L_p is implied from the Luzin theorem.

If $p > q$ then $L_p(0,1)$ is embedded densely into $L_q(0,1)$ using the identity embedding operator j. Indeed, when $p > q$ the Hölder inequality implies that

$$\|x\|_{L_q} = \left(\int_0^1 |x(t)|^q dt \right)^{1/q} \leq \left(\int_0^1 |x(t)|^{q \times \frac{p}{q}} dt \right)^{\frac{q}{p} \times \frac{1}{q}} = \|x\|_{L_p}.$$

Since $C(0,1) \subset L_p(0,1) \subset L_q(0,1)$ is an everywhere dense set in any of the spaces L_q, when $p > q$, and $L_p(0,1)$ is an everywhere dense set in $L_q(0,1)$. Also, the Hölder inequality implies that l_p is embedded into l_q when $q > p$; the density of the embedding l_p into l_q is implied from the fact that a linear manyfold M consisting from sequences having only finite number of non-zero elements belongs to any l_p and it is an everywhere dense set.

One more important example of embedding of the Banach spaces are the Sobolev space and the space $L_p(D)$, where D is a bounded region in $\mathbb{R}^n$. The space $W_p^{(l)}(D)$ consisting of the functions $x(t)$, $t = (t_1, t_2, \ldots, t_n) \in D \subset \mathbb{R}^n$ having pth-order summable generalized partial derivatives in D up to lth order inclusive is embedded into $L_q(D)$ using the operator of natural embedding. In addition,

$$\|x\|_{L_p(D)} = \left(\int_D |x(t)|^p dt \right)^{1/p}$$

$$\leq \left(\int_D |x(t)|^p dt \right)^{1/p} + \sum_{k_1, k_2, \ldots, k_l = 1}^n \left(\int_D \left| \frac{\partial^l x}{\partial t_{k_1} \partial t_{k_2} \ldots \partial t_{k_l}} \right|^p dt \right)^{1/p}$$

$$= \|x\|_{W_p^{(l)}(D)}.$$

The fact that the embedding $W_p^{(l)}(D)$ into $L_p(D)$ is dense follows from the fact that the set $C^{(l)}(D)$ consisting of all lth-order continuously differentiable functions in D is everywhere dense both in $W_p^{(l)}(D)$ and in $L_p(D)$ [29].

Chapter 2
The Simplest Schemes of Generalized Solution of Linear Operator Equation

Let E, F be Banach spaces and $\mathscr{L}$ be a linear operator with an everywhere dense domain $D(\mathscr{L}) \subset E$, which acts from E into F. Let us consider an operator equation

$$\mathscr{L}u = f, \qquad u \in D(\mathscr{L}), f \in F \tag{2.1}$$

and an adjoint equation

$$\mathscr{L}^* \varphi = l, \qquad \varphi \in D(\mathscr{L}^*), l \in E^*, \tag{2.2}$$

where E^* and F^* are conjugate Banach spaces to E and F, respectively, $\mathscr{L}^*$ is an adjoint operator to $\mathscr{L}$. Suppose that the range $R(\mathscr{L}) \subset F$ of $\mathscr{L}$ is an everywhere dense set in F and (2.1) is uniquely solvable over $R(\mathscr{L})$, i.e. the null space $\mathrm{Ker}(\mathscr{L})$ of $\mathscr{L}$ consists only of the zero element $\theta : \mathrm{Ker}(\mathscr{L}) = \theta$. Thus, $\mathscr{L}$ sets a one-to-one mapping between $D(\mathscr{L})$ and $R(\mathscr{L})$. Note that the continuity of $\mathscr{L}$ is not supposed.

The aim of this chapter is to give a "meaningful" definition of the solution of (2.1) when $f \notin R(\mathscr{L})$.

2.1 Strong Generalized Solution

Let us introduce one more norm on the linear set $D(\mathscr{L})$ in the space E. Since $\mathscr{L} : E \to F$ is a linear injective operator with a domain $D(\mathscr{L}) \subset E$, the function

$$D(\mathscr{L}) \ni u \to \|\mathscr{L}u\|_F \in \mathbb{R}$$

has all properties of a norm on $D(\mathscr{L})$. Hence, $D(\mathscr{L})$ with this norm turns into a normed space, which may be incomplete. Let $\bar{E}$ be a completion of this normed space.

The fact that $\|u\|_{\bar{E}} = \|\mathscr{L}u\|_F$ for all $u \in D(\mathscr{L})$ allows to extend $\mathscr{L}$ from $D(\mathscr{L})$ onto $\bar{E}$. Indeed, if u is an arbitrary element from $\bar{E}$, then the density of $D(\mathscr{L})$ in $\bar{E}$ implies that there is such a sequence $u_i \in D(\mathscr{L})$ that $u_i \to u$ in $\bar{E}$ as $i \to \infty$.

D.A. Klyushin et al., *Generalized Solutions of Operator Equations and Extreme Elements*, Springer Optimization and Its Applications 55, DOI 10.1007/978-1-4614-0619-8_2, © Springer Science+Business Media, LLC 2012

Since u_i is a convergent sequence in $\bar{E}$ and hence it is a Cauchy sequence in $\bar{E}$, and $\|u_i - u_j\|_E = \|\mathscr{L}u_i - \mathscr{L}u_j\|_F$ then $\mathscr{L}u_i$ is a Cauchy sequence in F. However, F is a complete normed space. Thus, there is such an element f in F that $\mathscr{L}u_i \to f$ in F as $i \to \infty$. Determine a value of the operator $\mathscr{L}$ on the element u in the following way: $\bar{\mathscr{L}}u = f$, where $\bar{\mathscr{L}} : \bar{E} \to F$ is an extended operator defined on the entire space $\bar{E}$. Note that the value $\bar{\mathscr{L}}u$ is defined correctly, i.e. the element $f = \bar{\mathscr{L}}u \in F$ does not depend on the selection of the sequence u_i. Thus, the operator $\bar{\mathscr{L}} : \bar{E} \to F$, $D(\bar{\mathscr{L}}) = \bar{E}$ is an extension of $\mathscr{L}$ on the whole space $\bar{E}$.

Definition 2.1. *A strong generalized solution* of (2.1) is such an element $u \in \bar{E}$ that equality (2.1) holds for the extended operator $\bar{\mathscr{L}}$.

Remark 2.1. If $E = F$ is a Hilbert space and $\mathscr{L}$ is a symmetric operator then the operator $\bar{\mathscr{L}}$ is called a self-adjoint extension of operator by Friedrichs.

As mentioned before, the concept of a strong generalized solution $\bar{u}$ arises when a right-hand side f of (2.1) does not belong to the range $R(\mathscr{L})$ of $\mathscr{L}$. In this case, the ordinary (classical) solution does not exist. The word "strong" means that the topology of the space $\bar{E}$ is normed.

Let us study some properties of $\bar{\mathscr{L}}$. It follows from the linearity of $\mathscr{L}$ that the operator $\bar{\mathscr{L}}$ is linear also. Let us prove that $\bar{\mathscr{L}}$ is an injective operator. Indeed, if $u \in \bar{E}$ is such an element that $\bar{\mathscr{L}}u = 0$, then selecting a sequence $u_i \in D(\mathscr{L})$ converging to u in $\bar{E}$ as $i \to \infty$ we have that $\mathscr{L}u_i \to \bar{\mathscr{L}}u = 0$ in F as $i \to \infty$. The last statement can be rewritten as $\|\mathscr{L}u_i\|_F \to 0$ or $\|u_i\|_{\bar{E}} \to 0$. Hence, $\|u\|_{\bar{E}} = 0$. Thus, the injectivity of the operator $\bar{\mathscr{L}}$ is proven. In addition, the equality $\|\mathscr{L}u\|_F = \|u\|_{\bar{E}}$, which holds for an arbitrary $u \in D(\mathscr{L})$, clearly holds for all $u \in \bar{E}$ (taking into account the replacement of $\mathscr{L}$ by $\bar{\mathscr{L}}$). From $\|\bar{\mathscr{L}}u\|_F = \|u\|_{\bar{E}}$, where $u \in \bar{E}$, it follows that the operator $\bar{\mathscr{L}}$ is continuous and coercive.

The properties of the operator $\bar{\mathscr{L}}$ can be proven in another way. Indeed, the operator $\mathscr{L}$ is a one-to-one map between $D(\mathscr{L})$ and $R(\mathscr{L})$. In addition, if ($D(\mathscr{L})$ is a normed space with the norm $\|u\|_{\bar{E}}$ and $R(\mathscr{L})$ is a normed space with the norm $\|f\|_F$) then the completion of $D(\mathscr{L})$ coincides with $\bar{E}$ and the completion $R(\mathscr{L})$ coincides with F (remember that $R(\mathscr{L})$ is a dense subset of F). On the other hand, granting the equality $\|\mathscr{L}u\|_F = \|u\|_{\bar{E}}$, which holds for all $u \in D(\mathscr{L})$, we have that the operator $\mathscr{L}$ is an isometry between the normed spaces $D(\mathscr{L})$ and $R(\mathscr{L})$. Hence, their completions are isometrical. This isometry defines the completion $\bar{\mathscr{L}}$ of the operator $\mathscr{L}$. Thus, the operator $\bar{\mathscr{L}}$ sets an isometry between $\bar{E}$ and F. This implies the above-mentioned properties of $\bar{\mathscr{L}}$. The foregoing implies the following theorem.

Theorem 2.1. *For any $f \in F$ there exists a unique strong generalized solution of* (2.1) *in the sense of Definition 2.1.*

If $f \in R(\mathscr{L})$, then a strong generalized solution $\bar{u}$ turns into a classic solution. It is also clear that the classic solution is strong, and it is classic if $\bar{u} \in D(\mathscr{L})$.

Let us clarify the relations between the spaces E and $\bar{E}$. Since $D(\mathscr{L})$ is a dense linear subset of E (of course, in the sense of the norm of the space E), then the set E may be obtained by completing $D(\mathscr{L})$ with respect to the norm $\|u\|_E$. Thus, the spaces E and $\bar{E}$ may be considered as completions of the same linear set $D(\mathscr{L})$

with respect to the two different norms: $\|u\|_E$ and $\|u\|_{\bar{E}}$. Unfortunately, in general case, elements of the spaces E and $\bar{E}$ are incomparable. It is explained by the fact that, on one hand, the operator $\mathscr{L} : E \to F$ can be unbounded and, from the other hand, it can be non-coercive, even though it is a linear injective operator. This means that in general case the norms $\|u\|_E$ and $\|\mathscr{L}u\|_F = \|u\|_{\bar{E}}$ can induce incomparable topologies on $D(\mathscr{L})$.

When $\mathscr{L} : E \to F$ is a linear continuous operator the case is more simple. Then the topology induced on $D(\mathscr{L})$ by the norm $\|u\|_{\bar{E}}$ is weaker than the topology of the space E.[1]

Consider another possibility. Let an operator $\mathscr{L} : E \to F$ be coercive, i.e. there exists such a constant $c > 0$ that

$$\|u\|_E \le c\|\mathscr{L}u\|_F = c\|u\|_{\bar{E}} \tag{2.3}$$

for all $u \in D(\mathscr{L})$. In this case, the norms $\|u\|_E$ and $\|\mathscr{L}u\|_F = \|u\|_{\bar{E}}$ are comparable over $D(\mathscr{L})$ (the topology of the space $\bar{E}$ is stronger than the topology of the space E on $D(\mathscr{L})$) and there is a relation between elements of E and $\bar{E}$.

Theorem 2.2. *Let $\mathscr{L}$ be a closable coercive operator. Then there exists a dense continuous embedding $\bar{E} \subset E$.*

Proof. Since the spaces $\bar{E}$ and E are the completions of the linear set $D(\mathscr{L})$ with respect to two norms and (2.3) holds, then in order to prove the theorem, it is enough to check the condition:

(π) if $u_i \in D(\mathscr{L})$ and $u_i \to u$ in $\bar{E}$, $u_i \to 0$ in E, then $u = 0$.

However, this condition can be rewritten in the following way:

(π) if $u_i \in D(\mathscr{L})$ and $\mathscr{L}u_i \to f$ in F, $u_i \to 0$ in E, then $f = 0$.

The last condition is clear, since the operator $\mathscr{L}$ is closable. $\qquad\square$

Thus, we ascertained that $\bar{E} \subset E$, i.e. an arbitrary strong generalized solution of (2.1) is an element of the space E.

2.2 Strong Near-Solution

Suppose that the right-hand side of (2.1), i.e. the element f, does not belong to the range $R(\mathscr{L})$ of an operator $\mathscr{L}$. Since $R(\mathscr{L})$ is everywhere dense in F and (2.1)

---

[1] Note that studying of a closable operator $\mathscr{L} : E \to F$ can be reduced (at least theoretically) to studying of a linear continuous operator $\mathscr{L}_1$ defined on the same set $D(\mathscr{L})$, but with respect to another norm. Indeed, introducing in $D(\mathscr{L})$ a graph norm

$$\|u\|_\Gamma = \|u\|_E + \|\mathscr{L}u\|_F,$$

with respect to which the linear set $D(\mathscr{L})$ is Banach, we have that the operator $\mathscr{L}_1 : D(\mathscr{L}) \to F$ is linear and continuous ($\mathscr{L}_1 u = \mathscr{L}u$, $u \in D(\mathscr{L}) = D(\mathscr{L}_1)$).

is uniquely solvable, then there exists a sequence $f_n \in R(\mathscr{L})$ such that $f_n \to f$ as $n \to \infty$, and a sequence $u_n = \mathscr{L}^{-1}(f_n)$ convergent to some element $\bar{u} \in \bar{E}$ in $\bar{E}$.

Definition 2.2. A sequence of elements $u_n \in D(\mathscr{L})$ is called a *strong near-solution* of the operator equation (2.1), if $f_n = \mathscr{L}u_n \to f$ as $n \to \infty$ in the metric of the space F. An element $\bar{u} \in \bar{E}$ is called the *strong limit element* of the near-solution.

The concept of a "near-solution" is justified by the following arguments. In many important practical cases, it is impossible or almost impossible to determine the right-hand side f of (2.1) absolutely exactly; therefore, we have to consider its ε-approximation, i.e. and element $f \in R(\mathscr{L})$ such that $\rho(f, f_\varepsilon) = \|f - f_\varepsilon\| < \varepsilon$. In this case, there exists an element $u_\varepsilon = \mathscr{L}^{-1}(f_\varepsilon)$ from the domain of the operator $\mathscr{L}$, which can be considered as "ε - approximation" of the solution of (2.1), i.e. the right-hand (2.1) is closely approximated by its image $\mathscr{L}u_\varepsilon = f_\varepsilon$. If the elements u_ε "become stabilize" as $\varepsilon \to 0$, i.e. if they converge in some topology (in $\bar{E}$) to the fixed element $\bar{u} \notin D(\mathscr{L})$, then it is naturally to consider the element u_ε as an "ε -solution" or "near-solution". Note that in many cases the "accuracy" of a solution u_ε is defined by the closeness of its image $\mathscr{L}u_\varepsilon = f_\varepsilon$ to the element f, i.e. by a norm of the space $\bar{E}$.

The definitions of a strong generalized solution and a near-solution of (2.1) imply that these concepts are equivalent, i.e. an element $u \in \bar{E}$ is a strong generalized solution of the operator equation (2.1) iff it is a strong limit element of a near-solution.

2.3 Weak Generalized Solution

Let us consider a definition of a generalized solution of an operator equation in a linear topological space with a topology which is not necessarily induced by a norm.

As before, suppose that $\mathscr{L} : E \to F$ is a linear injective operator, which acts between Banach spaces E, F with everywhere dense domain and range in E and F, respectively. In addition, suppose that $D(\mathscr{L}^*)$ is a total subset of F^* in a duality (F, F^*), and $R(\mathscr{L}^*)$ is a total subset of E^* in a duality (E, E^*). Note that the totality property of $R(\mathscr{L}^*)$ may be replaced by one of the following conditions:

(a) The space E is reflexive (if the space F is reflexive also, then the set $D(\mathscr{L}^*)$ is strongly dense in F^*).

(b) The operator $\mathscr{L}$ is continuous, i.e. $D(\mathscr{L}) = E$;

In condition (a), the totality of $R(\mathscr{L}^*)$ follows from [40], and in case (b) it follows from the formulae

$$R(\mathscr{L}^*)^\circ \cap D(\mathscr{L}) = \mathrm{Ker}(\mathscr{L}), \qquad (2.4)$$

where $R(\mathscr{L}^*)^\circ \subset E$ is a polar of the set $R(\mathscr{L}^*) \subset E^*$ in a duality (E, E^*). Let us prove (2.4) for an arbitrary linear operator. Since $R(\mathscr{L}^*)$ is a linear set, then

$$R(\mathscr{L}^*)^\circ \cap D(\mathscr{L}) = \{u \in E : u \in D(\mathscr{L}), l(u) = 0, \forall l \in R(\mathscr{L}^*)\}$$
$$= \{u \in E : u \in D(\mathscr{L}), \varphi(\mathscr{L}u) = 0, \forall \varphi \in D(\mathscr{L}^*)\}$$

Since $D(\mathscr{L}^*)$ is a total linear subspace, then

$$R(\mathscr{L}^*)^\circ \cap D(\mathscr{L}) = \{u \in E : u \in D(\mathscr{L}), \mathscr{L}u = 0\} = \mathrm{Ker}(\mathscr{L}).$$

Therefore, formulae (2.4) is proved. From (2.4) it is follows that

$$(R(\mathscr{L}^*)^\circ \cap D(\mathscr{L}))^\circ = (\mathrm{Ker}(\mathscr{L}))^\circ.$$

If $\mathscr{L}$ is a continuous injective operator, then $D(\mathscr{L}) = E$, $\mathrm{Ker}(\mathscr{L}) = \varnothing$. Therefore,

$$(R(\mathscr{L}^*))^{\circ\circ} = (\mathrm{Ker}(\mathscr{L}))^\circ = E^*.$$

So, a bipolar of the set $R(\mathscr{L}^*)$, i.e. a weak closure $R(\mathscr{L}^*)$ coincides with E^*; hence, $R(\mathscr{L}^*)$ is total in E^*.

Finally, we see that the set of functionals $R(\mathscr{L}^*) \subset E^*$ is a total linear manyfold with respect to the duality (E^*, E); the linear subspaces F and $D(\mathscr{L}^*)$ are in duality also.

Denote by $\widetilde{E}$ a completion of a space E with respect to a topology $\sigma(E, R(\mathscr{L}^*))$. Since the sets E and $R(\mathscr{L}^*)$ are in duality, then the space $\widetilde{E}$ is a Hausdorff locally convex topological vector space. Each of the functionals $l \in E^*$ which has the form $l = \mathscr{L}^*\varphi$, where $\varphi \in D(\mathscr{L}^*)$, allows a unique extension by continuity on the whole space $\widetilde{E}$, which we will denote as $\widetilde{l}$. A conjugate space to $\widetilde{E}$ is a space consisting of various functionals $\widetilde{l}$, where $l = \mathscr{L}^*\varphi$, $\varphi \in D(\mathscr{L}^*)$.

Let us consider an arbitrary continuous linear functional $\varphi \in D(\mathscr{L}^*)$. Then (2.1) implies that

$$\varphi(\mathscr{L}u) = \varphi(f), \quad l(u) = (\mathscr{L}^*\varphi)(u) = \varphi(f). \tag{2.5}$$

Definition 2.3. *A weak generalized solution of the operator equation (2.1) is an element $u \in \widetilde{E}$, which satisfies the relation*

$$\widetilde{l}(u) = \varphi(f) \qquad \text{for all } \varphi \in D(\mathscr{L}^*), \tag{2.6}$$

where $l = \mathscr{L}^*\varphi$.

A weak generalized solution $u \in \widetilde{E}$, as a strong generalized solution of (2.1) also arises when the right-hand side of (2.1), i.e. the element f, does not belong to the range $R(\mathscr{L})$ of the operator $\mathscr{L}$ and a classic solution does not exist.

Relations (2.5) imply that any classic solution is a weak solution also. On the other hand, if $f \in R(\mathscr{L})$, then a weak generalized solution u turns into a classic one. Indeed, let $f \in R(\mathscr{L})$ and $u \in \widetilde{E}$ be a weak generalized solution. Therefore, for all $\varphi \in D(\mathscr{L}^*)$ we have

$$\widetilde{l}(u) = \varphi(f), \quad l = \mathscr{L}^*\varphi.$$

Moreover, there exists such an element $u_1 \in D(\mathscr{L})$ that $\mathscr{L}u_1 = f$. This element u_1 is a weak generalized solution, i.e.

$$\widetilde{l}(u_1) = \varphi(f), \quad l = \mathscr{L}^*\varphi$$

for all $\varphi \in D(\mathscr{L}^*)$. Thus, for all $\varphi \in D(\mathscr{L}^*)$ the equality $\tilde{l}(u_1) = \tilde{l}(u)$ holds, where $l = \mathscr{L}^* \varphi$. Since the set of all functionals $\tilde{l}$, where $l = \mathscr{L}^* \varphi$, $\varphi \in D(\mathscr{L}^*)$, coincides with the space $\tilde{E}^*$, then $u = u_1$ in $\tilde{E}$, and therefore in E.

Analogously, if a weak generalized solution $\tilde{u}$ belongs to $D(\mathscr{L})$, then it is a classic solution.

2.4 Weak Near-Solution

Analogous to a strong near-solution let us introduce a weak near-solution.

Definition 2.4. A sequence $u_n \subset D(\mathscr{L})$ is called a *weak near-solution* of the operator equation (2.1) if $f_n = \mathscr{L}u_n \to f$ as $n \to \infty$ with respect to the metric of the space F and $u_n = \mathscr{L}^{-1}(f_n) \to \tilde{u} \in \tilde{E}$ as $n \to \infty$ with respect to the weak topology $\sigma(\tilde{E}, R(\mathscr{L}^*))$; an element $\tilde{u} \in \tilde{E}$ is called a *Limit element!weak*.

As it will be proved below, the effect of stabilizing of a sequence of elements u_n in the space $\tilde{E}$ is a corollary of the convergence of f_n to f; so, there exists an analogy between string and weak near-solutions.

Consider the relation between a weak generalized solution and a weak near-solution. Let us prove that u is a weak generalized solution of the operator equation (2.1) iff it is a weak limit element of a near solution. Indeed, let u be a limit element of a near-solution, then $u \in \tilde{E}$ and there exists a sequence $\{f_n\} \subset R(\mathscr{L})$ such that f_n tends to f as $n \to \infty$ with respect to the metric of F, and $u_n = \mathscr{L}^{-1}(f_n) \in D(\mathscr{L})$ tends to u as $n \to \infty$ in the topology $\sigma(\tilde{E}, R(\mathscr{L}^*))$. Therefore, for all $\varphi \in D(\mathscr{L}^*) \subset F^*$ we have that $\varphi(\mathscr{L}u_n) = \varphi(f_n)$, hence

$$l(u_n) = (\mathscr{L}^*\varphi)(u_n) = \varphi(f_n) \to \varphi(f)$$

as $n \to \infty$, where $l = \mathscr{L}^* \varphi \in R(\mathscr{L}^*)$.

In addition, since $l \in R(\mathscr{L}^*)$, then $l(u_n) = \tilde{l}(u_n) \to \tilde{l}(u)$ as $n \to \infty$. Thus, we have that

$$\tilde{l}(u) = \varphi(f)$$

for all $\varphi \in D(\mathscr{L}^*)$ such that $l = \mathscr{L}^* \varphi$, i.e. u is a weak generalized solution of (2.1).

Conversely, let us suppose that u is a weak solution of (2.1), i.e.

$$\tilde{l}(u) = \varphi(f) \qquad \text{for all } \varphi \in D(\mathscr{L}^*),$$

where $l = \mathscr{L}^* \varphi$.

Let $\{f_n\}$ be an arbitrary sequence from $R(\mathscr{L})$ convergent to f as $n \to \infty$ with respect to the norm of F. Denote $\mathscr{L}^{-1}(f_n)$ as u_n. Then for an arbitrary functional $\varphi \in D(\mathscr{L}^*)$ such that $l = \mathscr{L}^* \varphi$ we have

$$l(u_n) = (\mathscr{L}^*\varphi)(u_n) = \varphi(\mathscr{L}x_n) = \varphi(f_n) \to \varphi(f)$$

as $n \to \infty$.

Thus, for any functional $\in R(\mathscr{L}^*)$ we have that $l(u_n) \to \varphi(f) = \widetilde{l}(u)$ as $n \to \infty$. Therefore, the sequence u_n converges to u with respect to the topology $\sigma(E, R(\mathscr{L}^*))$, hence u is a limit element of a near-solution $\{u_n\}$.

2.5 Existence and Uniqueness of a Weak Generalized Solution of a Linear Operator Equation

In this section, we prove the theorem on existence and uniqueness of a weak generalized solution of the operator equation (2.1) on the assumptions stated above, i.e. if $\mathscr{L}$ is a linear operator with dense domain $D(\mathscr{L})$ and dense range $R(\mathscr{L})$, (2.1) is uniquely solvable, and the sets $D(\mathscr{L}^*)$ and $R(\mathscr{L}^*)$ are total in the spaces F^* and E^* with respect to the corresponding weak topologies.

Let us start with the relatively simple problem of uniqueness. Suppose that the operator equation (2.1) in addition to a weak generalized solution $u \in \widetilde{E}$ has another weak generalized solution $\widetilde{u} \in \widetilde{E}$ $(u \neq \widetilde{u})$, then

$$\widetilde{l}(u) = \varphi(f) = \widetilde{l}(\widetilde{u})$$

for all $\varphi \in D(\mathscr{L}^*)$, $l = \mathscr{L}^*\varphi$.

Since the set of the functionals $\widetilde{l}$ coincides with the conjugate space $\widetilde{E}^*$, then $u = \widetilde{u}$, and we have a contradiction. Thus, the operator equation (2.1) may not have more than one weak generalized solution.

Now, let us consider the problem of existence of a weak generalized solution. Suppose that the right-hand side of the operator equation (2.1), i.e. the element f does not belong to the range $R(\mathscr{L})$ of the operator $\mathscr{L}$. Since (2.1) is densely solvable, then there exists such a sequence of elements f_n from $R(\mathscr{L})$ that $f_n \to f$ as $n \to \infty$ with respect to the norm F. Let us prove that the sequence $u_n = \mathscr{L}^{-1}(f_n)$ is a weak near-solution, and its limit element u belongs to $\widetilde{E}$. For this purpose let us consider the inverse operator $u = \mathscr{L}^{-1}(f)$, which acts from the vector space $R(\mathscr{L})$ into E. Denote by T the topology induced in $R(\mathscr{L}) \subset F$ by the norm of the Banach space F, and denote by $(R(\mathscr{L}), T)$, $(E, \sigma(E, R(\mathscr{L}^*)))$ the vector spaces $R(\mathscr{L})$ and E endowed with the topologies T and $\sigma(E, R(\mathscr{L}^*))$, respectively. Let us prove that the inverse operator $B = \mathscr{L}^{-1}$ is a continuous linear operator, which acts from the normed space $(R(\mathscr{L}), T)$ into the Hausdorff topological vector space $(E, \sigma(E, R(\mathscr{L}^*)))$. Since the set

$$W(l_1, \ldots, l_n; \varepsilon) = \{u : u \in E, l_1(u) < \varepsilon, \ldots, l_n(u) < \varepsilon\},$$

where $\varepsilon \in \mathbb{R}$, $l_i \in R(\mathscr{L}^*)$, $i \in \{1, 2, \ldots, n\}$, form a fundamental system of neighborhoods of zero in $(E, \sigma(E, R(\mathscr{L}^*)))$, it is enough to prove that the following preimages $B^{-1}[W(l_1, \ldots, l_n; \varepsilon)] = \mathscr{L}[W(l_1, \ldots, l_n; \varepsilon)]$ are neighborhoods of zero in $(R(\mathscr{L}), T)$.

Indeed,

$$\mathscr{L}[W(l_1,\ldots,l_n;\varepsilon)] = \{\mathscr{L}u : l_1(u) < \varepsilon, \ldots, l_n(u) < \varepsilon\}$$
$$= \{\mathscr{L}u : \varphi_1(\mathscr{L}u) < \varepsilon, \ldots, \varphi_n(\mathscr{L}u) < \varepsilon\},$$

where $l_i(u) = (\mathscr{L}^*\varphi_i)(u) = \varphi_i(\mathscr{L}u)$, $\varphi_i \in D(\mathscr{L}^*)$, $i \in \{1,2\ldots,n\}$. Therefore,

$$\mathscr{L}[W(l_1,\ldots,l_n;\varepsilon)] = \{f \in R(\mathscr{L}) : \varphi_1(f) < \varepsilon, \ldots, \varphi_n(f) < \varepsilon\}$$
$$= W_{R(\mathscr{L})}(\varphi_1,\ldots,\varphi_n;\varepsilon),$$

where $W_{R(\mathscr{L})}(\varphi_1,\ldots,\varphi_n;\varepsilon)$ is a neighborhood that belongs to a fundamental system of neighborhoods of zero in the vector space $R(\mathscr{L})$ endowed with the topology $\sigma(R(\mathscr{L}),D(\mathscr{L}^*))$. Since the normed topology T is stronger than the weak topology $\sigma(R(\mathscr{L}),D(\mathscr{L}^*))$, then the set $W_{R(\mathscr{L})}(\varphi_1,\ldots,\varphi_n;\varepsilon)$ is a neighborhood of zero with respect to the topology T. Thus, the operator $B = \mathscr{L}^{-1} : (R(\mathscr{L}),T) \to (E,\sigma(E,R(\mathscr{L}^*)))$ is continuous.

Since the space $\widetilde{E}$ is complete, F,E are the Hausdorff topological vector spaces, and every continuous linear map B of the space $(R(\mathscr{L}),T)$ into E is uniquely extendable to a continuous linear map $\widetilde{B}$ from F into $\widetilde{E}$ [7], then the sequence $\{u_n = \mathscr{L}^{-1}(f_n) = \widetilde{B}(f_n)\}$ converges to some element $\widetilde{u} \in \widetilde{E}$, which is a limit element of the near-solution $\{u_n = \mathscr{L}^{-1}(f_n)\}$. As it was shown above, in this case $\widetilde{u}$ is a weak generalized solution of (2.1). Thus, the existence of a weak generalized solution of (2.1) is proved.

2.6 Relation Between Weak and Strong Solutions of a Linear Operator Equation

Let us establish the relation between the solvability in sense of Definitions 2.1 and 2.3.

Theorem 2.3. *The space $\bar{E}$ is densely embedded into the space $\widetilde{E}$.*

Proof. Let some network $\{u_\alpha\}_{\alpha\in\mathscr{A}}, u_\alpha \in E$ converges to 0 with respect to the topology of the space $\bar{E}$. Then $\mathscr{L}u_\alpha \to 0$ in F, hence $\varphi(\mathscr{L}u_\alpha) \to 0$ for any $\varphi \in F^*$. Thus, $l(u_\alpha) \to 0$ for all $l \in R(\mathscr{L}^*)$. Therefore, the topology $\widetilde{E}$ is weaker than the topology $\bar{E}$. It remained only to prove that if $u_\alpha \to u$ with respect to the topology of the space $\bar{E}$ and $u_\alpha \to 0$ with respect to the topology $\widetilde{E}$, then $u = 0$ (condition π)). Taking into account the fact that $u_\alpha \to u$ is a convergent sequence, we have that

$$l(u_\alpha) = \mathscr{L}^*\varphi(u_\alpha) = \varphi(\mathscr{L}u_\alpha) \to \varphi(\bar{\mathscr{L}}u)$$

for all $l \in R(\mathscr{L}^*)$. In addition, the fact that $u_\alpha \to 0$ implies that $l(u_\alpha) \to 0$ for all $l \in R(\mathscr{L}^*)$ also. Thus, we have that $\varphi(\bar{\mathscr{L}}u) = 0$ for any $\varphi \in D(\mathscr{L}^*)$. Since the set

$D(\mathcal{L}^*)$ is total and the operator $\tilde{\mathcal{L}}$ is injective, then $u = 0$. Thus, the embedding $\bar{E} \subset \tilde{E}$ is proved.

The fact that the embedding is dense follows from the fact that the spaces $\bar{E} \subset \tilde{E}$ are obtained as a result of completing of the set $D(\mathcal{L})$, i.e. $D(\mathcal{L})$ is a dense set both in $\bar{E}$ and $\tilde{E}$.   □

Theorem 2.4. *Definitions 2.1 and 2.3 are equivalent.*

Proof. Let $u \in \bar{E}$ be a strong generalized solution of the equation $\mathcal{L}u = f$. Taking into account the fact that the set $R(\mathcal{L})$ is dense in F, we have that there exists such a sequence $f_n \in R(\mathcal{L})$ that converges to f, or, in other words, there exists such an element $u_n \in D(\mathcal{L})$, that $u_n \to u$ in $\bar{E}$. By virtue of Theorem 2.3 the elements $u \in \bar{E}$ belongs to the space $\tilde{E}$, and in addition $u_n \to u$ in $\tilde{E}$. Now, it is easy to see that, from one hand, for all $l = \mathcal{L}^*\varphi \in R(\mathcal{L})$

$$l(u_n) = \mathcal{L}^*\varphi(u_n) = \varphi(\mathcal{L}u_n) \to \varphi(f),$$

and, from the other hand, $- l(u_n) \to l(u)$ as $n \to \infty$. Thus, $u -$ is a weak generalized solution.

Let us prove that the solution $u \in \tilde{E}$ in the sense of Definition 2.3 is a solution in the sense of Definition 2.1 (and vice versa). Indeed, there exists a solution $u^* \in \bar{E}$ of the equation $\mathcal{L}u = f$. It is clear that $\mathcal{L}^*\varphi(u) = \varphi(f) = \varphi(\bar{\mathcal{L}}u^*)$ for all $\varphi \in D(\mathcal{L}^*)$. Hence $u = Ou^*$, where O is an operator of embedding of the space $\bar{E}$ into the space $\tilde{E}$.   □

Finally, let us point out that the concept of a generalized solution of the operator equation $\mathcal{L}u = y$ is very different from various concepts u^* of such equations (for example, from the concept of a quasi-solution introduce by V. K. Ivanov), which are described in [47] and [112], as far as $\bar{\mathcal{L}}\bar{u} = y$ for the generalized solution $\bar{u}$ always, where $\bar{\mathcal{L}}$ is a natural extension of the operator $\mathcal{L}$, whereas the equality $\bar{\mathcal{L}}u^* = y$ for the generalized solutions u^* holds not always.

Chapter 3
A Priori Estimates for Linear Continuous Operators

In this chapter, we will study a linear continuous operator $\mathscr{L}$, which acts from a normed space E $(D_f = E)$ into a normed space F. We will suppose that the $\mathscr{L}$ is injective and has a dense range in F.

Strong and weak solutions considered in the previous chapter belong to spaces $\bar{E}$ and $\tilde{E}$, but constructive description of the spaces $\bar{E}$ and $\tilde{E}$ is a very difficult problem for various operators $\mathscr{L}$ which are important from the practical point of view. So, it is necessary to establish the existence of a dense embedding of $\bar{E}$ or $\tilde{E}$ into another well-studied Banach or locally convex linear topological space H. In this chapter, we describe such spaces H for some integral, differential and abstract Hilbert–Schmidt operators in Hilbert space. In addition, we will study the properties of generalized solutions in H.

3.1 A Priori Inequalities

Let us consider the case when the space $\bar{E}$ in embedded into a Banach space H. This embedding implies that $c_1 \|u\|_H \le \|u\|_{\bar{E}}$ for all $u \in \bar{E}$. Hence,

$$c_1 \|u\|_H \le \|\mathscr{L}u\|_F \le c_2 \|u\|_E, \qquad \forall u \in E, \tag{3.1}$$

where c_1, c_2 are positive constants.

Such estimations are common in applications. They are called a priori estimations [40, 62]. In addition (3.1), the following a priori estimates hold

$$c_1 \|u\|_{\bar{E}} \le \|\mathscr{L}u\|_F \le c_2 \|u\|_E, \qquad \forall u \in E,$$
$$c_1 \|u\|_H \le \|\bar{\mathscr{L}}u\|_F \le c_2 \|u\|_{\bar{E}}, \qquad \forall u \in \bar{E},$$

where $\bar{\mathscr{L}}$, as in Chap. 2, is an extension $\mathscr{L}$ onto the entire space $\bar{E}$ by continuity, i.e. $\bar{\mathscr{L}} : \bar{E} \to F$.

D.A. Klyushin et al., *Generalized Solutions of Operator Equations and Extreme Elements*, Springer Optimization and Its Applications 55, DOI 10.1007/978-1-4614-0619-8_3,

Note that inequalities (3.1) themselves do not guarantee the embedding $\bar{E} \subset H$. They only allow to compare topologies induced in E by the norms $\| \cdot \|_E$ and $\| \cdot \|_H$.

Further, let us prove that estimations (3.1) may be a basis for constructing the theory of generalized solvability of an operator equation

$$\mathscr{L}u = f \tag{3.2}$$

As in Chap. 2 we shall consider the following concepts.

Definition 3.5. *A strong generalized solution* of (3.2) is such an element $u \in \bar{E}$ that $\bar{\mathscr{L}}u = f$.

3.2 A Generalized Solution of an Operator Equation in Banach Spaces

The previous section implies that inequalities (3.1) are the necessary conditions for construction of the theory of generalized solvability of linear operator equations in a Banach space H. Let us prove that inequalities (3.1) are sufficient conditions for the solvability of $\mathscr{L}u = f$ also (in some generalized sense) $\mathscr{L}u = f$. Note that various approaches to construction of the theory of generalized solutions of differential equations are relevant to this scheme also (see [5]).

Let us suppose that the linear operator $\mathscr{L}$, $(D(\mathscr{L}) = E, \overline{R(\mathscr{L})} = F)$ satisfies inequalities (3.1), where $u \in E$, $c_1, c_2 > 0$, H is a completion of the space E with respect to the norm $\|u\|_H$. It is clear that the right-hand sides of inequalities (3.1) imply the continuity of the operator $\mathscr{L}$, and the left-hand sides imply its injectivity. In addition, by virtue of the density of the embedding $E \subset H$ the set H^* (conjugate to H) is total in the conjugate space E^* and, hence, the spaces E and H^* are dual to each other.

Lemma 3.1. *An operator equation*

$$\mathscr{L}^* \varphi = l \tag{3.3}$$

is solvable on a subset H^* of the space E^*.

Proof. Let us consider the operator $\widetilde{\mathscr{L}} : H \to F$ $(D(\widetilde{\mathscr{L}}) = E)$, defined in the following way: $\widetilde{\mathscr{L}}u = \mathscr{L}u, u \in E$. Then the left-hand side of inequalities (3.1) implies the correct solvability of the operator $\widetilde{\mathscr{L}}$. Let us consider also the adjoint operator $\widetilde{\mathscr{L}}^* : F^* \to H^*$, $D(\widetilde{\mathscr{L}}^*) \subset D(\mathscr{L}^*) = F^*$. It is clear that if $\varphi \in D(\widetilde{\mathscr{L}}^*)$, then $\widetilde{\mathscr{L}}^* \varphi|_E = \mathscr{L}^* \varphi$, where $\widetilde{\mathscr{L}}^* \varphi|_E$ is a restriction of the functional $\widetilde{\mathscr{L}}^* \varphi \in H^*$ from the set H onto the set E. As well-known, the correct solvability of the operator $\widetilde{\mathscr{L}}$ implies the solvability of the operator $\widetilde{\mathscr{L}}^*$ everywhere [40]; hence, taking into account the facts above, we have the solvability of the operator $\mathscr{L}^*$ over the set H^* (as a subspace of E^*). $\qquad\square$

Remark 3.2. If the operator $\mathscr{L}$ satisfies inequalities (3.1), then

$$H^* \subset R(\mathscr{L}^*) \subset E^*.$$

Definition 3.6. *A generalized solution* of (3.2) with a right-hand side $f \in F$ is such an element $u \in H$, that the equality

$$\widetilde{\mathscr{L}^*}\varphi(u) = \varphi(f), \tag{3.4}$$

holds for any $\varphi \in D(\widetilde{\mathscr{L}^*})$.

It is clear that equality (3.4) is equivalent to

$$\mathscr{L}^*\varphi(u) = \varphi(f), \qquad \forall \varphi \in F^*, \mathscr{L}^*\varphi \in H^*.$$

Theorem 3.1. *For any right-hand side $f \in F$ there exists a unique solution $u \in H$ of (3.2) in the sense of Definition 3.6.*

Proof. Let us choose a sequence $f_p \in R(\mathscr{L})$ such that $f_p \to f$ in the space F. Hence, if $u_p \in E$ is a solution of the equation $\mathscr{L}u = f_p$, then taking into account (3.1) and the fact that the sequence $\{f_p\}$ is Cauchy, we have

$$\|u_{p_1} - u_{p_2}\|_H \le c_1^{-1}\|\mathscr{L}u_{p_1} - \mathscr{L}u_{p_2}\|_F = c_1^{-1}\|f_{p_1} - f_{p_2}\|_F \to 0, \qquad p_1, p_2 \to \infty.$$

Thus, there exist such $u^* \in H$ that $u_p \to u^*$ in H. Further, we have

$$\mathscr{L}^*\varphi(u_p) = \varphi(\mathscr{L}u_p) = \varphi(f_p), \qquad \varphi \in F^*.$$

Passing to the limit in the last equality as $p \to \infty$, we have

$$\mathscr{L}^*\varphi(u^*) = \varphi(f), \qquad \varphi \in F^*, \mathscr{L}^*\varphi \in H^*.$$

Thus, u^* is a solution of (3.2) in the sense of Definition 3.6.

Since $H^* \subset R(\mathscr{L}^*)$, then the equality

$$l(u^*) = \mathscr{L}^*\varphi(u^*) = 0, \qquad \forall \varphi \in F^*, \mathscr{L}^*\varphi \in H^*$$

implies that $u^* = 0$, and hence the solution is unique. $\qquad \square$

Definition 3.7. *A generalized solution* of problem (3.2) with a right-hand side $f \in F$ is such an element $u \in H$ that there exists a sequence $u_i \in E$, which satisfies the conditions

$$\|u_i - u\|_H \to 0, \qquad \|\mathscr{L}u_i - f\|_F \to 0, \qquad i \to \infty.$$

Theorem 3.2. *Definitions 3.6 and 3.7 are equivalent.*

Proof. Let u be a solution of the equation $\mathscr{L}u = f$ in the sense of Definition 3.6, i.e. $\mathscr{L}^*\varphi(u) = \varphi(f)$. Repeating reasonings which were used for proving of Theorem 3.1 we conclude that $u = u^*$, and hence $\|u_p - u\|_H \to 0$. From the other hand, $\|\mathscr{L}u_p - f\|_F = \|f_p - f\|_F \to 0$. Thus, u is a solution of (3.2) in the sense of Definition 3.7.

Let us prove the inverse statement. Let u be a solution of (3.2) in the sense of Definition 3.7. Then

$$\mathscr{L}^*\varphi(u) = \mathscr{L}^*\varphi(u_i) + \mathscr{L}^*\varphi(u - u_i) = \varphi(\mathscr{L}u_i) + \mathscr{L}^*\varphi(u - u_i)$$
$$= \varphi(\mathscr{L}u_i - f) + \varphi(f) + \mathscr{L}^*\varphi(u - u_i),$$

for all $\varphi \in F^*, \mathscr{L}^*\varphi \in H^*$.

Let us estimate the first and third term in the right-hand side of the last equality

$$|\varphi(\mathscr{L}u_i - f)| \le \|\varphi\|_{F^*}\|\mathscr{L}u_i - f\|_F \to 0,$$
$$|\mathscr{L}^*\varphi(u - u_i)| \le \|\mathscr{L}^*\varphi\|_{H^*}\|u - u_i\|_H \to 0, \qquad i \to \infty.$$

Hence,

$$\mathscr{L}^*\varphi(u) = \varphi(f), \qquad \varphi \in F^*, \mathscr{L}^*\varphi \in H^*,$$

i.e. u is a solution of (3.2) in the sense of Definition 3.6.   □

Remark 3.3. It is clear that solution in the sense of Definitions 3.6 and 3.7 coincide with a classic solution $u \in E$ if $f \in R(\mathscr{L})$. Also, it is easy to prove that the classical solution is a generalized solution, and if the generalized solution u belongs to $D(\mathscr{L})$, it is a classic solution.

Theorem 3.3. *If the space $\bar{E}$ is embedded into H, then Definitions 3.5, 3.6 and 3.7 are equivalent.*

Proof. Let us prove that Definition 3.5 is equivalent to Definition 3.7. Let $u \in H$ be a solution of (3.2) in the sense of Definition 3.7. As it was mentioned above, for any right-hand $f \in F$ there exists a unique solution $u^* \in \bar{E}$ in the sense of Definition 3.5. Let us prove that $u^* = u$ considering u^* to be the element of H (by virtue of the embedding $\bar{E} \subset H$). Indeed,

$$\|u - u^*\|_H \le \|u - u_i\|_H + \|u_i - u^*\|_H \le \|u - u_i\|_H + c_1^{-1}\|u_i - u^*\|_{\bar{E}} =$$
$$= \|u - u_i\|_H + c_1^{-1}\|\bar{\mathscr{L}}u_i - \bar{\mathscr{L}}u^*\|_F \to 0, \qquad i \to \infty,$$

where $u_i \in E$ is a sequence that converges to the solution $u \in H$.

Thus, u is a solution of (3.2) in the sense of Definition 3.5.

And vice versa, let u be a solution of (3.2) in the sense of Definition 3.5. Let us choose an arbitrary sequence $u_i \in E$ such that $\|u_i - u\|_{\bar{E}} \to 0$. Then, by virtue of the embedding $\bar{E} \subset H$ we have

$$\|u - u_i\|_H \to 0, \qquad \|u_i - u\|_{\bar{E}} = \|\mathscr{L}u_i - \bar{\mathscr{L}}u\|_F = \|\mathscr{L}u_i - f\|_F \to 0, \qquad i \to 0.$$

                                                                                    □

Remark 3.4. The theorem implies that there exist a constant $c > 0$ such that

$$\|u\|_H \leq c\|f\|_F, \qquad \forall f \in F, (c > 0),$$

where u is a solution of (3.2) with a right-hand side f in the sense of Definitions 3.5–3.7.

Remark 3.5. The embedding $\bar{E} \subset H$ follows either from the density of $D(\widetilde{\mathscr{L}^*})$ in the space F^* with respect to the weak topology $\sigma(F^*, F)$, either from the fact that the operator $\mathscr{L}$ is closable.

In many cases, inequalities (3.1) for the direct operator $\mathscr{L}$ immediately imply similar inequalities for the adjoint operator

$$c_1\|\varphi\|_G \leq \|\mathscr{L}^*\varphi\|_{E^*} \leq c_2\|\varphi\|_{F^*}, \qquad \forall \varphi \in F^*, \tag{3.5}$$

where G is a completion of the set F^* with respect to some norm. Consider this case for reflexive Banach spaces E, F. In this case, $\mathscr{L}^{**} = \mathscr{L}$ and similarly to Lemma 3.1 we have that the operator equation (3.2) is solvable over $G^* \subset F$. In addition, analogues of Theorems 3.1–3.3 for solvability of the adjoint equation (3.3) hold.

Theorem 3.4. *There is such a constant $c > 0$ that for any $f \in G^* \subset F$ and for any $l \in H^* \subset E^*$ the following inequalities are satisfied:*

$$\|u\|_E \leq c\|f\|_{G^*}, \tag{3.6}$$
$$\|\varphi\|_{F^*} \leq c\|l\|_{H^*}, \tag{3.7}$$

where $u \in E, \varphi \in F^*$ are the solutions of the equation $\mathscr{L}u = f$ and $\mathscr{L}^*\varphi = l$.

Proof. Let us prove inequality (3.6) (inequality (3.7) can be proved in a similar way). Since (3.3) is solvable (in the sense of analogues of Definitions 3.6 and 3.7 for the adjoint operator) for any $l \in E^*$, then for any $u \in E : \mathscr{L}u \in G^*$ the following equality holds (the second conjugate space is identified with the original space)

$$\mathscr{L}u(\varphi) = u(l), \tag{3.8}$$

where $\varphi \in G$ is a solution of (3.3) with the right-hand side $l \in E^*$. Hence,

$$|u(l)| \leq \|\mathscr{L}u\|_{G^*} \times \|\varphi\|_G$$

or

$$\left|\left(\frac{u}{\|\mathscr{L}u\|_{G^*}}\right)(l)\right| \leq \|\varphi\|_G.$$

Thus, the set of functionals

$$\left\{\frac{u}{\|\mathscr{L}u\|_{G^*}} : \mathscr{L}(u) \in G^*, u \in E\right\} \subset E^{**} = E$$

is bounded in any point $l \in E^*$, and therefore, by virtue of Banach–Steinhaus Theorem, it is bounded with respect to the norm of the space $E^{**} = E$, and therefore inequality (3.6) is proved. □

Theorem 3.5. *There exists such a constant $c > 0$ that for any $f \in F$ and for any $l \in E^*$ the following inequalities are satisfied:*

$$\|u\|_H \leq c\|f\|_F, \tag{3.9}$$

$$\|\varphi\|_G \leq c\|l\|_{E^*}, \tag{3.10}$$

where $u \in H$ and $\varphi \in G$ are the solutions of the equations $\mathscr{L}u = f$ and $\mathscr{L}^*\varphi = l$ in the sense of Definition 3.6 and 3.7.

Proof. Reasoning from inequality (3.8), we have

$$|\varphi(\mathscr{L}u)| = |\mathscr{L}u(\varphi)| \leq \|u\|_E \times \|l\|_{E^*}.$$

Applying inequality (3.6) to the right-hand side of inequality, we have

$$|\varphi(\mathscr{L}u)| \leq c\|\mathscr{L}u\|_{G^*} \times \|l\|_{E^*}$$

or

$$\|\varphi\|_G = \|\varphi\|_{G^{**}} = \sup_{\mathscr{L}u \in G^*} \frac{|\varphi(\mathscr{L}u)|}{\|\mathscr{L}u\|_{G^*}} \leq c\|l\|_{E^*}.$$

Taking into account the fact that $G^* \subset R(\mathscr{L})$, we obtain inequality (3.10). Inequality (3.9) is proved in a similar way. □

Remark 3.6. Since $\|f\|_F = \|\mathscr{L}u\|_F = \|u\|_{\bar{E}}$ and inequality (3.9) holds for any $f \in F$, it may seem that (3.9) guarantees the embedding $\bar{E} \subset H$. But it is not true. Indeed, in the space $\bar{E}$ there exists an element u^* such that $\|u^*\|_{\bar{E}} = \|f\|_F$. But until the embedding is not proved $\bar{E} \subset H$, it is impossible to compare $u \in H$ as a solution of (3.2) in the sense of Definitions 3.6 and 3.7, and $u^* \in \bar{E}$ as a solution of (3.2) in the sense of Definition 3.5.

3.3 A Generalized Solution in Locally Convex Linear Topological Spaces

Let us introduce several more definitions of a generalized solution. As before, let us suppose that $\mathscr{L}$ is a linear continuous operator. Let us select in the space E^* a total linear set $M \subset R(\mathscr{L}^*) \subset E^*$. Let $\tilde{M}$ be a completion of the set E with respect to the topology $\sigma(E, M)$. By virtue of the Banach theorem on a weakly continuous linear functional, the functional $l = \mathscr{L}^*\varphi \in M$ admits a unique extension by continuity onto the entire space $\tilde{M}$.

Definition 3.8. *A generalized solution* of (3.2) *is an element* $u \in \tilde{M}$ *satisfying the relation*

$$\mathscr{L}^* \varphi(u) = \varphi(f), \qquad \forall \varphi \in F^*, \mathscr{L}^* \varphi \in M.$$

Definition 3.9. *A generalized solution* of (3.2) *is an element* $u \in \tilde{M}$ *for which there exists a sequence* $u_i \in E$ *such that*

$$u_i \to u \text{ in topology } \tilde{M}, \qquad \|\mathscr{L}u_i - f\|_F \to 0, \qquad i \to \infty.$$

If $M = R(\mathscr{L}^*)$, Definitions 3.8 and 3.9 turn into Definitions 2.3 and 2.2 of a weak solution and a near-solution, respectively.

It is easy to prove the following theorem.

Theorem 3.6. *Definitions 3.8 and 3.9 are equivalent and for any element* $f \in F$ *there exists a generalized solution* $u \in \tilde{M}$ *of* (3.2) *in the sense of Definitions 3.8 and 3.9.*

Remark 3.7. Definition 3.9 implies that if $u \in \tilde{M}$ is a generalized solution of (3.2), then the point (u, f) is an adherent point of graphs $\Gamma(\mathscr{L})$ of the operator $\mathscr{L}$, i.e. an adherent point of the set

$$\{(u, f) : \mathscr{L}u = f, u \in E\} \subset \tilde{M} \times F$$

with respect to the topology $\tilde{M} \times F$.

Moreover, the point (u, f) belongs to a sequential closure of the graph $\Gamma(\mathscr{L})$ with respect to the topology $\tilde{M} \times F$ and hence $u \in \tilde{M}_s$, where $\tilde{M}_s$ is a sequential closure of the set E with respect to the topology $\tilde{M}$.

Remark 3.8. Taking into account Remark 3.7, we can define a generalized solution in sequentially complete spaces $\tilde{M}_s$.

In addition, it is easy to prove that a classic solution (3.2) is a generalized solution in the sense of Definitions 3.8 and 3.9. If $f \in R(\mathscr{L})$ or a generalized solution belongs to $D(\mathscr{L})$, then the generalized solution turn into the classic solution.

Since $M \subset R(\mathscr{L}^*)$, then there exists such a set $M_F \subset F^*$, that $M = \mathscr{L}^*(M_F)$.

Theorem 3.7. *An embedding of the space* $\bar{E}_{\mathscr{L}}$ *into the space* $\tilde{M}$ *exists iff* M_F *is a total linear subset of* F^*. *In this case, Definitions 3.8, 3.9, and 3.5 are equivalent.*

Proof. Similar to Theorem 2.3, it can be proved that in set E the topology induced by the norm $\|\cdot\|_{\bar{E}_{\mathscr{L}}}$ is stronger that the topology $\sigma(E, M)$. The condition π is considered similarly. We have that $\varphi(\bar{\mathscr{L}}u) = 0$ for any $\varphi \in M_F$. The fact that the operator $\bar{\mathscr{L}}$ sets an isometric isomorphism between the spaces $\bar{E}_{\mathscr{L}}$ and F implies that the condition $u = 0$ is equivalent to the totality of the set M_F. The equivalence of the definitions is proved similar to Theorem 2.4. $\square$

Remark 3.9. Theorem 3.7 implies that the set $\tilde{M}$ can be defined as a completion of the set $\bar{E}_{\mathscr{L}}$ with respect to the topology $\sigma(E, M)$ (under the conditions of the theorem).

Remark 3.10. The condition of totality of M_F is of fundamental importance and does not holds always.

Let us consider an example of a linear continuous injective operator which maps not total sets into total sets. Indeed, the operator $\mathscr{A} : \ell_2 \to \ell_2$, which acts on the vector $x = (\xi_1, \xi_2, \ldots)$ by the rule

$$\mathscr{A}x = \left(\frac{\xi_1}{2^0}, \frac{\xi_2}{2^1}, \frac{\xi_3}{2^2}, \ldots \right).$$

is linear, continuous (and even totally continuous), and injective. Also, it maps the vector system

$$g_1 = (1, 2, 0, 0, 0, \ldots),$$
$$g_2 = (0, 2, 4, 0, 0, \ldots),$$
$$g_3 = (0, 0, 4, 8, 0, \ldots),$$

$$\cdots$$

into the system

$$e_1 = (1, 1, 0, 0, 0, \ldots),$$
$$e_2 = (0, 1, 1, 0, 0, \ldots),$$
$$e_3 = (0, 0, 1, 1, 0, \ldots),$$

$$\cdots$$

The vector system $\{g_i\}$ is orthogonal to the vector

$$x^* = \left(1, -\frac{1}{2}, \frac{1}{4}, -\frac{1}{8}, \ldots \right)$$

and, hence, it is not total, but from the other hand, the totality of the system $\{e_i\}$ in ℓ_2 is obvious.

In addition, if $\mathscr{L}$ satisfies a priori inequalities (3.1), then the following theorem holds.

Theorem 3.8. *The Banach space H is embedded into the space $\widetilde{H}^*$. Definitions 3.8, 3.9 (for H^*), and 3.6, 3.7 are equivalent.*

Proof. It is clear that the topology induced by the norm $\| \cdot \|_H$ is stronger than the topology $\sigma(E, H^*)$. Let us test the condition π. Let a sequence u_n be convergent to u with respect to the topology of the space H and $u_n \to 0$ with respect to the topology $\sigma(E, H^*)$. Then Hahn–Banach Theorem implies that the norm $\| \cdot \|_H$ can be represented as

$$\|u_n - u\|_H = \sup_{l \in H^*} \frac{|l(u_n - u)|}{\|l\|_{H^*}}.$$

Hence, $l(u_n - u) \to 0$ as $n \to \infty$; therefore $l(u_n) \to l(u)$ for all $l \in H^*$. On the other hand, $l(u_n) \to 0$ for all $l \in H^*$. Thus, $l(u) = 0$ for all $l \in H^*$, and therefore, $u = 0$. □

3.4 Relation Between Generalized Solutions in Banach and Locally Convex Spaces

Note that there is some kind of analogy between the schemes of the construction of generalized solutions in the sense of Definitions 3.6 and 3.7 and the schemes of the construction of generalized solutions in the sense of Definitions 3.8 and 3.9.

Just as we construct a Hausdorff locally convex linear topological space $\widetilde{M}$ by a total set $M \subset R(\mathscr{L}^*) \subset E^*$, we can construct a Banach space $\bar{M}$, which defines a generalized solution in the sense of Definitions 3.6 and 3.7, by the total set $M \subset R(\mathscr{L}^*) \subset E^*$. Namely, let $\bar{M}$ be a completion of the set E with respect to the norm

$$\|u\|_{\bar{M}} = \sup_{\mathscr{L}^*\varphi \in M} \frac{|\mathscr{L}^*\varphi(u)|}{\|\varphi\|_{F^*}}. \tag{3.11}$$

Norm (3.11) can be rewritten as

$$\|u\|_{\bar{M}} = \sup_{\varphi \in M_F} \frac{|\varphi(\mathscr{L}u)|}{\|\varphi\|_{F^*}}. \tag{3.12}$$

From the other hand, if $M = R(\mathscr{L}^*)$, then the norm $\|u\|_{\bar{M}}$ coincides with the norm of the space $\bar{E}_{\mathscr{L}}$ and, hence, $\bar{M} = \bar{E}_{\mathscr{L}}$ (an analogue of the space $\widetilde{E}_{\mathscr{L}}$). Indeed,

$$\|u\|_{\bar{M}} = \sup_{\mathscr{L}^*\varphi \in M} \frac{|\mathscr{L}^*\varphi(u)|}{\|\varphi\|_{F^*}} = \sup_{\varphi \in F^*} \frac{|\varphi(\mathscr{L}u)|}{\|\varphi\|_{F^*}} = \|\mathscr{L}u\|_F.$$

This equality holds because, by virtue of Hahn–Banach Theorem, for any element $\mathscr{L}u \in F$ there exists such a functional $\varphi \in F^*$ with unit norm that $\varphi(\mathscr{L}u) = \|\mathscr{L}u\|_F$.

In addition, we have the following lemma.

Lemma 3.2. *If $M \subset R(\mathscr{L}^*)$ and M_F is a total subset of the space F^*, then the space $\bar{E}_{\mathscr{L}}$ is embedded into the space $\bar{M}$.*

Proof. The totality of the set M_F and the injectivity of the operator $\mathscr{L}$ imply the totality of the set M. It is easy to see that the norm $\|\cdot\|_{\bar{E}_{\mathscr{L}}}$ is stronger that the norm $\|\cdot\|_{\bar{M}}$. It remains only to test the condition π. Let a sequence $u_n \in E$ be convergent to $u \in \bar{E}_{\mathscr{L}}$ with respect to the norm $\|u\|_{\bar{E}_{\mathscr{L}}}$ and u_n be convergent to zero with respect to the norm $\|u\|_{\bar{M}}$, then, from the one hand, $\varphi(\mathscr{L}u_n) \to \varphi(\bar{\mathscr{L}}u)$, and from the other hand, $\varphi(\mathscr{L}u_n) \to 0$ for any $\varphi \in M_F$. Hence, $\varphi(\bar{\mathscr{L}}u) = 0$ for all $\varphi \in M_F$. The totality of M_F and the injectivity of $\mathscr{L}$ imply that $u = 0$. □

Remark 3.11. Theorem implies that the space $\bar{M}$ can be constructed by completing the set $\bar{E}_{\mathscr{L}}$ with respect to (3.11).

Thus, we can obtain an a priori estimate by any total set $M \subset R(\mathscr{L}^*)$ (and if M_F is total then the embedding $\bar{E}_{\mathscr{L}} \subset \bar{M}$ also)

$$c_1 \|u\|_{\bar{M}} \leq \|\mathscr{L}u\|_F \leq c_2 \|u\|_E, \qquad \forall u \in E,$$

This embedding implies the statements on solvability of (3.2) (in the sense of analogues of Definitions 3.5–3.7) which are similar to Theorems 3.1–3.3.

It is also easy to see that a topology induced by the norm $\|\cdot\|_{\bar{M}}$ on the space E is stronger than a topology induced by a topology of the space $\tilde{M}$.

Thus, evolving the idea on relation between generalized solutions in Banach and locally convex linear topological spaces, we can describe the results of Sect. 3.2 in the style of Sect. 3.3 (proceeding from a total linear subset of the set $R(\mathscr{L}^*)$), and the results of Sect. 3.3 – in the style of Sect. 3.2 (using analogues of a priori inequalities, i.e. assuming that a Hausdorff locally convex topology defined on the set E is weaker than the norm $\|\cdot\|_{\bar{E}_{\mathscr{L}}}$).

Lemma 3.3. *The following equality holds:* $R(\mathscr{L}^*) = (\bar{E}_{\mathscr{L}})^*$, *where* $R(\mathscr{L}^*)$, $(\bar{E}_{\mathscr{L}})^*$ *are subsets of the space* E^*.

Proof. Let $\bar{\mathscr{L}} : \bar{E}_{\mathscr{L}} \to F$ be a completion of the operator $\mathscr{L}$ by continuity, then the operators $\bar{\mathscr{L}}^*$ and $\mathscr{L}^*$ set an isomorphism between the linear sets

$$\bar{\mathscr{L}}^* : F^* \longleftrightarrow (\bar{E}_{\mathscr{L}})^*, \qquad \mathscr{L}^* : F^* \longleftrightarrow R(\mathscr{L}^*)$$

Let us prove that the sets $R(\mathscr{L}^*)$ and $(\bar{E}_{\mathscr{L}})^*$ coincide each other as sets of the space E^*. Let $O : E \to \bar{E}_{\mathscr{L}}$ be a linear continuous operator, which defines a canonic embedding of the space E into $\bar{E}_{\mathscr{L}}$. For any functional $l \in (\bar{E}_{\mathscr{L}})^*$ there exists such an element $\varphi \in F^*$, that

$$O^*l(u) = l(Ou) = \bar{\mathscr{L}}^*\varphi(Ou) = \varphi(\bar{\mathscr{L}}(Ou)) = \varphi(\mathscr{L}u) = \mathscr{L}^*\varphi(u), \qquad \forall u \in E.$$

Hence, $R(O^*) = R(\mathscr{L}^*)$. Q.E.D. □

Remark 3.12. Identifying $R(\mathscr{L}^*)$ and $(\bar{E}_{\mathscr{L}})^*$, we can interpret the set $R(\mathscr{L}^*)$ as a Banach space $R_{\bar{E}_{\mathscr{L}}}(\mathscr{L}^*)$ with the norm

$$\|\bar{l}\|_{R_{\bar{E}_{\mathscr{L}}}(\mathscr{L}^*)} = \|\varphi\|_{F^*},$$

where $\mathscr{L}^*\varphi = \bar{l}$.

Indeed, using this identification, we have

$$\|\bar{l}\|_{R_{\bar{E}_{\mathscr{L}}}(\mathscr{L}^*)} = \|l\|_{(\bar{E}_{\mathscr{L}})^*} = \|\bar{\mathscr{L}}^*\varphi\|_{(\bar{E}_{\mathscr{L}})^*} = \sup_{u \in \bar{E}_{\mathscr{L}}} \frac{|\bar{\mathscr{L}}^*\varphi(u)|}{\|u\|_{\bar{E}_{\mathscr{L}}}} = \sup_{f \in F} \frac{|\varphi(f)|}{\|f\|_F} = \|\varphi\|_{F^*}.$$

Thus, the spaces $R_{\bar{E}_{\mathscr{L}}}(\mathscr{L}^*)$ and $(E_{\mathscr{L}})^*$ are isometrically isomorphic.

In conclusion of the theoretical part we note that many aspects of our analysis have the topological character. Therefore, we can study (3.2) in locally convex linear topological spaces (and, may be, in just topological) spaces E and F also. In this case, instead of $\bar{E}_{\mathscr{L}}$ we may consider a completion of E with respect to the topology induced by the system of semi-norms

$$p_{\alpha,\bar{E}_{\mathscr{L}}}(u) = p_{\alpha,F}(\mathscr{L}u), \qquad \alpha \in \mathfrak{A},$$

where $\{p_{\alpha,F}\}_{\alpha \in \mathfrak{A}}$ is a system of semi-norms, which induces the topology of the space F, and instead of estimations (3.1) we have a chain of dense embeddings

$$E \subset \bar{E}_{\mathscr{L}} \subset H, \qquad \forall u \in E,$$

where H is a completion of the set E with respect to some locally convex topology which is weaker than the norm of the space $\bar{E}_{\mathscr{L}}$.

Chapter 4
Applications of the Theory of Generalized Solvability of Linear Equations

4.1 Application to the Equations with Hilbert–Schmidt Operator

Let $L_2(-\pi, \pi)$ be the Hilbert space of measurable, square integrable, complex valued functions with the standard inner product $(\cdot, \cdot)_0$ and $\{e_k\}_{k=-\infty}^{\infty}$ be an orthonormal basis consisting of eigenvectors of an self-adjoint Hilbert–Schmidt operator

$$\mathscr{L}u = f, \qquad \mathscr{L} : L_2(-\pi, \pi) \to L_2(-\pi, \pi). \qquad (4.1)$$

Then

$$\mathscr{L}u = \sum_{k=-\infty}^{\infty} \lambda_k (u, e_k)_0 e_k, \qquad \sum_{k=-\infty}^{\infty} \lambda_k^2 < +\infty.$$

Let us denote by $\mathscr{E}$ the vector space of all infinitely differentiable numerical functions over $(-\pi, \pi)$. Let us consider on the set $\mathscr{E}$ a countable system of semi-norm

$$p_m(f) = \sup_{t \in [-\pi, \pi]} |f^{(m)}(t)|,$$

where $m \geq 0$ is an integer number, $f^{(m)}$ is the derivative of the function f of order m. Thus, the set $\mathscr{E}$ turns into a metrizable topological vector space.

Let us select in the space $\mathscr{E}$ a closed subset $\mathscr{E}_=$ which consists of functions satisfying the additional condition

$$f^{(m)}(-\pi) = f^{(m)}(\pi), \qquad \forall m \in \mathbb{N} \cup \{0\}.$$

Denote by $\mathscr{E}_=^*$ the conjugate space of $\mathscr{E}_=$ endowed with a weak-* topology $\sigma(\mathscr{E}_=^*, \mathscr{E}_=)$. $\bar{E}_{\mathscr{L}}$ means a completion of the space $L_2(-\pi, \pi)$ with respect to the norm

$$\|u\|_{\bar{E}_{\mathscr{L}}} = \|\mathscr{L}u\|_{L_2(-\pi, \pi)} = \sum_{k=-\infty}^{\infty} \lambda_k^2 |(u, e_k)_0|^2.$$

D.A. Klyushin et al., *Generalized Solutions of Operator Equations and Extreme Elements*, Springer Optimization and Its Applications 55, DOI 10.1007/978-1-4614-0619-8_4, © Springer Science+Business Media, LLC 2012

Lemma 4.1. *Let the basis e_k be a trigonometric system of functions $e_k = e^{ikt}$, $k \in \mathbb{Z}$ and the eigenvalues satisfy the estimation*

$$|\lambda_k| > \frac{c}{k^s},$$

for fixed constants $c > 0$, $s \geq 1$. Then the Banach space $\bar{E}_{\mathscr{L}}$ is densely embedded into the space of distributions $\mathscr{E}_=^*$ with the help of the operator of canonical embedding j (for an arbitrary function $x(t) \in \bar{E}_{\mathscr{L}}$ we have $(j(x))(f) = \int_{-\pi}^{\pi} f(t)x(t)\,dt$, where $j(x) \in \mathscr{E}_=^*$ and $f \in \mathscr{E}_=$).

Proof. To prove the lemma, it is enough to check whether the following statements are true:

1. The topology τ_0 induced by the norm $\|\cdot\|_{\bar{E}_{\mathscr{L}}}$ on the set $L_2(-\pi, \pi)$ is stronger than the topology τ_1 induced by the weak-* topology $\sigma(\mathscr{E}_=^*, \mathscr{E}_=)$ on the set $L_2(-\pi, \pi)$.
2. For the topologies τ_0, τ_1 the following condition is satisfied:

> π) if the sequence $\{u_n\}$, where $u_n \in L_2(-\pi, \pi)$, is a Cauchy sequence in the topology τ_0 and it converges to zero in the topology τ_1, then the sequence $\{u_n\}$ converges to zero in the topology τ_0 also.

Let us verify Condition 1. Let $\{u_n\}$ be a sequence of elements of the space $L_2(-\pi, \pi)$ which converges to zero in the topology τ_0, i.e.

$$\|u_n\|_{\bar{E}_{\mathscr{L}}} = \sum_{k=-\infty}^{\infty} \lambda_k^2 |(u_n, e_k)_0|^2 \to 0, \qquad n \to \infty.$$

Let us prove that u_n converges to zero in the topology τ_1 also:

$$\int_{-\pi}^{\pi} u_n(t)\varphi(t)\,dt = (u_n, \varphi)_0 \xrightarrow[n \to \infty]{} 0, \qquad \forall \varphi \in \mathscr{E}_=.$$

Indeed, expanding φ into a Fourier series and using the continuity of the inner product $(u, v)_0$, we have

$$|(u_n, \varphi)_0| = \left| \left(u_n, \sum_{k=-\infty}^{\infty} (\varphi, e_k)_0 e_k \right)_0 \right| = \left| \sum_{k=-\infty}^{\infty} (\varphi, e_k)_0 (u_n, e_k)_0 \right|$$

$$\leq \sum_{k=-\infty}^{\infty} |(\varphi, e_k)_0| \times |(u_n, e_k)_0|.$$

Further, we have

$$|(u_n, \varphi)_0| \leq \sum_{k=-\infty}^{\infty} \frac{|(\varphi, e_k)_0|}{\lambda_k} \times \lambda_k |(u_n, e_k)_0|$$

$$\leq \left(\sum_{k=-\infty}^{\infty} \frac{|(\varphi, e_k)_0|^2}{\lambda_k^2} \right)^{1/2} \times \left(\sum_{k=-\infty}^{\infty} \lambda_k^2 |(u_n, e_k)_0|^2 \right)^{1/2}$$

$$= \left(\sum_{k=-\infty}^{\infty} \frac{|(\varphi, e_k)_0|^2}{\lambda_k^2} \right)^{1/2} \times \|u_n\|_{\bar{E}_{\mathscr{L}}}.$$

If for a fixed φ

$$\sum_{k=-\infty}^{\infty} \frac{|(\varphi, e_k)_0|^2}{\lambda_k^2} < C < +\infty, \qquad (4.2)$$

then

$$0 \le |(u_n, \varphi)_0| \le \sqrt{C} \times \|u_n\|_{\bar{E}_{\mathscr{L}}} \xrightarrow[n\to\infty]{} 0.$$

Let us prove (4.2). Integrating by parts, it is easy to see that the coefficients

$$a_k = \left(\varphi(t), e^{ikt} \right)_0$$

of the Fourier series $\varphi \in \mathscr{E}_-$ satisfy the following relations

$$a_k = \int_{-\pi}^{\pi} \varphi(t) e^{ikt} \, dt = -\frac{1}{ik} \int_{-\pi}^{\pi} \varphi'(t) e^{ikt} \, dt = -\frac{1}{ik} b_k, \qquad (4.3)$$

where b_k are the Fourier coefficients of the function $\varphi'(t)$. It is well-known that $|b_k| \to 0$ as $k \to \infty$, and hence, $|ka_k| \to 0$.

Using the analogous formulae several times, it is easy to prove that

$$|k|^p |a_k| \xrightarrow[k\to\infty]{} 0, \qquad \forall p \in \mathbb{N}.$$

Hence,

$$\sum_{k=-\infty}^{\infty} k^{2p} |a_k|^2 < +\infty, \qquad \forall p \in \mathbb{N}.$$

Since $|\lambda_k| > c/k^s$,

$$\sum_{k=-\infty}^{\infty} \frac{|(\varphi, e_k)_0|^2}{\lambda_k^2} < \frac{1}{c^2} \sum_{k=-\infty}^{\infty} k^{2s} |a_k|^2 < C < +\infty.$$

Thus, Condition 1 is verified.

Let us verify the condition π). Let a sequence $\{u_n\} \subset L_2(-\pi, \pi)$ be a Cauchy sequence with respect to the norm $\|\cdot\|_{\bar{E}_{\mathscr{L}}}$, and in the topology τ_1 it converges to zero, i.e.

$$\mathscr{L} u_n \to f^* \text{ in the space } L_2(-\pi, \pi) \text{ and}$$
$$(u_n, \varphi)_0 \xrightarrow[n\to\infty]{} 0, \qquad \forall \varphi \in \mathscr{E}_-.$$

Since $\mathscr{L}$ is a Hermitian operator,

$$(\mathscr{L} u_n, \varphi)_0 = (u_n, \psi)_0, \qquad \psi = \mathscr{L} \varphi. \qquad (4.4)$$

Let us prove that $\psi = \mathscr{L} \varphi$ is an infinitely differentiable function. Having differentiated $\mathscr{L} \varphi$ formally and taking into account (4.3), we have

$$\psi'(t) = (\mathscr{L} \varphi)'(t) = \sum_{k=-\infty}^{\infty} \lambda_k a_k e_k'(t) = \sum_{k=-\infty}^{\infty} \lambda_k \frac{-b_k}{ik} e_k'(t) = -\sum_{k=-\infty}^{\infty} \lambda_k b_k e_k(t).$$

The correctness of differentiating follows from convergence of the series

$$\sum_{k=-\infty}^{\infty} \lambda_k b_k e_k(t).$$

Indeed, using the Hilbert–Schmidt condition

$$\left(\sum_{k=-\infty}^{\infty} \lambda_k^2 \right)^{1/2} < +\infty$$

and the Bessel inequality, we have

$$\left| \sum_{k=-\infty}^{\infty} \lambda_k b_k e_k(t) \right| \leq \sum_{k=-\infty}^{\infty} |\lambda_k b_k| \leq \left(\sum_{k=-\infty}^{\infty} \lambda_k^2 \right)^{1/2} \times \left(\sum_{k=-\infty}^{\infty} |b_k|^2 \right)^{1/2} < +\infty.$$

Thus, the function $\psi(t)$ is differentiable. Repeating these reasonings, we can prove that $\psi(t)$ is an infinitely differentiable function.

On the other hand, since the eigenvectors of the basis $\{e_k\}$ belong to the space $\mathscr{E}_=$, then

$$\psi = \mathscr{L}\varphi = \sum_{k=-\infty}^{\infty} \lambda_k (\varphi, e_k)_0 e_k \in \mathscr{E}_=.$$

Thus, proceeding to the limit in (4.4) as $n \to \infty$, we have

$$(f^*, \varphi)_0 = 0, \qquad \forall \varphi \in \mathscr{E}_=.$$

Taking into account the fact that the set $\mathscr{E}_=$ is dense in the space $L_2(-\pi, \pi)$, we conclude that $f^* = 0$, so the condition π) holds.   □

The embedding $\bar{E}_{\mathscr{L}} \subset \mathscr{E}_=^*$ implies the existence and uniqueness of a generalized solution.

Theorem 4.1. *It follows from Lemma 4.1 that there exists a unique generalized solution of (4.1) in the sense of Definitions 3.8, 3.9 ($M = \mathscr{E}_=$).*

Note that the theorem proved above can be interpreted in the sense of Sect. 3.3. Indeed, the space $\mathscr{E}_=^*$ is induced by the set $M = \mathscr{E}_= \subset (L_2(-\pi, \pi))^*$. It is just necessary to verify the embedding $\mathscr{E}_= \subset R(\mathscr{L}^*)$ or, taking into account the self-adjointness of the operator, the embedding $\mathscr{E}_= \subset R(\mathscr{L})$. To prove the latter inclusion, we should suppose that in $R(\mathscr{L})$ a complete system of functions exists a priori (in the example considered above this system consisted of the functions e^{ikt}). Selecting other systems we can get other results. For example, for the integral operator

$$\mathscr{L}u = \int_0^1 K(t, s)u(t)\,dt, \qquad (4.5)$$

which acts in the real Hilbert space $L_2(0, 1)$, we can prove the following theorem.

Theorem 4.2. *If*

(1)There exist functions $f_n \in L_2(0,1)$ such that

$$\mathscr{L}^* f_n = \int_0^1 K(t,s) f_n(s) \, ds = \sqrt{2} \sin \pi n t,$$

(2)There exist fixed constants $c > 0, p$ such that

$$\|f_n\|_{L_2(0,1)} \leq c n^p, \qquad \forall n = 0,1,\dots,$$

then (4.5) has a unique solution in the sense of Definitions 3.8 and 3.9 (M = $\mathscr{E}_-(0,1)$).

Remark 4.13. By virtue of the Mercer Theorem it is possible to construct integral Hilbert–Schmidt operators with the kernel

$$K(t,s) = \sum_{n=-\infty}^{\infty} \lambda_n e^{int} e^{-ins}, \qquad |\lambda_k| > \frac{c}{k^s},$$

which satisfy the conditions of Lemma 4.1. In the similar way we can construct kernels satisfying the conditions of Theorem 4.2. An example of such operator is

$$\mathscr{L}\varphi = \int_0^1 K(x,t)\varphi(t) \, dt,$$

where

$$K(x,t) = \begin{cases} (1-x)t, & 0 \leq t \leq x, \\ x(1-t), & x \leq t \leq 1. \end{cases}$$

The eigenvalues of this kernel are

$$\lambda_1 = \pi^2, \lambda_2 = (2\pi)^2, \dots, \lambda_n = (n\pi)^2, \dots,$$

and the corresponding eigenfunctions are

$$\varphi_1 = \sqrt{2} \sin \pi x, \varphi_2 = \sqrt{2} \sin 2\pi x, \dots, \varphi_n = \sqrt{2} \sin n\pi x, \dots.$$

Note that in general it is not a rule that the generalized solution of (4.1) is a generalized function even from $\mathscr{D}^*(-\pi, \pi)$ at least, where $\mathscr{D}^*(-\pi, \pi)$ is the conjugate space of $\mathscr{D}(-\pi, \pi)$, which is the space of finite and infinitely differentiable functions on $(-\pi, \pi)$ with a standard topology. Indeed, consider the integral operator

$$\mathscr{L}u = \int_{-\pi}^{\pi} K(t,s) u(t) \, dt,$$

with the kernel

$$K(t,s) = K(|t-s|) = K(\tau),$$
$$K(\tau) = \begin{cases} 0, & \tau \leq 0, \\ \tau, & \tau > 0. \end{cases}$$

This operator satisfies the Hilbert–Schmidt condition

$$\int_{-\pi}^{\pi}\int_{-\pi}^{\pi} K^2(t,s)\,dt\,ds < +\infty.$$

Let us prove that this operator is injective. We have

$$f = \mathscr{L}u = \int_{-\pi}^{s} K(|t-s|)u(t)\,dt + \int_{s}^{\pi} K(|t-s|)u(t)\,dt$$

$$= \int_{-\pi}^{s} K(s-t)u(t)\,dt + \int_{s}^{\pi} K(t-s)u(t)\,dt$$

$$= \int_{s}^{\pi} K(t-s)u(t)\,dt.$$

Differentiating the function f, we have $f'' = -u$, whence it follows that the operator $\mathscr{L}$ is injective.

On the other hand, it is easy to verify, that a solution of (4.1) with the right-hand side

$$f(s) = \begin{cases} s(\ln s - 1) + 1, & t \in (0,\pi], \\ 0, & t \in [-\pi, 0] \end{cases}$$

is the function

$$u(t) = \begin{cases} 1/t, & t \neq 0, t \in [-\pi, \pi], \\ 0, & t = 0, \end{cases}$$

which is not locally integrable, and hence it does not belong to the space $\mathscr{D}^*(-\pi, \pi)$.

Let us show that in general the space $S(-\pi, \pi)$ which consists of measurable functions on $(-\pi, \pi)$ does not cover the space $\bar{E}_{\mathscr{L}}$. Indeed, considering (4.1) with the first-order integral Fredholm operator with a kernel $K(t,s)$ which is square integrable, we have that the sequence of functions

$$\left\{ u_n(t) = e^{i(-1)^n[n/2]t} \right\}_{n=1}^{+\infty}$$

does not converge to zero with respect to the metric of space $S(-\pi, \pi)$, i.e. it does not converge to zero with respect to the Lebesgue measure. On the other hand, the sequence u_n converges to zero in the space $\bar{E}_{\mathscr{L}}$. The sequence $\{u_n(t)\}_{n=1}^{+\infty}$ forms an orthonormal basis in the space $L_2(-\pi, \pi)$, and hence it converges weakly in the space $L_2(-\pi, \pi)$. Since the function $k_s(t) = K(t,s)$ is square integrable for almost all s, then

$$f_n = \mathscr{L}u_n = \int_{-\pi}^{\pi} K(t,s)u_n(t)\,dt = \left(\bar{k}_s, u_n\right)_0 \to 0, \qquad n \to \infty.$$

for almost all $s \in (-\pi, \pi)$.

Further, we have

$$|f_n(s)|^2 = \left| \int_{-\pi}^{\pi} K(t,s)u_n(t)\,\mathrm{d}t \right|^2 \leq \left(\int_{-\pi}^{\pi} |K(t,s)|^2\mathrm{d}t \right) \times \left(\int_{-\pi}^{\pi} |u_n(t)|^2\mathrm{d}t \right)$$

$$= \int_{-\pi}^{\pi} |K(t,s)|^2\mathrm{d}t.$$

Since $K(t,s)$ is a square summable function on the set $(-\pi,\pi) \times (-\pi,\pi)$, then by virtue of the Fubini theorem the function

$$g(s) = \int_{-\pi}^{\pi} |K(t,s)|^2\mathrm{d}t,$$

is a summable function on $(-\pi,\pi)$. Thus, the sequence $|f_n|^2$ converges to zero almost everywhere, and $|f_n(s)|^2 \leq g(s)$, where $g(s)$ is a summable function. Then by virtue of Lebesgue theorem

$$\int_{-\pi}^{\pi} |f_n(s)|^2\mathrm{d}s \to 0, \qquad n \to \infty.$$

i.e. $u_n \to 0$ in $\bar{E}_{\mathscr{L}}$. Thus, the embedding $\bar{E}_{\mathscr{L}} \subset S(-\pi,\pi)$ does not exist and hence there exists a solution of (4.1), which does not belong to the space $S(-\pi,\pi)$.

Note that the development of the sufficiently general theory of solvability of (4.1) (including the proof of the embedding $\bar{E}_{\mathscr{L}} \subset H$, where H is a known space) is rather difficult. Indeed, let M be an arbitrary total linear subset of the space $L_2(-\pi,\pi)$. Let us consider the class $\mathscr{M}$ of operators of the form (4.1) which satisfy the condition $M \subset R(\mathscr{L}^*)$. Then it follows from Sect. 3.3 that it is reasonable to try to prove the embedding $\bar{E}_{\mathscr{L}}$ into $\tilde{M}$ for all $\mathscr{L} \in \mathscr{M}$. However, it can be shown that in M there exist linear total subsets M^+, M^-, such that for some operator $\mathscr{L}_M \in \mathscr{M}$ the embedding $\bar{E}_{\mathscr{L}_M} \subset \tilde{M}^+$ exists, but the embedding $\bar{E}_{\mathscr{L}_M} \subset \tilde{M}^-$ does not exist. (Note that, nevertheless, the space $\bar{E}_{\mathscr{L}_M}$ induces on $L_2(-\pi,\pi)$ a stronger topology than the topology in $\tilde{M}^-$). Thus, to prove the embedding of $\bar{E}_{\mathscr{L}}$ into $\tilde{M}$ it is necessary to use inherent properties of the set M.

Let us show how to select the sets M^-, M^- and the operator $\mathscr{L}_M$. The density of the set M in the space $L_2(-\pi,\pi)$ implies the fact that in the space $L_2(-\pi,\pi)$ it is possible to select an orthonormal basis $\{t_k\}_{k=1}^{\infty}$ from the vectors of M. Let us consider the operator $\mathscr{L}_M : L_2(-\pi,\pi) \to L_2(-\pi,\pi)$

$$\mathscr{L}_M u = \sum_{k=1}^{\infty} \lambda_k (u,t_k) \circ t_k, \qquad \sum_{k=1}^{\infty} \lambda_k^2 < +\infty,$$

which is an injective linear and continuous Hilbert–Schmidt operator. Extending the example from Remark 3.10, let us consider the system of vectors

$$g_1 = \frac{t_1}{\lambda_1} + \frac{t_2}{\lambda_2},$$

$$g_2 = \frac{t_2}{\lambda_2} + \frac{t_3}{\lambda_3},$$

$$g_3 = \frac{t_3}{\lambda_3} + \frac{t_4}{\lambda_4},$$

$$\dots$$

The operator $\mathscr{L}_M$ maps this system to the following system

$$s_1 = t_1 + t_2,$$
$$s_2 = t_2 + t_3,$$
$$s_3 = t_3 + t_4,$$

$$\dots$$

The system of vectors $\{g_i\}$ is orthogonal to the vector

$$g^* = \sum_{k=1}^{\infty} (-1)^{k+1} \lambda_k t_k$$

and hence it is not total (and the linear span of the system $G = \text{l.h.}\{g_i\}$ is not dense in $L_2(-\pi, \pi)$), and the totality of the system $\{s_i\}$ in the space L_2 is clear. Let $M^- = \text{l.s.}\{s_i\}$ be a linear span of the system of vectors s_i. So, it is obvious that $M^- \subset M$, M^- is dense in $L_2(-\pi, \pi)$ and $\mathscr{L}_M^{-1}(M^-) = G$. Theorem 3.7 implies that the embedding $\bar{E}_{\mathscr{L}_M} \subset \tilde{M}^-$ does not exist.

The set M^+ can be selected in the following way. Let $M^+ = \text{l.s.}\{t_i\}$, then the set $\mathscr{L}_M^{-1}(M^+) = M^+$ is total and hence, by virtue of Theorem 3.7, there exists the embedding $\bar{E}_{\mathscr{L}_M} \subset \tilde{M}^+$.

4.2 Generalized Solutions of an Infinite System of Linear Algebraic Equations

Let us consider several examples of generalized solutions of operator equations with a bounded linear operator $\mathscr{L}$, which acts in the Hilbert space

$$\ell_2 = \left\{ x = (x_1, \dots, x_k, \dots), \ \sum_{k=1}^{\infty} |x_k|^2 < \infty \right\}$$

with the inner product

$$(x, y) = \sum_{k=1}^{\infty} x_k y_k, \ y = (y_1, \dots, y_k, \dots).$$

As well-known [114], every such operator $\mathscr{L}$ is defined by an infinite matrix, which we will denote by the same letter $\mathscr{L}$ (and we will identify it with the operator $\mathscr{L}$):

$$\mathscr{L} = \begin{pmatrix} a_{11} & a_{12} & \cdots & a_{1n} & \cdots \\ a_{21} & a_{22} & \cdots & a_{2n} & \cdots \\ \cdot & \cdot & \cdots & \cdot & \cdots \\ a_{n1} & a_{n2} & \cdots & a_{nn} & \cdots \\ \cdot & \cdot & \cdots & \cdot & \cdots \end{pmatrix} \tag{4.6}$$

The element $f = \mathscr{L}u$ is represented as the product of the matrix $\mathscr{L}$ on the column-vector u from $\ell_2 : f = \mathscr{L}u$. If the matrix $\mathscr{L}$ satisfies the Hilbert–Schmidt condition

$$\sum_{i,j=1}^{\infty} |a_{ij}|^2 < \infty, \tag{4.7}$$

then $\mathscr{L}$ is a compact operator, which acts in the space ℓ_2 [114]. The domain $D(\mathscr{L})$ of a bounded linear operator $\mathscr{L} : \ell_2 \to \ell_2$ coincides with the space ℓ_2. In order to apply our theory of generalized solutions we have to clarify under which conditions on $\mathscr{L}$ its range $R(\mathscr{L})$ is everywhere dense in ℓ_2 and (2.1) is uniquely solvable (i.e the kernel $\mathrm{Ker}(\mathscr{L})$ of the operator $\mathscr{L}$ consists of only zero element $\mathrm{Ker}(\mathscr{L}) = \{\theta\}$). For this purpose, let us put $a_i = (a_{i1}, a_{i2}, \ldots, a_{in}, \ldots)$; so, inequality (4.7) implies $a_i \in \ell_2$, $i = 1, 2, \ldots$, therefore, each row a_i of the matrix $\mathscr{L}$ can be considered as the element of ℓ_2. Then

$$f = \mathscr{L}u, f = (f_1, \ldots, f_n, \ldots), u = (u_1, \ldots, u_n, \ldots),$$

$$f_i = \sum_{j=1}^{\infty} a_{ij} u_j = (a_i, u), (i = 1, 2, \ldots).$$

It is easy to see, that $\mathrm{Ker}(\mathscr{L}) = \{\theta\}$ iff the system of the elements

$$\mathfrak{R} = (a_1, a_2, \ldots, a_n, \ldots)$$

is total in ℓ_2 (or, saying in terms of Hilbert spaces theory, it is complete or closed): if $(a_i, u) = 0$ for any $i \in \mathbb{N}$, then $u = \theta$. To study whether $R(\mathscr{L})$ is dense in ℓ_2 let us introduce the following definition.

Definition 4.10. We will say that the infinite matrix (4.6) which satisfies the Hilbert–Schmidt condition (4.7) is a *matrix with sparse rows* if any row a_i of the matrix $\mathscr{L}$ does not belong to closed linear subspace

$$\bar{L}_i = \overline{L(a_1, \ldots, a_{i-1}, a_{i+1}, \ldots)},$$

induced by the other vectors $(a_1, \ldots, a_{i-1}, a_{i+1}, \ldots)$ of the system $\mathscr{L}$ with respect to the metric $\ell_2 : a_i \notin \bar{L}_i$ for all $i \in \mathbb{N}$.

Let us denote by $e_k (k \in \mathbb{N})$ the unit vectors in the space ℓ_2:

$$e_1 = (1, 0, 0, \ldots), e_2 = (0, 1, 0, \ldots), \ldots, e_k = (0, \ldots, 0, 1, 0, \ldots), \ldots$$

and let us put $B = \{e_k\}_{k=1}^{\infty}$. Let us show that the system of unit vectors B is contained in the range $R(\mathscr{L})$ of the operator $\mathscr{L}$ iff matrix (4.6) has sparse rows. Indeed, if $a_i \notin \bar{L}_i$ for all $i \in \mathbb{N}$, then $\bar{L}_i \neq \ell_2$ and there exists an element $u_i \neq \theta$, which is orthogonal to the closed subspace $\bar{L}_i : u_i \perp \bar{L}_i$. It is easy to see that $c_i = (a_i, u_i) \neq 0$, because otherwise $f_i = \mathscr{L}(u_i) = ((a_1, u_1), \ldots, (a_n, u_i), \ldots) = (0, 0, \ldots)$, but this contradicts the fact that the operator $\mathscr{L}$ is injective. Supposing that $\bar{u}_i = u_i/c_i$, we have $\mathscr{L}(u_i) = e_i$. Thus, $e_i \in R(\mathscr{L})$ for all $i \in \mathbb{N}$ and $B \subset R(\mathscr{L})$. On the other hand, if $B \subset R(\mathscr{L})$, then for any natural number i there exists such an element u_i that $e_i = \mathscr{L}u_i$. The latter inequality means that $(a_i, u_i) = 1$ and $(a_k, u_i) = 0$ for all $k \neq i$, and hence $(g, u_i) = 0$ for all $g \in \bar{L}(a_1, \ldots, a_{i-1}, a_{i+1}, \ldots) = \bar{L}_i$, i.i. $u_i \perp \bar{L}_i$. This implies that $a_i \notin \bar{L}_i$, therefore $\mathscr{L}$ is a matrix with sparse rows. It is easy to see, that every injective bounded linear operator $\mathscr{L}$ in ℓ_2 which is defined by a matrix $\mathscr{L}$ with sparse rows has an everywhere dense range $R(\mathscr{L})$. Indeed, for such operator we have $B \subset R(\mathscr{L})$; since $R(\mathscr{L})$ is a linear manyfold, then the vector space $L(B)$ induced by unit vectors is contained in $R(\mathscr{L})$ also, and since $L(B)$ is everywhere dense in ℓ_2, then the completion of $R(\mathscr{L})$ coincides with ℓ_2. The concept of a matrix with sparse rows is an extension of classical concept of nonsingular matrix in case of infinite matrices, since a finite matrix is non-singular iff it has sparse rows. There are many examples of matrices with sparse rows: diagonal matrices $\mathscr{L} = \{a_{ij}\}_{i,j=1}^{\infty}$, where $a_{ij} = 0$ for all $i \neq j$, $a_{ii} \neq 0$, triangular matrices $\mathscr{L} = \{a_{ij}\}_{i,j=1}^{\infty}$, where $a_{ij} = 0$ for $i > j$ (or $i < j$) with the non-zero diagonal elements, unitary matrices an so on.

The problem of solving an operator equation $\mathscr{L}u = f$ in ℓ_2 with a bounded linear operator $\mathscr{L}$ is equivalent to the problem of solving an infinite system of linear algebraic equations

$$\sum a_{ij}u_j = b_i, \quad u = (u_1, \ldots, u_n, \ldots), b = (b_1, \ldots, b_n, \ldots) \in \ell_2\, i = 1, 2, \ldots \quad (4.8)$$

In order to find a generalized solution of (4.8) let us consider the space s of all numerical sequences with the metric

$$\rho(x, y) = \sum_{n=1}^{\infty} \frac{1}{2^n} \frac{|x_n - y_n|}{1 + |x_n - y_n|},$$

where $x = (x_1, \ldots, x_n, \ldots)$, $y = (y_1, \ldots, y_n, \ldots) \in s$. As already known [114], s is a complete linear metric space, the conjugate space s^* consists of functionals $\varphi(x) = \sum_{k=1}^{n} \varphi_k x_k$, and the convergence in the metric of s is equivalent to coordinatewise convergence. This implies that the convergence of the sequence $\{x_n\} \subset s$ with respect to the weak topology $\sigma(s, s^*)$ is equivalent to the coordinatewise convergence also. Therefore, in s the weak convergence and the convergence in the metric ρ are equivalent. Remember [40] that a sequence $\{x_n\}$ is a weakly Cauchy sequence if for any $\varphi \in E^*$ there exists a finite limit $\lim \varphi(x_n)$. The space is weakly sequentially complete if every weakly Cauchy sequence weakly converges to an element of E. If $\{x_n = (x_1^{(n)}, \ldots, x_i^{(n)}, \ldots)\}$ is a weakly Cauchy sequence in s, then selecting φ as $\varphi_i(x) = x_i$, we have that the coordinate sequence $\varphi_i(x_n) = x_i^{(n)}$ is convergent, and hence, the sequence $\{x_n\}$ is coordinatewise convergent. Whence it follows that the sequence $\{x_n\}$ is weakly convergent to an element of s, and hence, s is a weakly sequentially complete space.

It is easy to see that the Hilbert space ℓ_2 is densely embedded into the space $s : \ell_2 \subset s$ [40]. The analysis of Definitions 1 and 2 of a generalized solution shows that it belongs to weak sequential closure E^S of a vector space E in a Hausdorff locally convex topological vector space $\bar{E}$ (i.e., in a completion of E with respect to the topology $\sigma(E, R(\mathscr{L}^*))$). If the range $R(\mathscr{L}^*)$ of the adjoint operator $\mathscr{L}^* : \ell_2 \to \ell_2$ contains the unit vectors e_k, then $s^* = L(B) \subset R(\mathscr{L})$; therefore, the topology $\sigma(\ell_2, R(\mathscr{L}))$ is stronger than the topology $\sigma(\ell_2, L(B)) = \sigma(\ell_2, s^*)$. Let $\bar{u}$ be an arbitrary element of weak sequential closure ℓ_2^s and $\{u_n\}$ be a sequence from ℓ_2, which converges to the element $\bar{u}$ with respect to the topology $\sigma(E, R(\mathscr{L}^*))$. This implies that the sequence $\{u_n\}$ is a Cauchy sequence in the topology $\sigma(\ell_2, R(\mathscr{L}^*))$, and hence it is a Cauchy sequence in the topology $\sigma(\ell_2, s^*)$, and hence in the topology $\sigma(s, s^*)$ also, i.e. the sequence $\{u_n\}$ is a weakly Cauchy sequence in s. Since the space s is weakly sequentially complete, then $\{u_n\}$ converges in s to some element $\tilde{u}$; in addition, the space ℓ_2 is embedded into a completion of s in the topology $\sigma(s, s^*)$, and the latter topology is Hausdorff, therefore $\bar{u} = \tilde{u} \in s$. Thus, any generalized solution of system (4.8), for which the matrix $\mathscr{L}$ and the transposed matrix $\mathscr{L}^*$ have sparse rows, belongs to the space s. This fact is non-trivial for various classes of matrices with sparse rows (e.g., for the triangular matrixes). It shows that a generalized solution is an element of s, which is a limit of an near-solution $\{u_n\}$ in the metric of this space.

Finally, consider an example of generalized solving of system (4.8). Let an infinite matrix $\mathscr{L} = \|a_{ij}\|$ is diagonal: $a_{ij} = 0$ for all $i \neq j$, $a_{ii} = \lambda_i \neq 0$ and $\sum_{i=1}^{\infty} |\lambda_i|^2 < \infty$; in this case the matrix $\mathscr{L}$ induces in ℓ_2 an injective Hermitian operator $\mathscr{L} = \mathscr{L}^*$ and the matrices $\mathscr{L}$ and $\mathscr{L}^*$ have sparse rows. Let us put $a_{nn} = \lambda_n = \frac{1}{n}$, then

$$\mathscr{L}u = \left(u_1, \frac{u_2}{2}, \dots, \frac{u_n}{n}, \dots\right), \quad u = (u_1, \dots, u_n, \dots) \in \ell_2. \tag{4.9}$$

It is easy to see that $y = (b_1, b_2, \dots, b_n, \dots) = (1, \frac{1}{2}, \dots, \frac{1}{n}, \dots) \notin R(\mathscr{L})$, therefore, system (4.8) does not have a classical solution, but the element $\bar{u} = (1, 1, \dots, 1, \dots) \in s$ is a generalized solution; this element is a limit of the near-solution

$$u_n = (1, 1, \dots, 1, 0, 0, \dots) = \mathscr{L}^{-1}(y_n)$$

in s, where $y_n = (1, \frac{1}{2}, \dots, \frac{1}{n}, 0, 0, \dots)$. Let us consider an infinite system of linear algebraic equations

$$\begin{aligned}
a_{11}x_1 + a_{12}x_2 + \dots + a_{1n}x_n + \dots &= b_1, \\
a_{21}x_1 + a_{22}x_2 + \dots + a_{2n}x_n + \dots &= b_2, \\
&\dotso \\
a_{n1}x_1 + a_{n2}x_2 + \dots + a_{nn}x_n + \dots &= b_n,
\end{aligned} \tag{4.10}$$

A numerical sequence $x = (x_1, x_2, \ldots, x_n, \ldots)$ is called *a classical solution* of system (4.10), if after substitution of these values into the left-hand side of equalities (4.10) we obtain convergent (in the usual sense) numerical series and all these equalities are satisfied. Denote by $\mathscr{L}$ the infinite matrix $\mathscr{L} = \|a_{ij}\|_{i,j=\overline{1,\infty}}$ generated by the coefficients if system (4.10).

The infinite matrix $\mathscr{L}$ may be considered as an linear operator mapping some vector space E, consisting of sequences, into a similar space F containing the element b, i.e. $\mathscr{L} : E \to F$. We shall call the matrix operator $\mathscr{L}$ an operator defining system (4.10). Thus, applying the matrix operator $\mathscr{L}$ defining system (4.10) to the element x we have

$$y = \mathscr{L}(x) = \left(\sum_{i=1}^{\infty} a_{1k}x_k, \sum_{i=1}^{\infty} a_{2k}x_k, \ldots, \sum_{i=1}^{\infty} a_{nk}x_k, \ldots \right) \tag{4.11}$$

The solution of an operator equation (4.11) is an element x from E such that $\mathscr{L}(x) = b \in F$. Passing from the system of linear algebraic equations to an operator equation allows to specify the concept of the solution of system (4.10); now by the solution of the operator equation (4.11) we will means an element x from the vector space of sequences E for which $\mathscr{L}x = b \in E$. Comparing this concept with the concept of the classical solution x of system (4.10), it would seem that this sequence does not belong to the space E, since the solution of the operator (4.11) corresponding to the element $b = (b_1, b_2, \ldots, b_n, \ldots)$ does not exist whereas the classical solution exists. However, the classical solution of system (4.10) is a partial case of the solution of equation (4.11) if both E and F are the vector space S consisting of all numerical sequences.

Let us rewrite system (4.10) in the following way:

$$x_i = \sum_{i=1}^{\infty} c_{ik}x_k + b_i \tag{4.12}$$

where $c_{ik} = \delta_{ik} - a_{ik}$, δ_{ik} is the Kroneker symbol.

Let us introduce the following notations: $C = \|c_{ik}\|_{i,k=\overline{1,\infty}}$ is an infinite matrix, which induces an operator C. If $\mathscr{F} \subset E$ then $Cx = (I - \mathscr{L})x$, where I is a unit operator in E. Affine operator T which acts from E to F such that $T(x) = (I - \mathscr{L})x + b$ allows to rewrite (4.11) in the following way:

$$x = Tx = (I - \mathscr{L})x + b \tag{4.13}$$

Therefore, solving system (4.12) is reduced to searching of a fixed point T.

It is known that [30] an infinite system like (4.12) is called *regular*, if

$$\sum_{k=1}^{\infty} |c_{ik}| < 1 \qquad i = 1, 2, \ldots, n \tag{4.14}$$

and *completely regular* if

$$\sum_{k=1}^{\infty} |c_{ik}| < 1 - \theta < 1 \qquad i = 1, 2, \ldots, n \tag{4.15}$$

Let us prove that for regular systems the operators $\mathscr{L}$ and T are continuous in the Banach space m consisting of all bounded sequences with the norm $\|x\| = \sup\limits_{k\in[1,\infty)} |x_k|$, $(x_1,...,x_k,...) \in m$. Indeed, if $x \in m$ then for a regular system

$$\|\mathscr{L}x\| = \|y\| = \sup_{i\in[1,\infty)} |y_i| = \sup_{i\in[1,\infty)} \sum_{k=1}^{\infty} |a_{ik}x_k| \leq \sup_{i\in[1,\infty)} \sum_{k=1}^{\infty} |a_{ik}| \sup_{k\in[1,\infty)} |x_k|$$

$$\leq \left(\sup_{i\in[1,\infty)} \sum_{k=1}^{\infty} |a_{ik}| \right) \|x\| \tag{4.16}$$

$$= \left(\sup_{i\in[1,\infty)} \sum_{k=1}^{\infty} |c_{ik}| + |c_{ii} - 1| = \right) \|x\| \tag{4.17}$$

$$\leq \left(\sup_{i\in[1,\infty)} \sum_{k=1}^{\infty} |c_{ik}| + 1 \right) \|x\| \leq 2\|x\| \tag{4.18}$$

So, the operator $\mathscr{L}$ is a bounded linear operator which acts in m and its norm does not exceed 2. By formula (4.12)

$$\|Tx\| = \|y\| = \|(I - \mathscr{L})x + b\| \leq \|\mathscr{L}x\| + \|x\| + \|b\| \tag{4.19}$$

$$\leq 2\|x\| + \|x\| + \|b\| \leq 3\|x\| + \|b\| \tag{4.20}$$

whence,

$$\|y\| = \sup_{k\in[1,\infty)} |y_i| \leq 3\|x\| + \|b\| < \infty,$$

therefore, $Tx \in m$.

When the operator T is completely regular in the space m, it is a contraction operator, as

$$\rho\left(Tx, Tx'\right) = \|Tx - Tx'\| = \| \sum_{k=1}^{\infty} c_{ik}x_k - \sum_{k=1}^{\infty} c_{ik}x_k' \|$$

$$= \| \sum_{k=1}^{\infty} c_{ik}\left(x_k - x_k'\right) \| \leq \sum_{k=1}^{\infty} |c_{ik}|\|x - x'\| \leq \|x - x'\| \sum_{k=1}^{\infty} |c_{ik}|$$

$$\leq (1 - \theta))\rho\left(x, x'\right) = q\rho\left(x, x'\right),$$

where $q = 1 - \theta < 1$. Therefore, a completely regular infinite system of linear algebraic equations in the space m has a unique solution and we can find this solution by the reduction method proposed by V.L. Kantorovich [30]. For solving finite systems of linear algebraic equations which occur in this method we will use the combined method described in Chap. 5.

Moreover, existence and uniqueness of a solution of a completely regular system and Banach inverse operator theorem imply that the operator $\mathscr{L}$ always has a bounded linear inverse operator $\mathscr{L}^{-1}$, which acts in the space m.

The solving of a regular systems is a more complex problem. In this case, the operator T is not a contraction operator and the concept of generalized solution must be considered.

Consider an arbitrary regular infinite system of linear algebraic equations with a bounded linear operator $\mathscr{L}$, which acts in the space m. Denote by $\mathscr{L}_1$ a restriction of the operator $\mathscr{L}$ on the space c_0 of all sequences convergent to zero with the norm $\|x\| = \sup\limits_{k\in[1,\infty)} |x_k|$. Since $c_0 \subset m$, then the operator $\mathscr{L}_1$ maps the space c_0 into m and it is a bounded linear operator. P.S.Bondarenko proved that the operator $\mathscr{L}_1$ is injective, as a regular system may have only one solution converging to zero [6]. Denote by $R(\mathscr{L})$ and $R(\mathscr{L}_1)$ the ranges of the operators $\mathscr{L}$ and $\mathscr{L}_1$, respectively, by $F = \overline{R(\mathscr{L})}$ and $F_1 = \overline{R(\mathscr{L}_1)}$ the closures of the ranges $R(\mathscr{L})$ and $R(\mathscr{L}_1)$ in the space m. Then $R(\mathscr{L}_1) \subset R(\mathscr{L})$, $F_1 \subset F$, and F, F_1 are closed Banach subspaces of m. Let us show that $c_0 \subset F$. Indeed, let $b = (b_1, b_2, ..., b_n, ...)$ be an arbitrary element of the space c_0 and $\tilde{b}_n = (b_1, b_2, ..., b_n, 0, ...)$. When the right-hand side of a regular system has only finite number of nonzero element, this regular system has a bounded solution [30], therefore there exists such an element $\tilde{x}_n \in m$ that $\mathscr{L}(\tilde{x}_n) = \tilde{b}_n \in R(\mathscr{L})$. Since $\tilde{b}_n \to b$ as $n \to \infty$ in the space m, then $c_0 \subset \overline{R(\mathscr{L})} = F$.

Assume that the matrix $\mathscr{L}$ defining system (4.10) is symmetrical. In this case the matrix C defining system (4.12) and all columns of the matrix C are elements of the space l_1, as $\sum_{i=1}^{\infty} |c_{ik}| < 1$. Since $l_1 \subset c_0$, the columns of the matrix C are elements of the space c_0. By virtue of condition (4.13), all columns of the matrix $\mathscr{L}$ are elements of the space c_0. Denote by $e_1 = (1,0,0,...,)$, $e_2 = (0,1,0,...,)$, ..., $e_n = (0,0,,...,1,0,...)$ the orts of the space c_0. Formula (4.11) implies that $y_1 = \mathscr{L}_1(e_1) = \mathscr{L}(e_1) = (a_{11}, a_{21}, ..., a_{k1}, ...)$, $y_2 = \mathscr{L}_1(e_2) = \mathscr{L}(e_2) = (a_{12}, a_{22}, ..., a_{k2}, ...)$, ..., $y_k = \mathscr{L}_1(e_k) = \mathscr{L}(e_k) = (a_{1k}, a_{2k}, ..., a_{kk}, ...)$, ... belong to the space c_0. Hence, $\forall u \in c_0$: $u = \sum_{i=1}^{\infty} \alpha_i e_k$ and $\mathscr{L}_1(u) = \mathscr{L}(u) = \sum_{i=1}^{\infty} \alpha_i \mathscr{L}_1(e_i) = \lim_{n\to\infty} \sum_{i=1}^{n} \alpha_i \mathscr{L}_1(e_i) \in c_0$. Thus, the operator $\mathscr{L}_1$ maps elements $u \in c_0$ to elements $\mathscr{L}_1(u) \in c_0$, i.e. the operator L_1 acts into c_0, so that $F_1 = \overline{R(\mathscr{L}_1)} \subset c_0$.

Thus, we conclude the following:

(1) If $b \in R(\mathscr{L}_1)$ then there exist a unique classical solution $x \in c_0$.

(2) If $b \in \overline{R(\mathscr{L}_1)}$, but $b \notin R(\mathscr{L}_1)$ then a classical solution does not exist, but the operator equation $\mathscr{L}_1(x) = b$, where $x \in c_0$, has a unique generalized solution $\tilde{x}$ in some space $\overline{E}$, which is a completion of c_0 with respect to the norm $\|x\|^* = \|\mathscr{L}_1(x)\|_{c_0}$.

(3) If $b \notin \overline{R(\mathscr{L}_1)}$ then there are no either classical either generalized solution of the equation $\mathscr{L}_1(x) = b$.

The most interesting case is that in which $R(\mathscr{L}_1)$ is everywhere dense in c_0 (with respect to the norm or in a weak topology). Then $\overline{R(\mathscr{L}_1)} = c_0$, as $R(\mathscr{L}_1) \subset c_0$. In this case, (4.10) always has a generalized solution if $\mathscr{L}$ is a symmetrical matrix, and system (4.12) is regular. If the operator $\mathscr{L}_1$ is injective, then $B = (e_1, e_2, ..., e_n, ...) \subset R(\mathscr{L}_1)$ iff the matrix $\mathscr{L}_1$ has sparse rows in ℓ_2. Since all rows of the matrix $\mathscr{L}$ of a regular system belong to l_1 and $l_1 \subset l_2$, then all rows of the matrix $\mathscr{L}$ are elements of l_2. From the other hand, the condition $B \subset R(\mathscr{L}_1)$ implies that $\overline{R(\mathscr{L}_1)} = c_0$, hence, $\overline{R(\mathscr{L}_1)} = c_0$ if the matrix $\mathscr{L}_1$ has sparse rows.

Let us consider a generalized solution $\bar{x}$ of system (4.10) when $b = (b_1, b_2, ..., b_n, ...) \in c_0$ and a classic solution in c_0 or m does not exist. In other words, let us try to give a constructive description of a space containing the generalized solution $\bar{x}$ of a system $\mathscr{L}_1(x) = b$, where x and b belong to c_0. In addition, this description must not depend on the kind of the operator $\mathscr{L}$.

The space c_0 is densely embedded into s. Definitions 1 and 2 imply that the generalized solutions belong to weak sequential closure E^s of the vector space E in a Hausdorff locally convex topological vector space $\overline{E}$, which is a completion of c_0 in topology $\sigma(E, R(\mathscr{L}_1^*))$. Let us show that the range $R(\mathscr{L}_1^*)$ of the adjoint operator $\mathscr{L}_1^* : l_1 \to l_1$ contains the orts $e_k = (0, ..., 0, 1, 0, ...)$, $k = 1, 2, ...$. By definition of the adjoint operator $f = \mathscr{L}_1^*(g)$, where $f, g \in c_0^* = l_1$, $f(x) = g(\mathscr{L}_1(x))$, $x = (x_1, x_2, ..., x_n, ...) \in c_0$. Put $y = \mathscr{L}_1(x)$. Then formula (4.11) implies that

$$y = \left(\sum_{k=1}^{\infty} a_{1k} x_k, \sum_{k=1}^{\infty} a_{2k} x_k, ..., \sum_{k=1}^{\infty} a_{ik} x_k, ... \right).$$

Thus,

$$f(x) = \sum_{i=1}^{\infty} g_i y_i = \sum_{i=1}^{\infty} g_i \left(\sum_{k=1}^{\infty} a_{ik} x_k \right) = \sum_{i=1}^{\infty} \left(\sum_{ki=1}^{\infty} a_{ik} g_i x_k \right)$$

$$= \sum_{i=1}^{\infty} \sum_{k=1}^{\infty} a_{ik} g_i x_k = \sum_{k=1}^{\infty} \left(\sum_{i=1}^{\infty} a_{ik} g_i x_k \right) = \sum_{k=1}^{\infty} \left(\sum_{i=1}^{\infty} a_{ki} g_i \right) x_k$$

since the matrix $\mathscr{L}$ is symmetrical. Swapping the indexes i and k, we obtain

$$f(x) = \sum_{k=1}^{\infty} f_i x_i = \sum_{i=1}^{\infty} \left(\sum_{k=1}^{\infty} a_{ik} g_k \right) x_i,$$

therefore,

$$f(x) = \sum_{i=1}^{\infty} a_{ik} g_k$$

This implies that the adjoint operator $\mathscr{L}_1^* : l_1 \to l_1$ is a matrix operator, and its inducing matrix $\mathscr{L}_1^*$ coincides with the matrix $\mathscr{L}$. Therefore, the matrix $\mathscr{L}$ has sparse rows in the space l_2 and hence $R(\mathscr{L}_1^*)$ contains the set of all orts in l_2. Therefore, $s^* = \mathscr{L}(B) \subset R(\mathscr{L}_1)$ and hence the topology $\sigma(c_0, R(\mathscr{L}_1))$ is stronger than $\sigma(c_0, \mathscr{L}(B)) = \sigma(c_0, s^*)$. Let $\bar{u}$ is an arbitrary element of weak sequential closure c_0^s and $\{u_n\}_{n=1}^{\infty}$ is a sequence from c_0, which converges to $\bar{u}$ in the topology $\sigma(c_0, R(\mathscr{L}_1^*))$. Therefore, the sequence $\{u_n\}_{n=1}^{\infty}$ is a Cauchy sequence in the topology $\sigma(c_0, R(\mathscr{L}_1^*))$ and hence it is a Cauchy sequence in the topology $\sigma(c_0, s^*)$ and, therefore, in the topology $\sigma(s, s^*)$. Hence, the sequence $\{u_n\}_{n=1}^{\infty}$ is a Cauchy sequence in the space s. Since the space s is sequentially complete, then the sequence $\{u_n\}_{n=1}^{\infty}$ converges to s to some element $\tilde{u}$. In addition, the space c_0 is embedded to the completion of the space s in the topology $\sigma(s, s^*)$ and this topology is Hausdorff. Therefore, $\bar{u} = \tilde{u} \in s$. Thus, any generalized solution of system (4.10) belongs to s, if inducing matrix $\mathscr{L}$ is symmetrical and has sparse rows, and $b \in c_0$. Hence, while the operator $\mathscr{L}_1$ acts in the space c_0 and the generalized solution $\bar{u}$ does not belong to c_0, the generalized solution is just an element of more wide space s and coincides with a classical solution of system (4.10).

Thus, a classical solution of the system $\mathscr{L}x = b$ and a classical solution of the operator equation $\mathscr{L}(x) = b$, where the operator $\mathscr{L} : E \to F$ does not coincide each other. Indeed, if x is the classical solution of the operator equation, then $x \in E$ and x is the classical solution of the system $\mathscr{L}x = b$. However, if x is the generalized solution of the system $\mathscr{L}x = b$, then $x \notin E$ and x is not a classical solution of the operator equation $\mathscr{L}(x) = b$, but x may be a classical solution of the system $\mathscr{L}x = b$.

The investigation of a infinite system of linear algebraic equations confirms that the main notion of functional analysis is an operator, but not a space [43].

4.3 Application to Volterra Integral Equation of the First Kind

Consider a Volterra integral operator of the first kind

$$\mathscr{L}u = \int_0^t K(t,s)u(s), ds, \qquad \mathscr{L} : L_2(0,1) \to L_2(0,1), \qquad (4.21)$$

where $L_2(0,1)$ is a space of measurable, square-integrable, real functions.

Let us denote by D the bounded region $\{(t,s) \,|\, 0 \le s \le t \le 1\}$.

Lemma 4.2. *Let a kernel $K(t,s)$ be bounded in D and have a square-integrable in D partial derivative (in the Sobolev sense) $\partial K / \partial t$ with respect to the variable t, the function $k_s(t) = K(t,s)$ be absolutely continuous with respect to t, $s \le t \le 1$, and the function $K(t,t)$ be a square-integrable on $[0,1]$, then the function*

$$f = \mathscr{L}u = \int_0^t K(t,s)u(s)\,ds$$

is absolutely continuous on $[0,1]$ and the following formulae is true:

$$\frac{df}{dt} = K(t,t)u(t) + \int_0^t \frac{\partial K(t,s)}{\partial t} u(s)\,ds.$$

Proof. At first, let us prove that the function f is absolutely continuous. Let us select on $[0,1]$ the points

$$0 \le a_1 < b_1 \le a_2 < b_2 \le \ldots \le a_n < b_n \le 1.$$

Then

$$\sum_{i=1}^{n} |f(b_i) - f(a_i)| = \sum_{i=1}^{n} \left| \int_0^{b_i} K(b_i,s)u(s)\,ds - \int_0^{a_i} K(a_i,s)u(s)\,ds \right|$$

$$\leq \sum_{i=1}^{n} \left| \int_{a_i}^{b_i} K(b_i,s)u(s)\,ds \right| + \sum_{i=1}^{n} \left| \int_0^{a_i} (K(b_i,s) - K(a_i,s))u(s)\,ds \right|$$

$$\leq \max_{(t,s)\in D} |K(t,s)| \times \sum_{i=1}^{n} \left| \int_{a_i}^{b_i} u(s)\,ds \right|$$

$$+ \sum_{i=1}^{n} \left| \int_0^{a_i} \left(\int_{a_i}^{b_i} \frac{\partial K(t,s)}{\partial t}\,dt \right) u(s)\,ds \right|$$

$$\leq C \sum_{i=1}^{n} \int_{a_i}^{b_i} |u(s)|\,ds + \sum_{i=1}^{n} \left| \int_{a_i}^{b_i} \left(\int_0^{a_i} \frac{\partial K(t,s)}{\partial t} u(s)\,ds \right) dt \right|$$

$$\leq \sum_{i=1}^{n} \int_{a_i}^{b_i} \left(C|u(t)| + \int_0^1 \left| \frac{\partial K(t,s)}{\partial t} u(s) \right| ds \right) dt.$$

By the conditions of the lemma, we have that the function

$$z(t) = C|u(t)| + \int_0^1 \left| \frac{\partial K(t,s)}{\partial t} u(s) \right| ds$$

is integrable on $[0,1]$, and the absolute continuity of the Lebesgue integral implies the absolute continuity of the function f.

Thus, the function $f(t)$ has an integrable derivative almost everywhere on $[0,1]$. On the other hand

$$\int_0^t \left(K(\tau,\tau)u(\tau) + \int_0^\tau \frac{\partial K(\tau,s)}{\partial t} u(s)\,ds \right) d\tau$$

$$= \int_0^t K(\tau,\tau)u(\tau)\,d\tau + \int_0^t \int_s^t \frac{\partial K(\tau,s)}{\partial t}\,d\tau u(s)\,ds$$

$$= \int_0^t K(\tau,\tau)u(\tau)\,d\tau + \int_0^t (K(t,s) - K(s,s))u(s)\,ds = f(t).$$

Hence, the lemma is proved. $\qquad\square$

Remark 4.14. Usually, the formula from Lemma 4.2 is cited under the stronger restrictions on $K(t,s)$.

Remark 4.15. Lemma 4.2 directly implies that the function $f'(t)$ is square integrable.

Lemma 4.3. *Let the conditions of Lemma 4.2 be satisfied and $K(t,t) \geq \varepsilon > 0$, then the range of operator (4.21) coincides with the set $W_{2,0}^1(0,1)$ of absolutely continuous functions that are equal to zero at the point $t = 0$ and have a square-integrable derivative.*

Proof. The fact that $f'(t)$ is square-integrable was probed in Remark 4.15. It is also easy to see that $f(0) = 0$. Let us prove that for any function $f \in W_{2,0}^1$ there exists $u \in L_2(0,1)$ such that $\mathscr{L}u = f$. To do this, let us consider the Volterra integral equation of the second type

$$K(t,t)u(t) + \int_0^t \frac{\partial K(t,s)}{\partial t} u(s)\, ds = f'(t), \tag{4.22}$$

or

$$u(t) + \int_0^t \frac{1}{K(t,t)} \frac{\partial K(t,s)}{\partial t} u(s)\, ds = \frac{f'(t)}{K(t,t)}.$$

As well-known, this equation has a solution in $L_2(0,1)$ for any right-hand side. Thus, for any function f there exists $u \in L_2(0,1)$, which is a solution of (4.22). Integrating the equality (4.22) from 0 to t and taking into account Lemma 4.2, we have that any function $f \in W_{2,0}^1$ belongs to the range of the operator (4.21). $\square$

Remark 4.16. In particular, the lemma implies that the set $R(\mathscr{L})$ is dense in $L_2(0,1)$.

Remark 4.17. Similarly, we can prove the following statement. Let us suppose that the following conditions are satisfied:

(a) A kernel $K(t,s)$ is bounded and absolutely continuous with respect to s for any fixed value t.
(b) $\partial K/\partial s$ is square-integrable in D.
(c) $K(t,t) \geq \varepsilon > 0$.
(d) The function $K(t,t)$ is square-integrable in $[0,1]$.

Then $R(\mathscr{L}^*) = W_{2,1}^1(0,1)$ and since $R(\mathscr{L}^*)$ is dense in $L_2(0,1)$; the operator $\mathscr{L}$ is injective, where $W_{2,1}^1(0,1)$ is a set of absolutely continuous functions on $[0,1]$, which are equal to zero at the point $t = 1$ and have square-integrable derivative on $(0,1)$.

Lemma 4.4. *Let the conditions of Remark 4.17 be satisfied. Then the space $R_{\bar{E}_{\mathscr{L}}}(\mathscr{L}^*)$ (in terms of Lemma 3.3) is isomorphic to the space $W_{2,1}^1$ (with respect to the structures of topological vector spaces) in the norm*

$$\|f\|_{W_{2,1}^1} = \left(\int_0^1 \left(\frac{\partial f}{\partial t} \right)^2 dt \right)^{1/2}.$$

Proof. It is sufficiently to prove that the norm $\| \cdot \|_{W_{2,1}^1}$ is equivalent to the norm $\| \cdot \|_{R_{\bar{E}_{\mathscr{L}}}(\mathscr{L}^*)}$. Indeed, the result of Lemma 4.2, the boundedness of the kernel $K(t,s)$ and the square-integrability of $\partial K/\partial s$ imply that

$$\|f\|_{W_{2,1}^1}^2 = \int_0^1 \left(\frac{\partial \mathscr{L}^* \varphi}{\partial s} \right)^2 ds = \int_0^1 \left(K(s,s)\varphi(s) + \int_0^s \frac{\partial K(t,s)}{\partial s} \varphi(t)\, dt \right)^2 ds$$

$$\leq 2 \int_0^1 (K(s,s)\varphi(s))^2\, ds + 2 \int_0^1 \left(\int_0^s \frac{\partial K}{\partial s} \varphi(t)\, dt \right)^2 ds$$

$$\leq 2 \int_0^1 K^2(s,s) \, ds \int_0^1 \varphi^2 ds + 2 \int_0^1 \left(\int_0^s \left(\frac{\partial K}{\partial s} \right)^2 dt \right) ds \times \int_0^1 \left(\int_0^s \varphi^2(t) \, dt \right) ds$$

$$\leq C \|\varphi\|^2_{L_2(0,1)} = C \|f\|^2_{R_{\bar{E}_{\mathscr{L}}}(\mathscr{L}^*)}.$$

Thus, we have proved that the norm $\| \cdot \|_{R_{\bar{E}_{\mathscr{L}}}(\mathscr{L}^*)}$ is stronger than the norm $\| \cdot \|_{W^1_{2,1}}$. In addition, the spaces $R_{\bar{E}_{\mathscr{L}}}(\mathscr{L}^*)$, $W^1_{2,1}$ are complete and hence by virtue of the Banach theorem on the inverse operator, these norms are equivalent. $\qquad \square$

Lemma 4.5. *Let the following conditions be satisfied:*

(a)$K(t,s)$ is a measurable and bounded function in D.
(b)The function $K(t,s)$ is absolutely continuous with respect to each variable t,s and have square-integrable partial derivatives $\partial K/\partial t$ and $\partial K/\partial s$.
(c)The function $K(t,t) \geq \varepsilon > 0$ is square-integrable on $[0,1]$.

Then the space $\bar{E}_{\mathscr{L}}$ is densely embedded into the space H, which is homeomorphic to the negative space $W^{-1}_{2,1}$, where the space $W^{-1}_{2,1}$ is constructed by the pair $W^1_{2,1} \subset L_2(0,1)$.

Proof. Lemmas 3.3 and 4.4 imply that the spaces $W^1_{2,1}$ and $(\bar{E}_{\mathscr{L}})^*$ are linear and topologically isomorphic. Therefore, the spaces $(W^1_{2,1})^* = W^{-1}_{2,1}$ and $(\bar{E}_{\mathscr{L}})^{**}$ are linear and topologically isomorphic also. Taking into account the fact that the embedding $\bar{E}_{\mathscr{L}} \subset \bar{E}^{**}_{\mathscr{L}}$ is dense, we prove the lemma. $\qquad \square$

Theorem 4.3. *If the conditions of the previous lemma are satisfied there exists a unique generalized solution of the Volterra equation in the sense of Definitions 3.6 and 3.7 (H is homeomorphic to $W^{-1}_{2,1}$).*

Remark 4.18. The conditions such that the kernel is absolute continuous with respect to t and its partial derivative $\partial K/\partial t$ is square-integrable are necessary only to prove that the range $R(\mathscr{L})$ is densely embedded into $L_2(0,1)$. If it is known a priori, then these conditions may be discarded.

4.4 Application to the Statistics of Random Processes

Let $x(t)$ be a random process with continuous time, whose trajectory is observed on the segment $[a,b]$, $r(t,s)$ be a correlation function of this process. By Mercer theorem a correlation function can be represented as a uniformly convergent series

$$r(t,s) = \sum_{k=1}^{\infty} \frac{\varphi_k(s)\varphi_k(t)}{\lambda_k}, \qquad (4.23)$$

if the eigenvalues λ_k and the eigenvectors φ_k of the integral operator

$$A\varphi = \int_a^b r(t,s)\varphi(s)\,ds$$

are known.

If $\lambda_k > 0$, then the function $r(t,s)$ defined by the formulae (4.23) is positive-definite, and by Khinchin's theorem $r(t,s)$ can be considered as a correlation function of some random (may be Gaussian)) process $x(t)$.

In the theory of testing of statistical hypotheses on random processes with continuous time and in the theory of estimation of unknown coefficients in a linear regression scheme, there is an important problem of solving Fredholm equation of the first type

$$\int_a^b r(t,s)\varphi(s)\,ds = a(t)$$

for the right-hand sides $a(t) \in L_2(a,b)$ [23, 96]. In particular, if the process $x(t)$ has a constant, but unknown mathematical expectation $Mx(t) = m$, then the best linear estimation m^* of this mathematical expectation is defined by the following formula:

$$m^* = \int_a^b f(t)x(t),dt, \qquad f \in L_2(a,b)$$

and it exists if the function $f \in L_2(a,b)$ is a solution of the integral equation

$$\int_a^b r(s,t)f(t)\,dt = 1. \tag{4.24}$$

Since (4.24) rarely has a classic solution, it is naturally to study a generalized solution of this equation and to search for the best linear estimation of the mathematical expectation as a functional from the space of generalized functions (see Theorems 4.1, 4.2).

4.5 Application to Parabolic Differential Equation in a Connected Region

A priori estimates and generalized solutions are the effective methods of qualitative analysis of partial differential equations (e.g. see [5, 62, 63]). As an example let us consider the application of the method of a priori estimates to the issue of existence and uniqueness of generalized solutions of parabolic equations.

4.5.1 Problem Definition

Let $\Omega \subset \mathbb{R}^n$ be a bounded region with a regular border $\partial\Omega$. Let us consider in the cylinder $Q = (0,T) \times \Omega$ the following parabolic equation

$$u_t + \mathscr{A}u = f(t,\xi), \tag{4.25}$$

where $\mathscr{A}$ is an elliptic differential expression

$$\mathscr{A}u = -\sum_{i,j=1}^{n} \left(a_{ij}u_{\xi_j}\right)_{\xi_i} + \sum_{i=1}^{n} (a_i u)_{\xi_i} + au.$$

Let us study the following boundary value problem: we need to find the function $u(t,\xi)$, which satisfies (4.25) in Q and the following conditions

$$u|_{t=0} = 0, \qquad u|_{\xi \in \partial\Omega} = 0. \tag{4.26}$$

Let $D \subset C^1(\bar{Q})$ be a linear manyfold of functions, which satisfy conditions (4.26), $D_+ \subset C^1(\bar{Q})$ be a linear manyfold of functions, which satisfy the conditions

$$v|_{t=T} = 0, \qquad v|_{\xi \in \partial\Omega} = 0.$$

To study boundary value problem (4.25), (4.26) let us consider the following Hilbert spaces. Let W be a Hilbert space, which is a completion of D with respect to the norm

$$\|u\|_W^2 = \int_Q \left(u_t^2 + \sum_{i=1}^{n} u_{\xi_i}^2\right) d\xi\, dt, \tag{4.27}$$

W_+ be a Hilbert space, which is a completion of D_+ with respect to the norm (4.27), and H be a Hilbert space, which is a completion of D with respect to the norm

$$\|u\|_H^2 = \int_Q \left(u^2 + \sum_{i=1}^{n} u_{\xi_i}^2\right) d\xi\, dt. \tag{4.28}$$

Note that the space H is a completion of D_+ with respect to the norm (4.28). In addition, there are dense and compact embeddings

$$W \subset H \subset L_2(Q), \qquad W_+ \subset H \subset L_2(Q).$$

Passing to the conjugate spaces, we have the following chains of dense and compact embeddings

$$L_2(Q) \subset H^- \subset W^-, \qquad L_2(Q) \subset H^- \subset W_+^-.$$

Denote by $\langle \cdot, \cdot \rangle_{W,W^-}$ an extension of the inner product $(\cdot, \cdot)_{L_2(Q)}$ by continuity on $W \times W^-$. The next denotations have similar sense $\langle \cdot, \cdot \rangle_{W_+^-, W_+}$, $\langle \cdot, \cdot \rangle_{H^-, H}$ and $\langle \cdot, \cdot \rangle_{H, H^-}$.

Let us suppose that the following conditions are satisfied:

(1) Functions $a_{ij} = a_{ji}$ belong to the space $C(\bar{\Omega})$ and satisfy the uniform ellipticity condition

$$\sum_{i,j=1}^{n} a_{ij}(\xi)\lambda_i\lambda_j \geq \alpha \sum_{i=1}^{n}\lambda_i^2, \quad \lambda_i \in \mathbb{R}, \quad \xi \in \Omega,$$

where α is a positive constant.

(2) Functions $a \in C(\bar{\Omega})$, $a_i \in C^1(\bar{\Omega})$ satisfy the inequalities

$$\sum_{i=1}^{n}\frac{\partial a_i}{\partial \xi_i}(\xi) + a(\xi) \geq 0; \qquad a(\xi) \geq 0$$

for all $\xi \in \Omega$.

4.5.2 Properties of Operators Associated with a Boundary Value Problem

Let the function $u \in C^2(\bar{Q})$ satisfy (4.25), (4.26) with a smooth right-hand side f, then the Gauss–Ostrogradsky formula implies, that for any function $v \in D_+$ the function u satisfies the identity

$$l(u,v) = \int_Q \left(u_t v + \sum_{i,j=1}^{n} a_{ij}u_{\xi_j}v_{\xi_i} - \sum_{i=1}^{n} a_i uv_{\xi_i} + auv \right) d\xi\, dt = (f,v)_{L_2(Q)} \qquad (4.29)$$

or the identity

$$\bar{l}(u,v) = \int_Q \left(-uv_t + \sum_{i,j=1}^{n} a_{ij}u_{\xi_j}v_{\xi_i} - \sum_{i=1}^{n} a_i uv_{\xi_i} + auv \right) d\xi\, dt = (f,v)_{L_2(Q)}. \qquad (4.30)$$

Also, the integral identities (4.29), (4.30) are meaningful for the pair of functions $(u,v) \in W \times H$, $(u,v) \in H \times W_+$, respectively (of course, all of the derivatives are regarded as generalized).

For the bilinear forms

$$W \times H \ni (u,v) \mapsto l(u,v), \qquad H \times W_+ \ni (u,v) \mapsto \bar{l}(u,v)$$

using the integral Cauchy–Bunyakovsky inequality it is easy to get the estimate

$$|l(u,v)| \leq c\|u\|_W\|v\|_H, \qquad |\bar{l}(u,v)| \leq c\|u\|_H\|v\|_{W_+}.$$

Thus, there exists such linear continuous operators

$$\mathscr{L}: W \to H^-, \quad \mathscr{L}^*: H \to W^-, \quad \bar{\mathscr{L}}: H \to W_+^-, \quad \bar{\mathscr{L}}^*: W_+ \to H^-,$$

that

$$l(u,v) = \langle \mathscr{L}u,v \rangle_{H-,H} = \langle u, \mathscr{L}^*v \rangle_{W,W-} \quad \forall (u,v) \in W \times H,$$

$$\bar{l}(u,v) = \langle \bar{\mathscr{L}}u,v \rangle_{W_+^-,W_+} = \langle u, \bar{\mathscr{L}}^*v \rangle_{H,H-} \quad \forall (u,v) \in H \times W_+.$$

Obviously,

$$\mathscr{L} \subset \bar{\mathscr{L}}, \quad \mathscr{L}^* \supset \bar{\mathscr{L}}^*,$$

i.e. the operator $\bar{\mathscr{L}}$ is a completion of the operator $\mathscr{L}$ by continuity and the operator $\bar{\mathscr{L}}^*$ is a contraction of the operator $\mathscr{L}^*$.

Let us obtain a priori estimates in negative norms for these operators.

Lemma 4.6. *The following inequalities hold:*

$$\|u\|_{L_2(Q)} \leq c_1 \|\bar{\mathscr{L}}u\|_{W_+^-} \leq c_2 \|u\|_H, \quad u \in H, \tag{4.31}$$

$$\|u\|_H \leq c_1 \|\mathscr{L}u\|_{H-} \leq c_2 \|u\|_W, \quad u \in W, \tag{4.32}$$

$$\|v\|_{L_2(Q)} \leq c_1 \|\mathscr{L}^*v\|_{W-} \leq c_2 \|v\|_H, \quad v \in H, \tag{4.33}$$

$$\|v\|_H \leq c_1 \|\bar{\mathscr{L}}^*v\|_{H-} \leq c_2 \|v\|_{W_+}, \quad v \in W_+. \tag{4.34}$$

Proof. It is sufficient to prove only left-hand sides of the double inequalities (4.31)–(4.34). Let us prove inequality (4.31). Let us put for an element $u \in D$

$$v(t,\xi) = \int_t^T e^{-N\tau} u(\tau,\xi)\,d\tau,$$

where N is a sufficiently large positive number. Its exact value we will determine later. It is clear that $v \in W_+$ and $u = -e^{Nt}v_t$. Let us consider the value of the functional $\bar{\mathscr{L}}u$ on the element v

$$\langle \bar{\mathscr{L}}u,v \rangle_{W_+^-,W_+} = \int_Q e^{Nt} \left(v_t^2 - \sum_{i,j=1}^n a_{ij}v_{\xi_j}v_{\xi_i t} + \sum_{i=1}^n a_i v_t v_{\xi_i} - avv_t \right) d\xi\,dt.$$

Using the formula of integration by parts and taking into account the boundary conditions and the coefficient conditions, we have

$$-\int_Q e^{Nt} \sum_{i,j=1}^n a_{ij}v_{\xi_j}v_{\xi_i t}\,d\xi\,dt = -\frac{1}{2}\int_\Omega \sum_{i,j=1}^n e^{Nt} a_{ij}v_{\xi_j}v_{\xi_i}\Big|_{t=0}^{t=T}\,d\xi$$

$$+\frac{N}{2}\int_Q e^{Nt} \sum_{i,j=1}^n a_{ij}v_{\xi_j}v_{\xi_i}\,d\xi\,dt$$

$$\geq \frac{\alpha N}{2}\int_Q e^{Nt} \sum_{i=1}^n v_{\xi_i}^2\,d\xi\,dt,$$

$$-\int_Q e^{Nt} avv_t\,d\xi\,dt = -\frac{1}{2}\int_{\Omega_k} e^{Nt} av^2\Big|_{t=0}^{t=T}\,d\xi + \frac{N}{2}\int_Q e^{Nt} av^2\,d\xi\,dt \geq 0,$$

$$\int_Q e^{Nt} \sum_{i=1}^n a_i v_t v_{\xi_i}\,d\xi\,dt \geq -c_0 \int_Q e^{Nt} \sum_{i=1}^n |v_t| \times |v_{\xi_i}|\,d\xi\,dt,$$

where $c_0 = \max\limits_{i=\overline{1,n}} \sup\limits_{\xi \in \Omega} |a_i(\xi)| < +\infty.$

Gathering all these estimates together gives the following inequality:

$$
\langle \mathcal{L}u, v \rangle_{W_+^-, W_+} \geq \int_Q e^{Nt} \left(v_t^2 + \frac{\alpha N}{2} \sum_{i=1}^n v_{\xi_i}^2 - c_0 \sum_{i=1}^n |v_t| \times |v_{\xi_i}| \right) d\xi \, dt
$$

$$
= \int_Q e^{Nt} \left(\frac{1}{2} v_t^2 + \frac{\alpha N}{4} \sum_{i=1}^n v_{\xi_i}^2 \right) d\xi \, dt
$$

$$
+ \int_Q e^{Nt} \sum_{i=1}^n \left(\frac{1}{2n} v_t^2 + \frac{\alpha N}{4} v_{\xi_i}^2 - c_0 |v_t| \times |v_{\xi_i}| \right) d\xi \, dt.
$$

Using the elementary inequalities $c_0 |v_t| \times |v_{\xi_i}| \leq \frac{1}{2n} v_t^2 + \frac{n}{2} c_0^2 v_{\xi_i}^2$, we have

$$
\langle \mathcal{L}u, v \rangle_{W_+^-, W_+} \geq \int_Q e^{Nt} \left(\frac{1}{2} v_t^2 + \frac{\alpha N}{4} \sum_{i=1}^n v_{\xi_i}^2 \right) d\xi \, dt + \int_Q e^{Nt} \sum_{i=1}^n \left(\frac{\alpha N}{4} - \frac{n}{2} c_0^2 \right) v_{\xi_i}^2 d\xi \, dt;
$$

hence, for a given $N > \frac{2n}{\alpha} c_0^2$, we have

$$
\langle \mathcal{L}u, v \rangle_{W_+^-, W_+} \geq c \|v\|_{W_+}^2.
$$

Applying the Schwarz inequality to the expression $\langle \mathcal{L}u, v \rangle_{W_+^-, W_+}$ and taking into account the estimates $\|u\|_{L_2(Q)} \leq e^{NT} \|v\|_{W_+}$, we get the inequality (4.31) for smooth functions $u \in D$. For reasons of the density, the inequalities (4.31) hold on the entire space H.

To prove (4.32), let us consider the value of the functional $\mathcal{L}u$ on an element $e^{-Nt} u$, where $u \in W$ and N is sufficiently large positive number. We have

$$
\langle \mathcal{L}u, e^{-Nt} u \rangle_{H^-, H} = \int_Q e^{-Nt} \left(u u_t + \sum_{i,j=1}^n a_{ij} u_{\xi_j} u_{\xi_i} - \sum_{i=1}^n a_i u u_{\xi_i} + a u^2 \right) d\xi \, dt
$$

$$
\geq \frac{N}{2} \int_Q e^{-Nt} u^2 d\xi \, dt + \int_Q e^{-Nt} \alpha \sum_{i=1}^n u_{\xi_i}^2 d\xi \, dt
$$

$$
- \int_Q e^{-Nt} \sum_{i=1}^n a_i u_{\xi_i} u \, d\xi \, dt.
$$

Further,

$$
\langle \mathcal{L}u, e^{-Nt} u \rangle_{H^-, H} \geq \int_Q e^{-Nt} \left(\frac{N}{4} u^2 + \frac{\alpha}{2} \sum_{i=1}^n u_{\xi_i}^2 \right) d\xi \, dt
$$

$$
+ \int_Q e^{-Nt} \sum_{i=1}^n \left(\frac{N}{4n} u^2 + \frac{\alpha}{2} u_{\xi_i}^2 - c_0 |u| |u_{\xi_i}| \right) d\xi \, dt.
$$

Using the inequality $c_0 |u| \cdot |u_{\xi_i}| \leq \frac{1}{2\alpha} c_0^2 u^2 + \frac{\alpha}{2} u_{\xi_i}^2$, we have

$$\langle \mathscr{L}u, e^{-Nt}u \rangle_{H^-,H} \geq \int_Q e^{-Nt} \left(\frac{N}{4} u^2 + \frac{\alpha}{2} \sum_{i=1}^n u_{\xi_i}^2 \right) d\xi\, dt$$

$$+ \int_Q e^{-Nt} \sum_{i=1}^n \left(\frac{N}{4n} - \frac{1}{2\alpha} c_0^2 \right) u^2\, d\xi\, dt;$$

hence, for a given $N > \frac{2n}{\alpha} c_0^2$, we obtain

$$\langle \mathscr{L}u, e^{-Nt}u \rangle_{H^-,H} \geq c \|u\|_H^2.$$

Applying the Schwarz inequality to $\langle \mathscr{L}u, e^{-Nt}u \rangle_{H^-,H}$, we get the estimate (4.32).

Estimates (4.33) and (4.34) can be proved similarly. In the first case, we have to estimate below the form

$$\left\langle \int_0^t e^{M\tau} v(\tau,\xi)\, d\tau, \mathscr{L}^* v \right\rangle_{W,W^-},$$

where $v \in D_+, M > 0$. It is clear that $u(t,\xi) = \int_0^t e^{M\tau} v(\tau,\xi)\, d\tau \in W$ and $v = e^{-Mt} u_t$. Let us consider the value of the functional $\mathscr{L}^* v$ on an element u

$$\langle u, \mathscr{L}^* v \rangle_{W,W^-} = \int_Q e^{-Mt} \left(u_t^2 + \sum_{i,j=1}^n a_{ij} u_{\xi_j} u_{\xi_i t} - \sum_{i=1}^n a_i u u_{t\xi_i} + a u u_t \right) d\xi\, dt.$$

Using the formula of integration by parts and taking into account the boundary conditions and the coefficient conditions, we have

$$\int_Q e^{-Mt} \sum_{i,j=1}^n a_{ij} u_{\xi_j} u_{\xi_i t}\, d\xi\, dt = \frac{1}{2} \int_\Omega \sum_{i,j=1}^n e^{-Mt} a_{ij} u_{\xi_j} u_{\xi_i} \Big|_{t=0}^{t=T} d\xi$$

$$+ \frac{M}{2} \int_Q e^{-Mt} \sum_{i,j=1}^n a_{ij} u_{\xi_j} u_{\xi_i}\, d\xi\, dt$$

$$\geq \frac{\alpha M}{2} \int_Q e^{-Mt} \sum_{i=1}^n u_{\xi_i}^2\, d\xi\, dt,$$

$$\int_Q e^{-Mt} a u u_t\, d\xi\, dt - \int_Q e^{-Mt} \sum_{i=1}^n a_i u u_{t\xi_i}\, d\xi\, dt = \int_Q e^{-Mt} \left(\sum_{i=1}^n \frac{\partial a_i}{\partial \xi_i} + a \right) u u_t\, d\xi\, dt$$

$$+ \int_Q e^{-Mt} \sum_{i=1}^n a_i u_{\xi_i} u_t\, d\xi\, dt$$

$$\geq -c_0 \int_Q e^{-Mt} \sum_{i=1}^n |u_t| |u_{\xi_i}|\, d\xi\, dt.$$

Gathering all these estimates together gives the following inequality:

$$\langle u, \mathscr{L}^* v \rangle_{W,W^-} \geq \int_Q e^{-Mt} \left(u_t^2 + \frac{\alpha M}{2} \sum_{i=1}^n u_{\xi_i}^2 - \sum_{i=1}^n c_0 |u_t| |u_{\xi_i}| \right) d\xi \, dt$$

$$= \int_Q e^{-Mt} \left(\frac{1}{2} u_t^2 + \frac{\alpha M}{4} \sum_{i=1}^n u_{\xi_i}^2 \right) d\xi \, dt$$

$$+ \int_Q e^{-Mt} \sum_{i=1}^n \left(\frac{1}{2n} u_t^2 + \frac{\alpha M}{4} u_{\xi_i}^2 - c_0 |u_t| |u_{\xi_i}| \right) d\xi \, dt.$$

Using the inequality $c_0 |u_t| \cdot |u_{\xi_i}| \leq \frac{1}{2n} u_t^2 + \frac{n}{2} c_0^2 u_{\xi_i}^2$, we have

$$\langle u, \mathscr{L}^* v \rangle_{W,W^-} \geq \int_Q e^{-Mt} \left(\frac{1}{2} u_t^2 + \frac{\alpha M}{4} \sum_{i=1}^n u_{\xi_i}^2 \right) d\xi \, dt$$

$$+ \int_Q e^{-Mt} \sum_{i=1}^n \left(\frac{\alpha M}{4} - \frac{n}{2} c_0^2 \right) u_{\xi_i}^2 \, d\xi \, dt,$$

hence, for a given $M > \frac{2n}{\alpha} c_0^2$ we have $\langle u, \mathscr{L}^* v \rangle_{W,W^-} \geq c \|u\|_W^2$. Applying the Schwarz inequality to $\langle u, \mathscr{L}^* v \rangle_{W,W^-}$ and taking into account $\|v\|_{L_2(Q)} \leq c \|u\|_W$, we obtain the inequality (4.33) for smooth functions $v \in D_+$. For reasons of the density, the inequalities (4.33) hold on the entire space H.

To prove (4.34) let us consider the value of the functional $\tilde{\mathscr{L}}^* v \in H^-$ on the element $e^{Mt} v$, where $v \in W_+$ and M is a sufficiently large positive number.   □

4.5.3 Generalized Solvability of the Boundary Value Problem

Using the inequalities proved above we will prove that the operator equations $\mathscr{L}u = f$, $\mathscr{L}^* v = g$ and $\tilde{\mathscr{L}}u = f$, $\tilde{\mathscr{L}}^* v = g$ are correctly and densely solvable. It is naturally to call the solutions of these equations the generalized solutions of the boundary value problem (4.25), (4.26) and of the adjoint problem.

Theorem 4.4. *For any function $f \in L_2(Q)$ there exists a unique solution $u \in W$ of the operator equation $\mathscr{L}u = f$, and the following estimate holds*

$$\|u\|_W \leq c \|f\|_{L_2(Q)}.$$

Proof. Let us consider $f \in L_2(Q)$. By virtue of inequalities (4.33) the following estimate holds

$$(f, v)_{L_2(Q)} \leq \|f\|_{L_2(Q)} \|v\|_{L_2(Q)} \leq c \|\mathscr{L}^* v\|_{W^-}, v \in H.$$

Therefore, the expression $l(w) = (f, v)_{L_2(Q)}$ $(w = \mathscr{L}^* v)$ specifies a linear bounded functional on $R(\mathscr{L}^*) \subset W^-$, moreover $\|l\| \leq c \|f\|_{L_2(Q)}$. Let us extend the functional l linearly and with preservation of norm to a linear continuous functional

$\bar{l} \in (W^-)^*$. By the Riesz representation theorem for a linear continuous functional in W^- there exists such an element $u \in W$ that $\bar{l}(w) = \langle u, w \rangle_{W,W^-}$ and $\|u\|_W = \|\bar{l}\|$. Since for an arbitrary element $v \in H$

$$(f, v)_{L_2(Q)} = \bar{l}(\mathscr{L}^* v) = \langle u, \mathscr{L}^* v \rangle_{W,W^-} = \langle \mathscr{L} u, v \rangle_{H^-,H},$$

then $\mathscr{L} u = f$. The uniqueness of the solution follows from the inequality (4.32) and the embedding $W \subset H$.   □

Corollary 4.1. Problem (4.25), (4.26) for any right-hand side $f \in L_2(Q)$ has a unique weak solution $u \in W$, i.e. an element $u \in W$ for an arbitrary function $v \in H$ that satisfies the integral identity

$$\int_Q \left(u_t v + \sum_{i,j=1}^n a_{ij} u_{\xi_j} v_{\xi_i} - \sum_{i=1}^n a_i u v_{\xi_i} + a u v \right) d\xi \, dt = (f, v)_{L_2(Q)}.$$

Similar to Theorem 4.4 we can prove the following theorems.

Theorem 4.5. *For any element $f \in H^-$ there exists a unique solution $u \in H$ of the operator equation $\mathscr{L} u = f$, and the following estimate holds*

$$\|u\|_H \leq c \|f\|_{H^-}.$$

Corollary 4.2. For any right-hand side $f \in H^-$ the boundary value problems (4.25) and (4.26) has a unique weak solution $u \in H$, i.e. an element that for any function $v \in W_+$ satisfies the integral identity

$$\int_Q \left(-u v_t + \sum_{i,j=1}^n a_{ij} u_{\xi_j} v_{\xi_i} - \sum_{i=1}^n a_i u v_{\xi_i} + a u v \right) d\xi \, dt = \langle f, v \rangle_{H^-,H}.$$

Theorem 4.6. *For an arbitrary function $g \in L_2(Q)$ there exists a unique solution $v \in W_+$ of the operator equation $\mathscr{L}^* v = g$, and the following estimate holds*

$$\|v\|_{W_+} \leq c \|g\|_{L_2(Q)}.$$

Theorem 4.7. *For any element $g \in H^-$ there exists a unique solution $v \in H$ of the operator equation $\mathscr{L}^* v = g$, and the following estimate holds*

$$\|v\|_H \leq c \|g\|_{H^-}.$$

To prove the solvability of problems (4.25) and (4.26) for the right-hand sides which are more singular, let us extend the class of generalized solutions.

Definition 4.11. A function $u \in L_2(Q)$ is called a *generalized solution* of the operator equation $\mathscr{L} u = f$ ($f \in W_+^-$) if it satisfies the identity

$$(u, g)_{L_2(Q)} = \langle f, v \rangle_{W_+^-, W_+} \quad \forall g \in L_2(Q),$$

where $v \in W_+$ is a solution of the operator equation $\mathscr{L}^* v = g$.

Remark 4.19. The generalized solutions in $L_2(Q)$ are nothing else than ultra-weak solutions in the method of transposition of isomorphism defined by J.-L.Lions.

The naturalness of this definition is based on the following properties of a generalized solution.

1. A classic solution of the equation $\mathscr{L}u = f$ is a generalized solution. Indeed, since $H \subset L_2(Q)$, then such an element $u \in H$, that $\mathscr{L}u = f$ ($f \in W_+^-$), satisfies the identity

$$\langle f, v \rangle_{W_+^-, W_+} = \langle \mathscr{L}u, v \rangle_{W_+^-, W_+} = \langle u, \mathscr{L}^*v \rangle_{H, H^-} = (u, \mathscr{L}^*v)_{L_2(Q)}$$

   for all $v \in W_+:$ $\mathscr{L}^*v \in L_2(Q)$, i.e. u is a generalized solution.
2. If a generalized solution $u \in L_2(Q)$ of the equation $\mathscr{L}u = f$ belongs to the space H, than it is a classic solution. Indeed, let $u \in H$ be a generalized solution. Then the following equality holds

$$\langle f, v \rangle_{W_+^-, W_+} = (u, \mathscr{L}^*v)_{L_2(Q)} = \langle u, \mathscr{L}^*v \rangle_{H, H^-} = \langle \mathscr{L}u, v \rangle_{W_+^-, W_+}$$

   for all $v \in W_+:$ $\mathscr{L}^*v \in L_2(Q)$.
   The density of the set $\{v \in W_+ : \mathscr{L}^*v \in L_2(Q)\}$ in the space W_+ implies that $\mathscr{L}u = f$.
3. If $u \in L_2(Q)$ is a generalized solution of the equation $\mathscr{L}u = f$ with the right-hand side $f \in R(\mathscr{L})$, than $u \in H$ is a classic solution also.

Using the inequality (4.31) we can prove the following theorem

Theorem 4.8. *For an arbitrary element $f \in W_+^-$ there exists a unique generalized solution $u \in L_2(Q)$ of the operator equation $\mathscr{L}u = f$, and the following estimate holds*

$$\|u\|_{L_2(Q)} \leq c \|f\|_{W_+^-}.$$

Remark 4.20. Similarly, we can give the definition of a generalized solution for the adjoint problems in $L_2(Q)$ and prove its existence.

4.6 Application to Parabolic Differential Equation in a Disconnected Region

Let us consider once more example of application of the method of a priori estimates to a boundary value problem for parabolic equation in a disconnected region with a contact condition. Such boundary value problems arises in the theory of heat and mass transport in heterogeneous media Ω_1, Ω_2, when media contact each other via a thin three-layered inclusion $\gamma = \gamma_1 \cup \gamma_2 \cup \gamma_3$ (see, e.g. [67]).

In such cases, a zone of foreign layer γ is excluded from a region, where the process is carried, and the influence of an inclusion is described by conditions of conjugation Ω_1, Ω_2. In order to define the problem correctly, it is necessary to pose the contact conditions in every region Ω_1 and Ω_2 in addition to initial and classical boundary conditions. In the case of three-layered inclusion, we obtain a problem with heterogeneous non-ideal contact conditions [11,55–57,81] (γ_1, γ_3 – poorly permeable inclusions where the coefficient of filtration is much less than in other media Ω_1, Ω_2, γ_2 is a highly permeable inclusion, for example, thin tectonic break, thin layer of highly permeable media, formed as a result of dissolution of thin salt layers and so on). Thus, we come to a boundary condition in disconnected region.

This process can be considered from the other point of view [55–57,77,83]. An excluded layer is returned to the region of solution, but coefficients of equations become generalized functions. In this section, we develop this approach for a parabolic system with heterogeneous non-ideal contact conditions of conjugation.

4.6.1 Main Definitions

Let the state of a system be described by the function $u(t, \xi_1, \xi_2, \ldots, \xi_n)$ defined in a cylindrical region $Q = (0, T) \times \Omega$, where $\Omega = \Omega_1 \cup \gamma \cup \Omega_2 \subset \mathbb{R}^n$ is a bounded simply connected domain of changes of space variables $\xi = (\xi_1, \ldots, \xi_n)$ with a regular bound $\partial \Omega$ which is broken by a smooth hypersurface $\bar{\gamma} = \overline{\Omega}_1 \cap \overline{\Omega}_2 \subset \mathbb{R}^n$ of dimension $(n-1)$ onto two simply connected domains Ω_1 and Ω_2 ($\Omega_1 \cap \Omega_2 = \varnothing$). Let us introduce the following notations: $Q_i = (0, T) \times \Omega_i$, $i = \overline{1,2}$, $Q_3 = (0, T) \times \gamma$.

Let us consider the diffusion process that evolves in two heterogeneous regions Q_1 and Q_2 Hausdorff by a three-layered inclusion Q_3

$$\frac{\partial u}{\partial t} + q(\xi)u - \sum_{i,j=1}^{n} \frac{\partial}{\partial \xi_i}\left(k_{ij}(\xi)\frac{\partial u}{\partial \xi_j}\right) = f(t, \xi), \quad (t, \xi) \in Q_1 \cup Q_2, \quad (4.35)$$

$$u|_{t=0} = 0, \qquad u|_{\xi \in \partial \Omega} = 0, \qquad (4.36)$$

$$[(\omega, \mathbf{n})_{\mathbb{R}^n}] = f_0(t, \xi), \qquad (t, \xi) \in Q_3, \qquad (4.37)$$

$$[u] + R_1(\omega, \mathbf{n})_{\mathbb{R}^n}^- + R_3(\omega, \mathbf{n})_{\mathbb{R}^n}^+ = 0, \qquad (t, \xi) \in Q_3, \qquad (4.38)$$

where $\omega = -\mathbf{K}\,\mathrm{grad}\,u$ in $Q_1 \cup Q_2$, $\mathbf{K} = \{k_{ij}\}_{i,j=1}^{n}$ is a non-degenerated matrix, $\mathrm{grad}\,u = \left(u_{\xi_1}, \ldots, u_{\xi_n}\right)$, $[u]$ is a jump of function $u(t, \xi)$ on Q_3, i.e.

$$[u](t, \xi_0) = u^+(t, \xi_0) - u^-(t, \xi_0), \qquad \xi_0 \in \gamma,$$
$$u^+(t, \xi_0) = \lim_{\xi^+ \to \xi_0} u(t, \xi^+),$$
$$u^-(t, \xi_0) = \lim_{\xi^- \to \xi_0} u(t, \xi^-), \qquad \xi^+ \in \Omega_2, \xi^- \in \Omega_1,$$

$[(\omega,\mathbf{n})_{\mathbb{R}^n}], (\omega,\mathbf{n})^+_{\mathbb{R}^n}, (\omega,\mathbf{n})^-_{\mathbb{R}^n}$ are defined similarly, $R_1(\xi) \geq 0$, $R_3(\xi) \geq 0$ are the functions, continuous on $\bar{\gamma}$ and describe physical parameters of inclusions γ_1, γ_3 $(R_1(\xi) + R_3(\xi) > 0)$, $\mathbf{n} = (n_{\xi_1}, \ldots, n_{\xi_n})$ is the normal to the surface γ, which is outward to Ω_1, $q(\xi), k_{ij}(\xi)$ have a simple disconnection on the surface Q_3.

Note that when $f_0 = 0$ the conditions of conjugation (4.37), (4.38) turn into heterogeneous conditions of conjugation of non-ideal contact type

$$[(\omega,\mathbf{n})_{\mathbb{R}^n}] = 0, \quad \alpha[u] + (\omega,\mathbf{n})_{\mathbb{R}^n} = 0, \quad (t,\xi) \in Q_3,$$

where $\alpha = 1/(R_1 + R_3)$.

According to [56, 57], let us pass from (4.35) to (4.38) to a first order system of linear differential equations (with respect to (u, ω)), which allow to take into account the conditions of conjugation (4.37) and (4.38) in the equations of the system.

Let $C^k(\overline{Q}_1, \overline{Q}_2)$ be a set of functions of the class $C^k(Q_1 \cup Q_2)$, which allow extensions keeping the smoothing from Q_1 into $\overline{Q}_1$ and from Q_2 into $\overline{Q}_2$.

Let us define on the set $C^0(\overline{Q}_1, \overline{Q}_2)$ the linear functionals $\delta^-(\gamma), \delta^+(\gamma)$ in the following way:

$$\delta^-(\gamma)(u) = \int_{Q_3} u^-(t,\xi)\,dQ_3, \quad \delta^+(\gamma)(u) = \int_{Q_3} u^+(t,\xi)\,dQ_3, \quad u \in C^0(\overline{Q}_1, \overline{Q}_2).$$

By a generalized left (right) derivative of the function $f \in C^1(\overline{Q}_1, \overline{Q}_2)$ we mean a functional defined on the functions $u \in C^0(\overline{Q}_1, \overline{Q}_2)$ by the rule

$$\frac{\partial_l f}{\partial \xi_i} = f^*_{\xi_i} + [f]n_{\xi_i}\delta^+(\gamma), \quad \left(\frac{\partial_r f}{\partial \xi_i} = f^*_{\xi_i} + [f]n_{\xi_i}\delta^-(\gamma)\right),$$

where $f^*_{\xi_i}$ is a classic derivative of the function f in $Q_1 \cup Q_2$.

Then for the function $u(t,\xi) \in C^2(\overline{Q}_1, \overline{Q}_2)$ the following equalities hold:

$$\frac{R_3}{R_1 + R_3}\operatorname{grad}_l u = \frac{R_3}{R_1 + R_3}(u^*_{\xi_1}, \ldots, u^*_{\xi_n}) + \frac{R_3[u]\mathbf{n}\delta^+(\gamma)}{R_1 + R_3},$$

$$\frac{R_1}{R_1 + R_3}\operatorname{grad}_r u = \frac{R_1}{R_1 + R_3}(u^*_{\xi_1}, \ldots, u^*_{\xi_n}) + \frac{R_1[u]\mathbf{n}\delta^-(\gamma)}{R_1 + R_3}.$$

Taking into account the equality $\omega = -\mathbf{K}\operatorname{grad} u$ in $Q_1 \cup Q_2$ and the condition (4.38) on the surface Q_3, we get

$$\frac{R_3}{R_1 + R_3}\operatorname{grad}_l u = -\frac{R_3\mathbf{K}^{-1}\omega}{R_1 + R_3} - \frac{R_3\mathbf{n}\delta^+(\gamma)(R_1(\omega,\mathbf{n})^-_{\mathbb{R}^n} + R_3(\omega,\mathbf{n})^+_{\mathbb{R}^n})}{R_1 + R_3},$$

$$\frac{R_1}{R_1 + R_3}\operatorname{grad}_r u = -\frac{R_1\mathbf{K}^{-1}\omega}{R_1 + R_3} - \frac{R_1\mathbf{n}\delta^-(\gamma)(R_1(\omega,\mathbf{n})^-_{\mathbb{R}^n} + R_3(\omega,\mathbf{n})^+_{\mathbb{R}^n})}{R_1 + R_3},$$

or

$$\frac{R_3\operatorname{grad}_l u + R_1\operatorname{grad}_r u}{R_1 + R_3} = -\mathbf{K}^{-1}\omega - \frac{(R_1\mathbf{n}\delta^-(\gamma) + R_3\mathbf{n}\delta^+(\gamma))^2\,\omega}{R_1 + R_3}, \quad (t,\xi) \in Q.$$

$$(4.39)$$

Similarly, generalized derivatives of the vector ω may be written as

$$\frac{R_1 \operatorname{div}_l \omega}{R_1 + R_3} = \sum_{i=1}^{n} \frac{R_1(\omega_i)^*_{\xi_i}}{R_1 + R_3} + \frac{R_1 \delta^+(\gamma)([\omega], \mathbf{n})_{\mathbb{R}^n}}{R_1 + R_3},$$

$$\frac{R_3 \operatorname{div}_r \omega}{R_1 + R_3} = \sum_{i=1}^{n} \frac{R_3(\omega_i)^*_{\xi_i}}{R_1 + R_3} + \frac{R_3 \delta^-(\gamma)([\omega], \mathbf{n})_{\mathbb{R}^n}}{R_1 + R_3}.$$

Hence, taking into account (4.37) we have

$$\frac{R_1 \operatorname{div}_l \omega + R_3 \operatorname{div}_r \omega}{R_1 + R_3} = \sum_{i=1}^{n} (\omega_i)^*_{\xi_i} + \frac{R_1 f_0 \delta^+(\gamma) + R_3 f_0 \delta^-(\gamma)}{R_1 + R_3}.$$

Thus, (4.35) may be rewritten as

$$\frac{\partial u}{\partial t} + q(\xi) u + \frac{R_1 \operatorname{div}_l \omega + R_3 \operatorname{div}_r \omega}{R_1 + R_3} = f + \frac{R_1 f_0 \delta^+(\gamma) + R_3 f_0 \delta^-(\gamma)}{R_1 + R_3}, \qquad (4.40)$$

where $(t, \xi) \in Q$ and the derivatives are meant in a the generalized sense.

Thus, instead of (4.35) and conditions (4.37) and (4.38) we obtain the first order system of linear partial derivative equations (4.39) and (4.40).

Let us consider this problem in the generalized sense.

Let $C^1_{bd}(\overline{Q}_1, \overline{Q}_2)$ be a subset of $C^1(\overline{Q}_1, \overline{Q}_2)$ which consists of functions satisfying initial and boundary conditions (4.36). Similarly, let $C^1_{bd*}(\overline{Q}_1, \overline{Q}_2)$ be a subset of functions of $C^1(\overline{Q}_1, \overline{Q}_2)$ satisfying the adjoint conditions

$$v|_{t=T} = 0, \quad v|_{\xi \in \partial \Omega} = 0. \qquad (4.41)$$

Let us denote by C_{bd} the set of pairs of functions $x = (u, \omega) \in C^1_{bd}(\overline{Q}_1, \overline{Q}_2) \times (C^0(\overline{Q}_1, \overline{Q}_2))^n$ which satisfy adjoint condition (4.38). Similarly, let C_{bd*} be a set of pairs of functions $y = (v, \eta) \in C^1_{bd*}(\overline{Q}_1, \overline{Q}_2) \times (C^0(\overline{Q}_1, \overline{Q}_2))^n$, satisfying on Q_3 the condition

$$[v] = R_1(\eta, \mathbf{n})^-_{\mathbb{R}^n} + R_3(\eta, \mathbf{n})^+_{\mathbb{R}^n}.$$

In addition, let $W_2^{1,1/1}(Q)$ be a completion of the set $C^1_{bd}(\overline{Q}_1, \overline{Q}_2)$ with respect to the norm

$$\|u\|^2_{W_2^{1,1/1}(Q)} = \sum_{k=1}^{2} \int_{Q_k} u_t^2 + \sum_{i=1}^{n} u_{\xi_i}^2 \, dQ_k, \qquad (4.42)$$

and $W_2^{1,1}(Q)$ be a completion of $C^1(\overline{Q})$ with respect to the same Sobolev norm (4.42).

It is clear that elements of the space $W_2^{1,1/1}(Q)$ may be interpreted as pairs of functions $(u_1, u_2) \in W_2^{1,1}(Q_1) \times W_2^{1,1}(Q_2)$ satisfying condition (4.36) on corresponding parts of bound. Similarly, $W_{2,*}^{1,1/1}(Q)$ is a completion of $C^1_{bd*}(\overline{Q}_1, \overline{Q}_2)$ with respect to the norm (4.42). Denote as $W_2^{-1,1/1}(Q)$ and $W_{2,*}^{-1,1/1}(Q)$ the spaces which are conjugate to $W_2^{1,1/1}(Q)$ and $W_{2,*}^{1,1/1}(Q)$, respectively.

By the theorem on traces, functions from $(u_1, u_2) \in W_2^{1,1}(Q_1) \times W_2^{1,1}(Q_2)$ leave the traces $(u^-, u^+) \in L_2(Q_3) \times L_2(Q_3)$ on the surface Q_3, and the trace operator is continuous. That is why for all $u \in W_2^{1,1/1}(Q)$ the following inequality holds:

$$\int_{Q_3} [u]^2 dQ_3 \leq c \|u\|^2_{W_2^{1,1/1}(Q)},$$

where c hereinafter is some positive constant. The inequality $\|[v]\|_{L_2(Q_3)} \leq c\|v\|_{W_2^{1,1/1}(Q)}$ for all $v \in W_{2,*}^{1,1/1}(Q)$ can be proved in the same way.

Let us introduce the space X (Y) as a completion of the set C_{bd} (C_{bd*}, respectively) in the norm

$$\|x\|^2 = \|u\|^2_{W_2^{1,1/1}(Q)} + \|\omega\|^2_{L_2^n(Q)}.$$

In the pair $x = (u, \omega) \in X$ the vector ω leaves a trace like $R_1(\omega, \mathbf{n})_{\overline{\mathbb{R}^n}}^- + R_3(\omega, \mathbf{n})_{\overline{\mathbb{R}^n}}^+$ on the surface Q_3, defined by the equality

$$R_1(\omega, \mathbf{n})_{\overline{\mathbb{R}^n}}^- + R_3(\omega, \mathbf{n})_{\overline{\mathbb{R}^n}}^+ = -[u].$$

The relations between the vector-function ω and its trace $R_1(\omega, \mathbf{n})_{\overline{\mathbb{R}^n}}^- + R_3(\omega, \mathbf{n})_{\overline{\mathbb{R}^n}}^+$ on Q_3 is seen better when we consider on C_{bd} the norm

$$\|x\|^2_0 = \|u\|^2_{W_2^{1,1/1}(Q)} + \|\omega\|^2_{L_2^n(Q)} + \|R_1(\omega, \mathbf{n})_{\overline{\mathbb{R}^n}}^- + R_3(\omega, \mathbf{n})_{\overline{\mathbb{R}^n}}^+\|^2_{L_2(Q_3)}$$

which is equivalent to the norm of the space X. Now, let us denote as $L_{2,\gamma}^n(Q)$ the completion of the set $(C^0(\overline{Q}_1, \overline{Q}_2))^n$ with respect to the norm

$$\|\omega\|^2_{L_{2,\gamma}^n(Q)} = \|\omega\|^2_{L_2^n(Q)} + \|R_1(\omega, \mathbf{n})_{\overline{\mathbb{R}^n}}^- + R_3(\omega, \mathbf{n})_{\overline{\mathbb{R}^n}}^+\|^2_{L_2(Q_3)},$$

then, element of the space $L_{2,\gamma}^n(Q)$ is a set of functions ω from $L_2^n(Q)$, whose traces $R_1(\omega, \mathbf{n})_{\overline{\mathbb{R}^n}}^- + R_3(\omega, \mathbf{n})_{\overline{\mathbb{R}^n}}^+ \in L_2(Q_3)$ are meaningful. More precisely, the set $L_{2,\gamma}^n(Q)$ is isometric $L_2^n(Q) \times L_2(Q_3)$, where the operator of isometry $O : L_{2,\gamma}^n(Q) \to L_2^n(Q) \times L_2(Q_3)$ is set as a completion by continuity of the operator

$$(C^0(\overline{Q}_1, \overline{Q}_2))^n \ni \omega \to O\omega = (\omega, R_1(\omega, \mathbf{n})_{\overline{\mathbb{R}^n}}^- + R_3(\omega, \mathbf{n})_{\overline{\mathbb{R}^n}}^+) \in L_2^n(Q) \times L_2(Q_3),$$

on the entire space $L_{2,\gamma}^n(Q)$.

Similarly, in the pair $y = (v, \eta) \in Y$ the vector η leaves the trace $R_1(\eta, \mathbf{n})_{\overline{\mathbb{R}^n}}^- + R_3(\eta, \mathbf{n})_{\overline{\mathbb{R}^n}}^+ = [v]$ on Q_3.

Let the natural bilinear form $\langle \cdot, \cdot \rangle_{X \times X^*}$ be defined on Cartesian product of the original space and its conjugate space (for example, X and X^*).

Let us consider a system describing heat and mass transport in two heterogeneous media with heterogeneous conditions of conjugation of non-ideal contact type:

$$\mathscr{L}x = F, \tag{4.43}$$

where the operator $\mathscr{L}$ is defined as a symbolic matrix

$$\mathscr{L} = \left(\begin{array}{c|c} \dfrac{\partial}{\partial t} + q & \dfrac{R_1 \operatorname{div}_l + R_3 \operatorname{div}_r}{R_1 + R_3} \\ \hline \dfrac{R_3 \operatorname{grad}_l + R_1 \operatorname{grad}_r}{R_1 + R_3} & \mathbf{M} \end{array}\right), \quad x = \begin{pmatrix} u \\ \omega \end{pmatrix}.$$

The function $u(t,\xi)$ describes heat and mass transport, $\omega = (\omega_1, \ldots, \omega_n)$ is a vector of specific flux of substance. The operator $\mathscr{L}$ acts from X into Y^*, the domain of $\mathscr{L}$ is the set $D(\mathscr{L}) = C_{bd}$.

The coefficients of the system satisfy the following conditions: $q(\xi) \in C^0(\overline{\Omega}_1, \overline{\Omega}_2)$, $q \geq 0$, the coefficient matrix $\mathbf{M} = \{\sigma_{ij}\}_{i,j=1}^n$ has the form

$$\mathbf{M} = \mathbf{K}^{-1} + \frac{(R_1 \mathbf{n}\delta^-(\gamma) + R_3 \mathbf{n}\delta^+(\gamma))^2}{R_1 + R_3},$$

where $\mathbf{K}^{-1} = \{\bar{k}_{ij}\}_{i,j=1}^n$ is an inverse matrix to the coefficient matrix $\mathbf{K} = \{k_{ij}\}_{i,j=1}^n$ of the original parabolic equation ($\bar{k}_{ij}(\xi) \in C^0(\overline{\Omega}_1, \overline{\Omega}_2)$). We suppose that the matrix $\mathbf{K}$ is symmetric $k_{ij} = k_{ji}$ and uniformly positive defined in $\Omega_1 \cup \Omega_2$

$$\sum_{i,j=1}^n k_{ij}(\xi)\lambda_i\lambda_j \geq c^{-1} \sum_{i=1}^n \lambda_i^2, \quad \lambda_i \in \mathbb{R}, \xi \in \Omega_1 \cup \Omega_2,$$

where c is a positive constant which does not depend on ξ, λ_i.

Under $(R_1 \operatorname{div}_l \omega + R_3 \operatorname{div}_r \omega)/(R_1 + R_3)$ in the equations (4.43) we mean a linear continuous functional over $v \in W_{2,*}^{1,1/1}$ which acts by the rule

$$\left\langle \frac{R_1 \operatorname{div}_l \omega + R_3 \operatorname{div}_r \omega}{R_1 + R_3}, v \right\rangle_{W_{2,*}^{-1,1/1} \times W_{2,*}^{1,1/1}} = -\sum_{k=1}^2 \sum_{i=1}^n \int_{Q_k} \omega_i \frac{\partial v}{\partial \xi_i} dQ_k$$

$$- \int_{Q_3} \frac{R_1(\omega, \mathbf{n})_{\mathbb{R}^n}^- + R_3(\omega, \mathbf{n})_{\mathbb{R}^n}^+}{R_1 + R_3} [v] dQ_3.$$

Note, that for smooth functions this equality corresponds to the formula of integration by parts.

By $(R_3 \operatorname{grad}_l u + R_1 \operatorname{grad}_r u)/(R_1 + R_3)$ we mean a linear continuous functional over $\eta \in L_{2,\gamma}^n(Q)$ (or over $y \in Y$):

$$\left\langle \frac{R_3 \operatorname{grad}_l u + R_1 \operatorname{grad}_r u}{R_1 + R_3}, \eta \right\rangle_{(L_{2,\gamma}^n)^* \times L_{2,\gamma}^n} = \sum_{k=1}^2 \sum_{i=1}^n \int_{Q_k} \frac{\partial u}{\partial \xi_i} \eta_i dQ_k$$

$$+ \int_{Q_3} \frac{R_1(\eta, \mathbf{n})_{\mathbb{R}^n}^- + R_3(\eta, \mathbf{n})_{\mathbb{R}^n}^+}{R_1 + R_3} [u] dQ_3.$$

By $\mathbf{M}\omega$ we mean a functional over $\eta \in L_{2,\gamma}^n(Q)$:

$$\langle \mathbf{M}\omega, \eta \rangle_{(L_{2,\gamma}^n)^* \times L_{2,\gamma}^n} = \sum_{k=1}^{2} \sum_{i,j=1}^{n} \int_{Q_k} \bar{k}_{ij} \omega_j \eta_i \, dQ_k$$

$$+ \int_{Q_3} \frac{\left(R_1(\omega, \mathbf{n})_{\mathbb{R}^n}^- + R_3(\omega, \mathbf{n})_{\mathbb{R}^n}^+ \right) \left(R_1(\eta, \mathbf{n})_{\mathbb{R}^n}^- + R_3(\eta, \mathbf{n})_{\mathbb{R}^n}^+ \right)}{R_1 + R_3} \, dQ_3.$$

Thus, taking into account (4.38), we have

$$\langle \mathscr{L}x, y \rangle_{Y^* \times Y} = \sum_{k=1}^{2} \int_{Q_k} \frac{\partial u}{\partial t} v + quv + \sum_{i,j=1}^{n} \bar{k}_{ij} \omega_j \eta_i \, dQ_k$$

$$+ \sum_{k=1}^{2} \sum_{i=1}^{n} \int_{Q_k} \frac{\partial u}{\partial \xi_i} \eta_i - \frac{\partial v}{\partial \xi_i} \omega_i \, dQ_k + \int_{Q_3} \frac{[u][v]}{R_1 + R_3} \, dQ_3. \qquad (4.44)$$

By $\mathscr{L}^+$ we denote an adjoint operator

$$\mathscr{L}^+ y = G, \qquad \mathscr{L}^+ : Y \to X^*, \quad y = (v, \eta).$$

Let us write a symbolic matrix which defines the operator $\mathscr{L}^+$:

$$\mathscr{L}^+ = \left(\begin{array}{c|c} -\dfrac{\partial}{\partial t} + q & -\dfrac{R_1 \operatorname{div}_l + R_3 \operatorname{div}_r}{R_1 + R_3} \\ \hline -\dfrac{R_3 \operatorname{grad}_l + R_1 \operatorname{grad}_r}{R_1 + R_3} & \mathbf{M} \end{array} \right),$$

where values of symbolic operators are defined similarly to the matrix of the operator $\mathscr{L}$. For the moment, let take as a domain of the operator $\mathscr{L}^+$ the set $D(\mathscr{L}^+) = C_{\mathrm{bd}*}$. Then

$$\langle \mathscr{L}^+ y, x \rangle_{X^* \times X} = \sum_{k=1}^{2} \int_{Q_k} -\frac{\partial v}{\partial t} u + quv + \sum_{i,j=1}^{n} \bar{k}_{ij} \omega_j \eta_i \, dQ_k$$

$$+ \sum_{k=1}^{2} \sum_{i=1}^{n} \int_{Q_k} \frac{\partial u}{\partial \xi_i} \eta_i - \frac{\partial v}{\partial \xi_i} \omega_i \, dQ_k$$

$$+ \int_{Q_3} \frac{[v][u]}{R_1 + R_3} \, dQ_3$$

$$= \langle y, \mathscr{L}x \rangle_{Y \times Y^*}, \qquad (4.45)$$

for all $x \in D(\mathscr{L}), y \in D(\mathscr{L}^+)$.

4.6.2 Properties of Operators Associated with a Boundary Value Problem

By means (4.44) and (4.45), we can prove that the operators $\mathscr{L}$ and $\mathscr{L}^+$ are continuous in their domains. The density of the set $D(\mathscr{L})$ in X ($D(\mathscr{L}^+)$ in Y, respectively) allows to extend $\mathscr{L}$ ($\mathscr{L}^+$, respectively) by continuity on the entire space X (Y, respectively). Extended operators we denote by $\bar{\mathscr{L}}$, $\bar{\mathscr{L}}^+$. Thus, the following lemma is true.

Lemma 4.7. *There exists such a positive constant $c > 0$ that for all $x \in X, y \in Y$ the following inequalities hold*

$$\|\bar{\mathscr{L}}x\|_{Y^*} \leq c\|x\|_X, \qquad \|\bar{\mathscr{L}}^+y\|_{X^*} \leq c\|y\|_Y. \tag{4.46}$$

Remark 4.21. Passing to the limit in (4.45) and taking into account (4.46), it is easy to prove that the operators $\bar{\mathscr{L}}, \bar{\mathscr{L}}^+$ satisfy the relation

$$\langle \bar{\mathscr{L}}x, y \rangle_{Y^* \times Y} = \langle x, \bar{\mathscr{L}}^+y \rangle_{X \times X^*}, \qquad \forall x \in X, y \in Y,$$

i.e. $\bar{\mathscr{L}}^+$ is adjoint operator to $\bar{\mathscr{L}}$.

Let us show that a solution of the equation $\bar{\mathscr{L}}x = F$ is connected with the classical solvability of the problems (4.35)–(4.38).

Theorem 4.9. *Let the coefficients k_{ij} of the operator $\bar{\mathscr{L}}$ and the solution $x = (u, \omega) \in X$ of the equation*

$$\bar{\mathscr{L}}x = \left(\bar{f} + \bar{f}_0 \frac{R_3 \delta^-(\gamma) + R_1 \delta^+(\gamma)}{R_1 + R_3}, 0 \right) \in Y^*, \qquad \bar{f} \in L_2(Q), \bar{f}_0 \in L_2(Q_3),$$

are smooth sufficiently to guarantee the classical solvability of the problem (4.35)–(4.38):

(1)$u_t \in C(Q_1 \cup Q_2), u_{\xi_i \xi_j} \in C(Q_1 \cup Q_2), k_{ij} \in C^1(\Omega_1 \cup \Omega_2), i, j = \overline{1, n}.$
(2)There exist one-sided pointwise limits:
$$\lim_{\xi_1 \to \xi_0} (\mathbf{K} \operatorname{grad} u, \mathbf{n})_{\mathbb{R}^n}, \lim_{\xi_2 \to \xi_0} (\mathbf{K} \operatorname{grad} u, \mathbf{n})_{\mathbb{R}^n}, \quad \xi_0 \in \gamma, \xi_k \in \Omega_k, k = \overline{1, 2}.$$

Then there exist such $f \in C(Q_1 \cup Q_2), f_0 \in C(Q_3)$, that the function $u(t, \xi)$ pointwise satisfies relations (4.35)–(4.38), $\omega = -\mathbf{K} \operatorname{grad} u$ in $Q_1 \cup Q_2$, and the equalities $\bar{f} = f$, $\bar{f}_0 = f_0$ hold almost everywhere.

Proof. The conditions (4.36) hold because $x = (u, \omega) \in X$ and the norm of $W_2^{1,1/1}(Q)$ preserves the corresponding limit values.

On the other hand, for any $y = (v, \eta) \in Y$ the following inequality holds.

$$\langle \bar{\mathscr{L}}x, y \rangle_{Y^* \times Y} = (\bar{f}, v)_{L_2(Q)} + \left(\frac{R_3 \bar{f}_0}{R_1 + R_3}, v^- \right)_{L_2(Q_3)} + \left(\frac{R_1 \bar{f}_0}{R_1 + R_3}, v^+ \right)_{L_2(Q_3)}. \tag{4.47}$$

At the same time, let $y = (0, \eta) \in Y$, then we can rewrite (4.47) as

$$\langle \mathscr{L}x, y \rangle_{Y^* \times Y} = \sum_{k=1}^{2} \sum_{i=1}^{n} \int_{Q_k} \sum_{j=1}^{n} \bar{k}_{ij} \omega_j \eta_i + u_{\xi_i} \eta_i \, dQ_k = 0, \qquad \forall \eta_i \in L_2(Q).$$

Hence, $\omega = -\mathbf{K}\operatorname{grad} u$ in the sense of equality in $L_2(Q)$, and taking into account the smoothing of $u(t, \xi)$ we can consider this equality in the pointwise sense in $Q_1 \cup Q_2$ also. Thus, $\omega \in (C^1(Q_1 \cup Q_2))^n$, and $(\omega, \mathbf{n})^-_{\mathbb{R}^n}, (\omega, \mathbf{n})^+_{\mathbb{R}^n}$ make sense.

Let us substitute into (4.47) such $y = (v, \eta) \in Y$ that $v \in C^1(\overline{Q})$ and $v = 0$ over Q_3. Then $[v] = 0$ over Q_3. Integrating by parts we have

$$\langle \mathscr{L}x, y \rangle_{Y^* \times Y} = \sum_{k=1}^{2} \int_{Q_k} \frac{\partial u}{\partial t} v + quv + \sum_{i=1}^{n} \frac{\partial \omega_i}{\partial \xi_i} v \, dQ_k = (\bar{f}, v)_{L_2(Q)}.$$

Hence, by virtue of the density of the considered set of functions $v(t, \xi)$ in $L_2(Q)$, we have

$$\frac{\partial u}{\partial t} + qu + \sum_{i=1}^{n} \frac{\partial \omega_i}{\partial \xi_i} = \frac{\partial u}{\partial t} + qu - \sum_{i,j=1}^{n} \frac{\partial}{\partial \xi_i} \left(k_{ij} \frac{\partial u}{\partial \xi_j} \right) = \bar{f}$$

in $L_2(Q)$. Denoting the left-hand side by $f \in C(Q_1 \cup Q_2)$, we have that $f = \bar{f}$ almost everywhere in $Q_1 \cup Q_2$.

Let us substitute into (4.47) such $y = (v, \eta) \in Y$ that $v \in C^1(\overline{Q}_1)$ and $v = 0$ in Q_2. Integrating by parts we have

$$\langle \mathscr{L}x, y \rangle_{Y^* \times Y} - (f, v)_{L_2(Q)} =$$

$$= \int_{Q_3} \left(\frac{[u]}{R_1 + R_3} + (\omega, \mathbf{n})^-_{\mathbb{R}^n} \right) [v] \, dQ_3 = - \int_{Q_3} \frac{R_3 \bar{f}_0}{R_1 + R_3} [v] \, dQ_3.$$

Whence it follows that

$$\frac{[u]}{R_1 + R_3} + (\omega, \mathbf{n})^-_{\mathbb{R}^n} = - \frac{R_3 \bar{f}_0}{R_1 + R_3}, \qquad (t, \xi) \in Q_3.$$

Similarly, we can prove that

$$\frac{[u]}{R_1 + R_3} + (\omega, \mathbf{n})^+_{\mathbb{R}^n} = \frac{R_1 \bar{f}_0}{R_1 + R_3},$$

This guarantees the holding of the conditions (4.37) and (4.38).   □

Remark 4.22. The theorem remains true if $\bar{f} \in \overset{\circ}{W}{}^{-1,1}_{2,*}$, where $\overset{\circ}{W}{}^{-1,1}_{2,*}$ is a completion of the set of the functions from the space $C^1(\overline{Q})$, satisfying conditions (4.41) and vanishing on the surface Q_3 in the norm (4.42), and $\overset{\circ}{W}{}^{-1,1}_{2,*}$ is the conjugate space to $\overset{\circ}{W}{}^{1,1}_{2,*}$.

Lemma 4.8. *There exists such a positive constant $c > 0$ that for all $x = (u, \omega) \in X$ the following inequality holds:*

$$c^{-1}\|u\|_{L_2(Q)} \leq \|\mathscr{L}x\|_{Y^*}.$$

Proof. Let us consider the value of functional $\mathscr{L}x \in Y^*$ on the element $y = Ix \in Y$, where

$$v = -\int_T^t e^{-\tau} u(\tau, \xi) \, d\tau, \quad \eta = \mathbf{K} \operatorname{grad} v.$$

It is clear, that $y = Ix$ belongs to the space Y.

By definition of the operator $\mathscr{L}x$, we have

$$\langle \mathscr{L}x, y \rangle_{Y^* \times Y} = \sum_{k=1}^{2} (u_t + qu, v)_{L_2(Q_k)} + \sum_{k=1}^{2} \sum_{i,j=1}^{n} (\bar{k}_{ij}\omega_j, \eta_i)_{L_2(Q_k)}$$

$$+ \sum_{k=1}^{2} \sum_{i=1}^{n} (u_{\xi_i}, \eta_i)_{L_2(Q_k)} - \sum_{k=1}^{2} \sum_{i=1}^{n} (\omega_i, v_{\xi_i})_{L_2(Q_k)}$$

$$+ \left(\frac{1}{R_1 + R_3} [u], [v] \right)_{L_2(Q_3)}.$$

Let us consider every item separately. Integrating by parts and taking into account (4.36), we have

$$(u_t + qu, v)_{L_2(Q)} = -(u, v_t)_{L_2(Q)} - (qe^t v_t, v)_{L_2(Q)}$$

$$= \int_Q e^{-t} u^2 \, dQ + \frac{1}{2} \int_\Omega qe^t v^2 \big|_{t=0} \, d\Omega + \frac{1}{2} \int_Q qe^t v^2 \, dQ$$

$$\geq c^{-1} \|u\|_{L_2(Q)}^2.$$

Let us pass to the second item.

$$\sum_{k=1}^{2} \sum_{i,j=1}^{n} \int_{Q_k} \bar{k}_{ij} \omega_i \eta_j \, dQ_k = \sum_{k=1}^{2} \sum_{i=1}^{n} \int_{Q_k} \omega_i \sum_{j=1}^{n} \bar{k}_{ij} \eta_j \, dQ_k$$

$$= \sum_{k=1}^{2} \sum_{i=1}^{n} \int_{Q_k} \omega_i \frac{\partial v}{\partial \xi_i} \, dQ_k.$$

Let us consider the third item. Integrating by parts and taking into account the fact that the matrix $\{k_{ij}\}_{i,j=1}^{n}$ is positively defined, we have

$$\sum_{k=1}^{2}\sum_{i=1}^{n}\int_{Q_k} u_{\xi_i}\eta_i\,dQ_k = \sum_{k=1}^{2}\sum_{i,j=1}^{n}\int_{Q_k} u_{\xi_i}k_{ij}v_{\xi_j}\,dQ_k$$

$$= -\sum_{k=1}^{2}\sum_{i,j=1}^{n}\int_{Q_k} k_{ij}e^{t}v_{t\xi_i}v_{\xi_j}\,dQ_k$$

$$= \sum_{k=1}^{2}\sum_{i,j=1}^{n}\int_{\Omega_k}\frac{1}{2}k_{ij}e^{t}v_{\xi_i}v_{\xi_j}|_{t=0}\,d\Omega_k + \sum_{k=1}^{2}\sum_{i,j=1}^{n}\int_{Q_k}\frac{1}{2}e^{t}k_{ij}v_{\xi_i}v_{\xi_j}\,dQ_k$$

$$\geq c^{-1}\sum_{k=1}^{2}\sum_{i=1}^{n}\int_{Q_k}\left(\int_{T}^{t}e^{-\tau}u_{\xi_i}\,d\tau\right)^{2}dQ_k.$$

Let us estimate the last item.

$$\int_{Q_3}\frac{[u][v]}{R_1+R_3}\,dQ_3 = -\int_{Q_3}\frac{e^{t}[v_t][v]}{R_1+R_3}\,dQ_3$$

$$= \int_{\gamma}\frac{e^{t}[v]^2}{2(R_1+R_3)}|_{t=0}\,d\gamma + \int_{Q_3}\frac{e^{t}[v]^2}{2(R_1+R_3)}\,dQ_3$$

$$\geq 0.$$

Let $\|x\|_{X_1}$ be a semi-norm over X:

$$\|x\|_{X_1}^2 = \int_{Q} u^2\,dQ + \sum_{k=1}^{2}\sum_{i=1}^{n}\int_{Q_k}\left(\int_{T}^{t}e^{-\tau}u_{\xi_i}\,d\tau\right)^2 dQ_k.$$

Thus, we conclude that $\langle \mathscr{L}x,y\rangle_{Y^*\times Y} \geq c^{-1}\|x\|_{X_1}^2$. Applying the Schwarz inequality, we have $\|\mathscr{L}x\|_{Y^*}\cdot\|y\|_{Y} \geq c^{-1}\|x\|_{X_1}^2$.

Let us show that $\|y\|_Y \leq c\|x\|_{X_1}$. Indeed, since $\eta = \mathbf{K}\,\mathrm{grad}\,v$, then

$$\|y\|_Y^2 = \|v_t\|_{L_2(Q)}^2 + \sum_{k=1}^{2}\sum_{i=1}^{n}\|v_{\xi_i}\|_{L_2(Q_k)}^2 + \|\eta\|_{L_2^n(Q)}^2$$

$$\leq \|v_t\|_{L_2(Q)}^2 + c\sum_{k=1}^{2}\sum_{i=1}^{n}\|v_{\xi_i}\|_{L_2(Q_k)}^2$$

$$\leq c\int_{Q} u^2\,dQ + c\sum_{k=1}^{2}\sum_{i=1}^{n}\int_{Q_k}\left(\int_{T}^{t}e^{-\tau}u_{\xi_i}\,d\tau\right)^2 dQ_k$$

$$= c\|x\|_{X_1}^2.$$

Thus, we proved the inequality

$$\|\mathscr{L}x\|_{Y^*} \geq c^{-1}\|x\|_{X_1} \geq c^{-1}\|u\|_{L_2(Q)}, \qquad \forall x \in X.$$

□

Similarly, we can prove the following lemma for an adjoint operator.

Lemma 4.9. *There exists such a positive constant $c > 0$, that for all $y = (v, \eta) \in Y$ the following inequality holds*

$$c^{-1}\|v\|_{L_2(Q)} \leq \|\bar{\mathscr{L}}^+ y\|_{X^*}.$$

To prove this, we have to consider the operator $\bar{\mathscr{L}}^+ y$ on the element $x = (u, \omega) = \bar{I}y$, where

$$u = \int_0^t e^\tau v(\tau, \xi)\,d\tau, \qquad \omega = -\mathbf{K}\operatorname{grad} u.$$

Note that in the left-hand side of the inequalities of Lemmas 4.8 and 4.9 semi-norms of the element x and y appear, and not the norms as in the previous chapters.

Lemma 4.10. *The operators $\mathscr{L}$ and $\bar{\mathscr{L}}^+$ are injective.*

Proof. Let us suppose that there exists such $x = (u, \omega) \in X$ that $\mathscr{L}x = 0$ in Y^*. Then $\langle \mathscr{L}x, y \rangle_{Y^* \times Y} = 0$ for all $y \in Y$, including $y = Ix$ defined in Lemma 4.8. Applying the inequality from Lemma 4.8, we have $0 = \langle \mathscr{L}x, y \rangle_{Y^* \times Y} \geq c^{-1}\|x\|_{X_1}^2$. Whence it follows that, $u = 0$ in $L_2(Q)$; hence, in $W_2^{1,1/1}(Q)$ also. Then the equality $\langle \mathscr{L}x, y \rangle_{Y^* \times Y} = 0$ can be rewritten as

$$\sum_{k=1}^{2}\sum_{i=1}^{n}\int_{Q_k}\omega_i\left(\sum_{j=1}^{n}\bar{k}_{ij}\eta_j - \frac{\partial v}{\partial \xi_i}\right)dQ_k = 0, \qquad \forall y = (v, \eta) \in Y.$$

If $y = (0, \mathbf{K}\omega)$ this inequality takes the form $\|\omega\|_{L_2^n(Q)}^2 = 0$, whence it follows that $\omega = \mathbf{0}$ in $L_2^n(Q)$.

The injectivity of the operator $\bar{\mathscr{L}}^+$ can be prove similarly.   □

4.6.3 Generalized Solution of a Parabolic System with Discontinuous Coefficients and Solutions

Theorem 4.10. *For any right-hand side $F \in S_1 = \{(f, 0)\,|\,f \in L_2(Q)\} \subset Y^*$ there exists the unique element $x \in X$ such that $\mathscr{L}x = F$ in Y^*.*

Proof. In view of Lemma 4.9, for any $y \in Y$ we have

$$|\langle F, y \rangle_{Y^* \times Y}| = |(f, v)_{L_2(Q)}| \leq \|f\|_{L_2(Q)}\|v\|_{L_2(Q)} \leq c\|\bar{\mathscr{L}}^+ y\|_{X^*}.$$

By virtue of the injectivity of the operator $\bar{\mathscr{L}}^+$ the expression $\langle F, y \rangle_{Y^* \times Y}$ can be considered as a linear continuous functional of $\mu = \bar{\mathscr{L}}^+ y$ in X^*. Applying The Hahn–Banach Theorem on on the extension of linear functionals, let us extend the functional from the set $R(\bar{\mathscr{L}}^+)$ on the entire space X^*. By the Riesz Representation Theorem on the general form of a linear continuous functional in X^* there exists such an element $x \in X$ that $\langle x, \bar{\mathscr{L}}^+ y \rangle_{X \times X^*} = \langle F, y \rangle_{Y^* \times Y}$ for all $y \in Y$. Hence, $\langle \mathscr{L} x, y \rangle_{Y^* \times Y} = \langle F, y \rangle_{Y^* \times Y}$ or $\mathscr{L} x = F$ in Y^*. The uniqueness of the solution follows from the injectivity of the operator $\mathscr{L}$.   □

Corollary 4.3. The parabolic system (4.35)–(4.38) with homogeneous conjugation conditions of non-ideal contact type ($f_0 = 0$) has the unique solution $u \in W_2^{1,1/1}(Q)$ for any right-hand side $f \in L_2(Q)$.

Corollary 4.4. The following equality holds $\{g \in L_2(Q) \,|\, (g, 0) \in R(\bar{\mathscr{L}}^+)\} = L_2(Q)$.

In order to study the problem in the case if $f_0 \neq 0$, let us introduce the concept of a generalized solution.

Definition 4.12. The function $u \in L_2(Q)$ is called a *generalized solution* of the equation $\mathscr{L} x = F$, if there exists such a sequence $x_k = (u_k, \omega_k) \in X$ that

$$\|u - u_k\|_{L_2(Q)} \to 0, \quad \|F - \mathscr{L} x_k\|_{Y^*} \to 0, \quad k \to \infty.$$

By Lemmas 4.7 and 4.8, it is easy to see that if $x = (u, \omega) \in X$ satisfies the equation $\mathscr{L} x = F$, then u is a generalized solution also. In addition, if u is a generalized solution and $F \in R(\mathscr{L})$ then $u \in W_2^{1,1/1}(Q)$ and there exists ω such that $\mathscr{L} x = F$, where $x = (u, \omega)$.

Theorem 4.11. *For any right-hand side $F \in S_2 = \{(f, 0) \,|\, f \in W_{2,*}^{-1,1/1}(Q)\} \subset Y^*$ there exists the unique generalized solution $u \in L_2(Q)$ in the sense of Definition 4.12.*

Proof. By the density of the set S_1 in S_2 in the sense of the convergence in the space Y^*, there exists such a sequence $F_k \in S_1$ that $F_k \to F$ in Y^* as $k \to \infty$. By Theorem 4.10 there exists such a sequence $x_k = (u_k, \omega_k) \in X$ that $\mathscr{L} x_k = F_k$ and by Lemma 4.8 the sequence u_k is a Cauchy with respect to the norm $\| \cdot \|_{L_2(Q)}$. Thus, there exists such an element $u \in L_2(Q)$ that $\|u - u_k\|_{L_2(Q)} \to 0$, i.e. u is a generalized solution in the sense of Definition 4.12.

Let us suppose that there exists one more solution $\bar{u} \in L_2(Q)$. Then the following inequalities hold

$$\|u - \bar{u}\|_{L_2(Q)} \leq \|u_k - \bar{u}_k\|_{L_2(Q)} + o(1) \leq c\|\mathscr{L} x_k - \mathscr{L} \bar{x}_k\|_{Y^*} + o(1) = o(1),$$

since $\mathscr{L} x_k \to F$, $\mathscr{L} \bar{x}_k \to F$ in Y^*.   □

Corollary 4.5. The parabolic system (4.35)–(4.38) in a region with inclusions has the unique generalized solution $u \in L_2(Q)$ for any right-hand side $f \in W_{2,*}^{-1,1/1}(Q)$ and $f_0 \in L_2(Q_3)$.

Corollary 4.6. There exists such a constant $c > 0$ that for all $F \in S_2$ the inequality $\|u\|_{L_2(Q)} \leq c\|F\|_{Y^*}$ holds, where u is a solution $\mathscr{L} x = F$ in the sense of Definition 4.12.

Theorem 4.12. *In order for the function $u \in L_2(Q)$ to be a generalized solution of the equation $\mathscr{L} x = F$ in the sense of Definition 4.12 it is necessary (and when*

$F \in S_2$ it is *sufficiently too) for all* $y \in Y$ *such that* $\bar{\mathscr{L}}^+ y = (g, \mathbf{0}), g \in L_2(Q)$ *the equality* $(u, g)_{L_2(Q)} = \langle F, y \rangle_{Y^* \times Y}$ *to be held.*

Proof. Let u be a solution of the equality $\bar{\mathscr{L}} x = F$ in the sense of Definition 4.12 and $x_k = (u_k, \omega_k) \in X$ be a sequence determining this solution. Then

$$(u_k, g)_{L_2(Q)} = \langle x_k, \bar{\mathscr{L}}^+ y \rangle_{X \times X^*} = \langle \bar{\mathscr{L}} x_k, y \rangle_{Y^* \times Y}$$

for all $y \in Y : \bar{\mathscr{L}}^+ y = (g, \mathbf{0})$, $g \in L_2(Q)$.

Passing to the limit as $k \to \infty$, we get the desirable equality.

Vice versa, let us suppose that for all $y \in Y$ such that $\bar{\mathscr{L}}^+ y = (g, \mathbf{0}), g \in L_2(Q)$ the equality $(u, g)_{L_2(Q)} = \langle F, y \rangle_{Y^* \times Y}$ holds. By Theorem 4.11 the equality $\bar{\mathscr{L}} x = F$ has a solution $u^* \in L_2(Q)$. Whence it follows that $(u - u^*, g)_{L_2(Q)} = 0$. By Corollary 4.4, we have that functions $g \in L_2(Q)$ run over the entire space $L_2(Q)$, i.e. $u = u^*$. □

Remark 4.23. Similar statements also hold for an adjoint operator.

4.6.4 Approximate Method for Solving the Boundary Value Problem for a Parabolic Equation with Inhomogeneous Transmission Conditions of Non-ideal Contact Type

In this section, we consider a new approximate method for solving the boundary value problem for a parabolic equation with inhomogeneous transmission conditions of non-ideal contact type which is an analogue of the Galerkin method, and the stability of the method is investigated. Note that in this section (contrary to previous ones) a parabolic equation is investigated in a direct statement (in a disconnected domain without considering generalized functions in coefficients). This approach allows to compare results obtained for a parabolic equation in a disconnected domain under different conditions of simulation of the diffusion because theorems on convergence of the numerical method allow to prove the existence of a unique solution.

Let us consider the diffusion process (4.35)–(4.38) in inhomogeneous media that are in contact with each other through a thin three-layer region.

Assume that D_{bd} is the set of $C^1(\overline{Q}_1, \overline{Q}_2)$ consisting of the functions satisfying conditions (4.36) and transmission conditions (4.38). Similarly, $D_{\mathrm{bd}*}$ is the set of functions in $C^1(\overline{Q}_1, \overline{Q}_2)$ satisfying the boundary conditions

$$v|_{t=T} = 0, \qquad v|_{\xi \in \partial \Omega} = 0$$

and the transmission conditions

$$[v] = R_1(\eta, \mathbf{n})_{\mathbb{R}^n}^- + R_3(\eta, \mathbf{n})_{\mathbb{R}^n}^+, \quad [(\eta, \mathbf{n})_{\mathbb{R}^n}] = 0, \qquad (t, \xi) \in Q_3,$$

where $\eta = \mathbf{K} \operatorname{grad} v$ in $Q_1 \cup Q_2$.

Denote by $W_{2,0}^{1,1/1}(Q)$ and $W_{2,T}^{1,1/1}(Q)$ the completions of D_{bd} and D_{bd*}, respectively, with respect to the norm

$$\|u\|^2 = \sum_{k=1}^{2} \int_{Q_k} u_t^2 + \sum_{m=1}^{n} u_{\xi m}^2 \, dQ_k.$$

The space $W_2^{0,1/1}(Q)$ is a completion of D_{bd} with respect to the norm

$$\|u\|_{W_2^{0,1/1}(Q)}^2 = \sum_{k=1}^{2} \int_{Q_k} u^2 + \sum_{m=1}^{n} u_{\xi m}^2 \, dQ_k. \tag{4.48}$$

Note that the completion of D_{bd*} with respect to norm (4.48) coincides with $W_2^{0,1/1}(Q)$.

Let $W_{2,0}^{-1,1/1}(Q)$, $W_{2,T}^{-1,1/1}(Q)$, and $W_2^{-0,1/1}(Q)$ are the conjugate spaces to $W_{2,0}^{1,1/1}(Q)$, $W_{2,T}^{1,1/1}(Q)$, and $W_2^{0,1/1}(Q)$ (with respect to $L_2(Q)$). Obviously, the following continuous dense embeddings hold:

$$W_{2,0}^{1,1/1}(Q) \subset W_2^{0,1/1}(Q) \subset L_2(Q) \subset W_2^{-0,1/1}(Q) \subset W_{2,0}^{-1,1/1}(Q),$$

$$W_{2,T}^{1,1/1}(Q) \subset W_2^{0,1/1}(Q) \subset L_2(Q) \subset W_2^{-0,1/1}(Q) \subset W_{2,T}^{-1,1/1}(Q).$$

Denote the bilinear form over $W_2^{-0,1/1}(Q) \times W_2^{0,1/1}(Q)$ by $\langle \cdot, \cdot \rangle$, over $W_{2,T}^{-1,1/1}(Q) \times W_{2,T}^{1,1/1}(Q)$ by $\langle \cdot, \cdot \rangle_T$ and over $W_{2,0}^{1,1/1}(Q) \times W_{2,0}^{-1,1/1}(Q)$ by $\langle \cdot, \cdot \rangle_0$.

It is assumed as before that the coefficients (4.35) satisfy $q(\xi) \in C^0(\overline{\Omega}_1, \overline{\Omega}_2)$ and $q \geqslant 0$; and the coefficient matrix $\mathbf{K} = \{k_{ml}(\xi)\}_{m,l=1}^{n}$ is symmetric, i.e. $k_{ml}(\xi) = k_{lm}(\xi) \in C^0(\overline{\Omega}_1, \overline{\Omega}_2)$ and uniformly positive definite in $\Omega_1 \cup \Omega_2$

$$\sum_{m,l=1}^{n} k_{ml}(\xi) \lambda_m \lambda_l \geqslant \alpha \sum_{m=1}^{n} \lambda_m^2, \qquad \lambda_m \in \mathbb{R}, \ \xi \in \Omega_1 \cup \Omega_2,$$

where α is a positive constant independent of λ_m or ξ.

By applying the integration-by-parts formula, it can easily be shown that, if $u \in D_{bd} \cap C^2(Q_1 \cup Q_2)$ satisfies (4.35) and (4.37) for continuous f and f_0, then the relation

$$\sum_{k=1}^{2} (u_t + qu, v)_{L_2(Q_k)} + \sum_{k=1}^{2} \sum_{m,l=1}^{n} (k_{ml} u_{\xi l}, v_{\xi m})_{L_2(Q_k)} + \left(\frac{[u]}{R_1 + R_3}, [v] \right)_{L_2(Q_3)}$$

$$= (f, v)_{L_2(Q)} + \left(\frac{R_3 f_0}{R_1 + R_3}, v^- \right)_{L_2(Q_3)} + \left(\frac{R_1 f_0}{R_1 + R_3}, v^+ \right)_{L_2(Q_3)} \tag{4.49}$$

is satisfied for all $v \in W_2^{0,1/1}(Q)$.

Thus, the left-hand side of (4.49) can be considered as a definition of the operator $\mathscr{S}: W_{2,0}^{1,1/1}(Q) \to W_2^{-0,1/1}(Q)$. The operator $\mathscr{S}$ is defined for all $u \in W_{2,0}^{1,1/1}(Q)$, and it is easily shown that it is linear and continuous.

Analogously, the left-hand side of (4.49) can be considered as a definition of the adjoint operator $\mathscr{S}^*: W_2^{0,1/1}(Q) \to W_{2,0}^{-1,1/1}(Q)$, which is also linear and continuous.

The right-hand side of (4.49) can be considered the value of the functional

$$F = f + \frac{R_1 f_0 \delta^+(\gamma) + R_3 f_0 \delta^-(\gamma)}{R_1 + R_3} \in W_2^{-0,1/1}(Q)$$

on the element $v \in W_2^{0,1/1}(Q)$, where $\delta^+(\gamma), \delta^-(\gamma) \in W_2^{-0,1/1}(Q)$ are the Dirac delta functions supported on the different sides γ^+ and γ^- of the hypersurface γ.

The functional $F \in W_2^{-0,1/1}(Q)$ makes sense for arbitrary $f \in W_2^{-0,1/1}(Q)$, $f_0 \in L_2(Q_3)$. Thus, taking into account (4.49), the following equation can be considered: $\mathscr{S}u = F, F \in W_2^{-0,1/1}(Q)$.

If in (4.49) the integration-by-parts formula is applied once more to the item

$$\sum_{k=1}^{2} (u_t, v)_{L_2(Q_k)} = \sum_{k=1}^{2} (u, -v_t)_{L_2(Q_k)}, \qquad \forall v \in W_{2,T}^{1,1/1}(Q),$$

the obtained relation defines a linear continuous operator $\mathscr{S}_1: W_2^{0,1/1}(Q) \to W_{2,T}^{-1,1/1}(Q)$ that is an extension of $\mathscr{S}: W_{2,0}^{1,1/1}(Q) \to W_2^{-0,1/1}(Q)$, and, respectively, the adjoint operator $\mathscr{S}_1^*: W_{2,T}^{1,1/1}(Q) \to W_2^{-0,1/1}(Q)$ that is a restriction of the operator $\mathscr{S}^*: W_2^{0,1/1}(Q) \to W_{2,0}^{-1,1/1}(Q)$. In this case, the following equation can be considered: $\mathscr{S}_1 u = F, F \in W_{2,T}^{-1,1/1}(Q)$ ($f \in W_{2,T}^{-1,1/1}(Q)$, $f_0 \in L_2(Q_3)$).

Lemma 4.11. For all $u \in W_2^{0,1/1}(Q)$, the following inequalities are satisfied:

$$c^{-1}\|u\|_{L_2(Q)} \leqslant \|\mathscr{S}_1 u\|_{W_{2,T}^{-1,1/1}(Q)} \leqslant c\|u\|_{W_2^{0,1/1}(Q)}. \tag{4.50}$$

Here and below, c is a sufficiently large positive constant independent of u and v.

Proof. The definition of $\mathscr{S}_1$ and the integral form of the Cauchy–Schwarz inequality imply the right-hand side of (4.50). In order to prove the left-hand side, it is necessary to consider the value of the functional $\mathscr{S}_1 u$ on the element

$$v(t, \xi) = -\int_T^t e^{-\tau} u(\tau, \xi) d\tau.$$

It is clear that $v \in W_{2,T}^{1,1/1}(Q)$, since the norm of $W_{2,T}^{1,1/1}(Q)$ "does not hold" the conditions $[(\eta, \mathbf{n})_{\mathbb{R}^n}] = 0$, $[v] = R_1(\eta, \mathbf{n})_{\mathbb{R}^n}^- + R_3(\eta, \mathbf{n})_{\mathbb{R}^n}^+$.

Applying the formula $avv_t = \frac{1}{2}(av^2)_t - \frac{1}{2}a_t v^2$, going to surface integrals, and taking into account the conditions on the coefficients of $\mathscr{S}_1$ and the Schwarz inequality, we obtain

$$c^{-1}\|v\|^2_{W^{1,1/1}_{2,T}(Q)} \leqslant \langle \mathscr{S}_1 u, v\rangle_T \leqslant \|\mathscr{S}_1 u\|_{W^{-1,1/1}_{2,T}(Q)}\|v\|_{W^{1,1/1}_{2,T}(Q)}.$$

To complete the proof, it is sufficient to take into account that $\|u\|_{L_2(Q)} \leqslant c\|v\|_{W^{1,1/1}_{2,T}(Q)}$. $\square$

Corollary 4.7. The operator $\mathscr{S}_1$ (and, therefore, $\mathscr{S}$) is injective.

Lemma 4.12. For all $u \in W^{1,1/1}_{2,0}(Q)$ the following inequalities are satisfied

$$c^{-1}\|u\|_{W^{0,1/1}_2(Q)} \leqslant \|\mathscr{S}u\|_{W^{-0,1/1}_2(Q)} \leqslant c\|u\|_{W^{1,1/1}_{2,0}(Q)}. \tag{4.51}$$

Proof. The right-hand side of (4.51) is proved by applying the integral form of the Cauchy–Schwarz inequality. In order to prove the left-hand side, it is necessary, as in Lemma 4.11, to consider the value of $\mathscr{S}u$ on the element $v = e^{-t}u$. $\square$

Remark 4.24. 1. The analogous of inequalities (4.50) and (4.51) for the adjoint operators $\mathscr{S}^*$ and $\mathscr{S}_1^*$ can be proved in a similar way

$$c^{-1}\|v\|_{L_2(Q)} \leqslant \|\mathscr{S}^*v\|_{W^{-1,1/1}_{2,0}(Q)} \leqslant c\|v\|_{W^{0,1/1}_2(Q)} \qquad \forall v \in W^{0,1/1}_2(Q), \tag{4.52}$$

$$c^{-1}\|v\|_{W^{0,1/1}_2(Q)} \leqslant \|\mathscr{S}_1^*v\|_{W^{-0,1/1}_2(Q)} \leqslant c\|v\|_{W^{1,1/1}_{2,T}(Q)} \qquad \forall v \in W^{1,1/1}_{2,T}(Q). \tag{4.53}$$

2. On the basis of the proved inequalities (4.50), (4.51), and (4.52), (4.53), we can prove the following statements:

(1) For any right-hand side of $F \in L_2(Q)$ (for example, $f \in L_2(Q)$, $f_0 = 0$) there exists a unique element $u \in W^{1,1/1}_{2,0}(Q)$ such that $\mathscr{S}u = F$.

(2) For any right-hand side of $F \in W^{-0,1/1}_2(Q)$ (for example, $f \in W^{-0,1/1}_2(Q)$, $f_0 = L_2(Q_3)$) there exists a unique element $u \in W^{0,1/1}_2(Q)$ such that $\mathscr{S}_1 u = F$.

(3) For any right-hand side of $F \in W^{-1,1/1}_{2,T}(Q)$ (for example, $f \in W^{-1,1/1}_{2,T}(Q)$, $f_0 \in L_2(Q_3)$) there exists a unique element $u \in L_2(Q)$ such that the equality

$$(\mathscr{S}^*v, u)_{L_2(Q)} = \langle F, v\rangle_T \tag{4.54}$$

is valid for an arbitrary function $v \in W^{1,1/1}_{2,T}(Q)$ satisfying the condition $\mathscr{S}^*v \in (L_2(Q))^* = L_2(Q)$.

3. Similar consideration for the adjoint operator imply that the set $\{\mathscr{S}^*v | v \in W^{1,1/1}_{2,T}(Q)\}$ covers the entire space $L_2(Q)$.

Assume that $\phi_i(\xi) \in C^1(\overline{\Omega}_1, \overline{\Omega}_2)$, $i \in \mathbb{N}$. Assume also that, for all $i \in \mathbb{N}$, the condition $\phi_i|_{\xi \in \partial\Omega} = 0$ is satisfied and the set of functions

$$\{\varphi(t)\phi_i(\xi) | i \in \mathbb{N}, \varphi(t) \in C([0,T]), \varphi(T) = 0\}$$

forms a total set in $W^{0,1/1}_2(Q)$.

Assume that $f \in L_2(Q)$ and $f_0 \in L_2(Q_3)$. An approximate solution to the equation $\mathscr{S}u = F$ is sought in the form

$$u_s(t, \xi) = \sum_{i=1}^{s} g_i^s(t) \phi_i(\xi),$$

where the functions $g_i^s(t)$ are the solutions to the Cauchy problem for the following system of ordinary differential equations with constant coefficients:

$$\sum_{i=1}^{s} \sum_{k=1}^{2} \left(\frac{dg_i^s}{dt} (\phi_i, \phi_j)_{L_2(\Omega_k)} + g_i^s (q\phi_i, \phi_j)_{L_2(\Omega_k)} + g_i^s \sum_{m,l=1}^{n} \left(k_{ml} (\phi_i)_{\xi_l}, (\phi_j)_{\xi_m} \right)_{L_2(\Omega_k)} \right)$$

$$+ \sum_{i=1}^{s} g_i^s \left(\frac{[\phi_i]}{R_1 + R_3}, [\phi_j] \right)_{L_2(\gamma)} = \sum_{k=1}^{2} (f, \phi_j)_{L_2(\Omega_k)} + \left(\frac{R_3 f_0}{R_1 + R_3}, \phi_j^- \right)_{L_2(\gamma)}$$

$$+ \left(\frac{R_1 f_0}{R_1 + R_3}, \phi_j^+ \right)_{L_2(\gamma)}, \qquad g_m^s(0) = 0, \quad m = \overline{1,s}, j = \overline{1,s}. \qquad (4.55)$$

Due to the well-known solvability theorems for systems of ordinary differential equations with constant coefficients, the solution to the Cauchy problem for system (4.55) exists and $g_i^s(t) \in W_2^1(0,T)$, where $W_2^1(0,T)$ is a Sobolev space.

Consider the set of functions $u(t, \xi) \in D_{\mathrm{bd}}$ whose derivatives $u_t(t, \xi)$ regarded as functions of ξ belong to the space $C^1(\overline{\Omega}_1, \overline{\Omega}_2)$. Assume that H is a completion of this space with respect to the norm

$$\|u\|_H^2 = \sum_{k=1}^{2} \int_{Q_k} u_t^2 + \sum_{m=1}^{n} u_{t\xi_m}^2 \, dQ_k.$$

It is easily seen that $u_s \in H$ and $H \subset W_{2,0}^{1,1/1}(Q)$, and this inclusion is continuous and dense.

Lemma 4.13. *For all $u \in H$ the following inequalities are satisfied:*

$$c^{-1} \|u\|_{W_{2,0}^{1,1/1}(Q)}^2 \leqslant \langle \mathscr{S}u, e^{-t} u_t \rangle \leqslant c \|u\|_H^2.$$

Proof. Obviously, $e^{-t} u_t \in W_2^{0,1/1}(Q)$. The right-hand side of the inequality is proved by applying to $\langle \mathscr{S}u, e^{-t} u_t \rangle$ the integral Cauchy–Schwarz inequality, the Friedrichs inequality, and the trace theorem.

Consider the proof of the left-hand side of the inequality. Transform the expression $\langle \mathscr{S}u, e^{-t} u_t \rangle$. Applying the formula $auu_t = \frac{1}{2} \left(au^2 \right)_t - \frac{1}{2} a_t u^2$, going to surface integrals, and taking into account the condition $u|_{t=0} = 0$, we obtain

$$\sum_{k=1}^{2} (qu, e^{-t} u_t)_{L_2(Q_k)} = \sum_{k=1}^{2} \frac{1}{2} \int_{\Omega_k} q e^{-T} u^2 \big|_{t=T} \, d\Omega_k + \sum_{k=1}^{2} \frac{1}{2} \int_{Q_k} q e^{-t} u^2 \, dQ_k \geqslant 0.$$

Similarly, we obtain the inequalities

$$\sum_{k=1}^{2}\sum_{m,l=1}^{n}\left(k_{ml}u_{\xi_l},e^{-t}u_{t\xi_m}\right)_{L_2(Q_k)} = \sum_{k=1}^{2}\sum_{m,l=1}^{n}\frac{1}{2}\int_{\Omega_k}k_{ml}e^{-T}u_{\xi_l}u_{\xi_m}\big|_{t=T}\,d\Omega_k$$

$$+\sum_{k=1}^{2}\sum_{m,l=1}^{n}\frac{1}{2}\int_{Q_k}k_{ml}e^{-t}u_{\xi_l}u_{\xi_m}\,dQ_k \geqslant c^{-1}\sum_{k=1}^{2}\sum_{m=1}^{n}\int_{Q_k}u_{\xi_m}^2\,dQ_k,$$

$$\left(\frac{[u]}{R_1+R_3},e^{-t}[u_t]\right)_{L_2(Q_3)} = \int_\gamma\frac{e^{-T}[u]^2\big|_{t=T}}{2(R_1+R_3)}\,d\gamma+\int_{Q_3}\frac{e^{-t}[u]^2}{2(R_1+R_3)}\,dQ_3 \geqslant 0.$$

This proves the left-hand side of the inequality in lemma.   □

Assume that $W_{2,0}^{1,0}(Q)$ is the completion of the set of functions $f \in C^1(\overline{Q}_1,\overline{Q}_2)$ satisfying $f(0,\xi) = 0$ ($\xi \in \overline{\Omega}$) with respect to the norm

$$\|f\|_{W_{2,0}^{1,0}(Q)}^2 = \sum_{k=1}^{2}\int_{Q_k}(f_t)^2\,dQ_k.$$

Assume that $W_{2,0}^1(Q_3)$ is the completion of the set of functions $f \in C(\overline{Q}_3)$ satisfying $f(0,\xi) = 0$ ($\xi \in \overline{\gamma}$) and having continuous t-derivatives with respect to the norm

$$\|f\|_{W_{2,0}^1(Q_3)}^2 = \int_{Q_3}(f_t)^2\,dQ_3. \tag{4.56}$$

Lemma 4.14. *Assume that $f \in W_{2,0}^{1,0}(Q)$, $f_0 \in W_{2,0}^1(Q_3)$. Then, the following inequality is satisfied:*

$$\|u_s\|_H \leqslant c\|f\|_{W_{2,0}^{1,0}(Q)} + c\|f_0\|_{W_{2,0}^1(Q_3)}. \tag{4.57}$$

Proof. It is easy to show that, if the conditions of the theorem are satisfied, the solutions $g_i^s(t)$ to system (4.55) belong to the Sobolev space $W_2^2(0,T)$.

Differentiating each equality in system (4.55) with respect to t, multiplying it by $e^{-t}(g_j^s)_t$, summing over j from 1 to s, and integrating with respect to t from 0 to T, we obtain

$$A = \sum_{k=1}^{2}\left((u_s)_{tt}+q(u_s)_t,e^{-t}(u_s)_t\right)_{L_2(Q_k)} + \sum_{k=1}^{2}\sum_{m,l=1}^{n}\left(k_{ml}(u_s)_{t\xi_l},e^{-t}(u_s)_{t\xi_m}\right)_{L_2(Q_k)}$$

$$+\left(\frac{[(u_s)_t]}{R_1+R_3},e^{-t}[(u_s)_t]\right)_{L_2(Q_3)} = \sum_{k=1}^{2}\left(f_t,e^{-t}(u_s)_t\right)_{L_2(Q_k)}$$

$$+\left(\frac{R_3(f_0)_t}{R_1+R_3},e^{-t}(u_s)_t^-\right)_{L_2(Q_3)} + \left(\frac{R_1(f_0)_t}{R_1+R_3},e^{-t}(u_s)_t^+\right)_{L_2(Q_3)}.$$

Applying the integral Cauchy–Schwarz inequality to the right-hand side, we have

$$A \leqslant \|f_t\|_{L_2(Q)}\|(u_s)_t\|_{L_2(Q)} + c\|(f_0)_t\|_{L_2(Q_3)}\|(u_s)_t^-\|_{L_2(Q_3)}$$
$$+c\|(f_0)_t\|_{L_2(Q_3)}\|(u_s)_t^+\|_{L_2(Q_3)} \leqslant c_1 \left(\|f_t\|_{L_2(Q)} + \|(f_0)_t\|_{L_2(Q_3)}\right)\|u_s\|_H. \quad (4.58)$$

On the other hand, let us prove that $A \geqslant c^{-1}\|u_s\|_H^2$. Applying the formula $a u_t u_{tt} = \frac{1}{2}\left(a u_t^2\right)_t - \frac{1}{2}a_t u_t^2$ and going to surface integrals, we obtain

$$\sum_{k=1}^{2} \left((u_s)_{tt}, e^{-t}(u_s)_t\right)_{L_2(Q_k)} = \sum_{k=1}^{2}\frac{1}{2}\int_{\Omega_k} e^{-t}(u_s)_t^2\Big|_{t=0}^{t=T}\mathrm{d}\Omega_k + \sum_{k=1}^{2}\frac{1}{2}\int_{Q_k} e^{-t}(u_s)_t^2\mathrm{d}Q_k.$$

Thus,

$$A \geqslant -\frac{1}{2}\sum_{k=1}^{2}\int_{\Omega_k}(u_s)_t^2\Big|_{t=0}\mathrm{d}\Omega_k + c^{-1}\sum_{k=1}^{2}\int_{Q_k}(u_s)_t^2 + \sum_{m=1}^{n}(u_s)_{t\xi_m}^2\mathrm{d}Q_k.$$

Multiplying each inequality of system (4.55) by $(g_j^s)_t$, summing it over j from 1 to s, setting $t = 0$, and taking into account the lemma conditions and the equality $g_m^s(0) = 0$, $m = \overline{1,s}$, we obtain

$$\sum_{k=1}^{2}\int_{\Omega_k}(u_s)_t^2\Big|_{t=0}\mathrm{d}\Omega_k = 0.$$

Thus, the inequality $A \geqslant c^{-1}\|u_s\|_H^2$ has been proved. Taking into account (4.58) completes the proof of the lemma. $\square$

Corollary 4.8. Assume that the conditions of the lemma are satisfied. Then inequality (4.57) implies that the sequence u_s is bounded in the Hilbert space H and, therefore, there exists a weakly converging subsequence $u_{s_k} \overset{w}{\to} u^*$ in H, such that the sequences $(u_{s_k})_t$, $(u_{s_k})_{\xi_m}$, and $(u_{s_k})_{\xi_{mt}}$ converge weakly in $L_2(Q)$ and the sequences $u_{s_k}^+$, $u_{s_k}^-$, $(u_{s_k})_t^+$, and $(u_{s_k})_t^-$ converge weakly in $L_2(Q_3)$ to the corresponding value of the generalized Sobolev derivative of the function $u^* \in H$.

Lemma 4.15. *Assume that $f \in W_{2,0}^{1,0}(Q)$, $f_0 \in W_{2,0}^{1}(Q_3)$. Then, for an arbitrary function $v \in W_2^{0,1/1}(Q)$, the following equality is satisfied:*

$$\langle \mathscr{S}u^*, v\rangle = \langle F, v\rangle,$$

where $u^* \in H$ is the function defined on Corollary 4.8.

Proof. Multiplying each equality in system (4.55) as $s = s_k$ by an arbitrary function $\varphi_j(t) \in C([0,T])$: $\varphi_j(T) = 0$, summing the result over j from 1 to p ($p = \overline{1,s_k}$) and integrating it with respect to t from 0 to T, we obtain

$$\left\langle \mathscr{S}u_{s_k}, \sum_{j=1}^{p}\varphi_j\phi_j\right\rangle = \left\langle F, \sum_{j=1}^{p}\varphi_j\phi_j\right\rangle, \qquad p = \overline{1,s_k}.$$

Pass to the limit as $k \to \infty$ and assume that $v_p = \sum_{j=1}^{P} \varphi_j \phi_j$. Due to Corollary 4.8, we obtain

$$\langle \mathscr{S} u^*, v_p \rangle = \langle F, v_p \rangle.$$

Since the system

$$\{\varphi(t)\phi_i(\xi) | i \in \mathbb{N}, \ \varphi(t) \in C([0,T]), \ \varphi(T) = 0\}$$

is total in $W_2^{0,1/1}(Q)$, a closure of its span coincides with the entire space $W_2^{0,1/1}(Q)$, which implies the lemma. $\square$

Corollary 4.9. For arbitrary functions $f \in W_{2,0}^{1,0}(Q)$ and $f_0 \in W_{2,0}^{1}(Q_3)$ there exists a unique solution $u^* \in H$ to the equation $\mathscr{S} u = F$.

Theorem 4.13. *Assume that $f \in W_{2,0}^{1,0}(Q)$ and $f_0 \in W_{2,0}^{1}(Q_3)$. Then the sequence of approximations u_s converges to the solution to the equation $\mathscr{S} u = F$ in the norm of $W_{2,0}^{1,1/1}(Q)$ and $\mathscr{S} u_s \to F$ in $W_2^{-0,1/1}(Q)$ as $s \to \infty$.*

Proof. Multiplying each equality of system (4.55) at $s = s_k$ by $e^{-t}(g_j^{s_k})_t$, summing then over j from 1 to s_k, and integrating with respect to t from 0 to T, we obtain

$$\langle \mathscr{S} u_{s_k}, e^{-t}(u_{s_k})_t \rangle = \langle F, e^{-t}(u_{s_k})_t \rangle.$$

Taking into account the properties of the subsequence s_k (see Corollary 4.8), we pass to the limit as $k \to \infty$ to obtain

$$\lim_{k \to \infty} \langle \mathscr{S} u_{s_k}, e^{-t}(u_{s_k})_t \rangle = \langle F, e^{-t}(u^*)_t \rangle. \tag{4.59}$$

Taking into account Lemma 4.13, we have

$$c^{-1} \| u_{s_k} - u^* \|^2_{W_{2,0}^{1,1/1}(Q)} \leqslant \langle \mathscr{S}(u_{s_k} - u^*), e^{-t}(u_{s_k} - u^*)_t \rangle$$

$$= \langle \mathscr{S} u_{s_k}, e^{-t}(u_{s_k})_t \rangle - \langle \mathscr{S} u^*, e^{-t}(u_{s_k})_t \rangle - \langle \mathscr{S} u_{s_k} - \mathscr{S} u^*, e^{-t}(u^*)_t \rangle.$$

Since u_{s_k} converges weakly to u^* in the corresponding spaces (Corollary 4.8), we conclude that $\langle \mathscr{S} u_{s_k} - \mathscr{S} u^*, e^{-t}(u^*)_t \rangle$ tends to zero as $k \to \infty$ and $\langle \mathscr{S} u^*, e^{-t}(u_{s_k})_t \rangle$ tends to $\langle \mathscr{S} u^*, e^{-t}(u^*)_t \rangle$. Therefore, due to (4.59) and Lemma 4.15, we obtain

$$\lim_{k \to \infty} c^{-1} \| u_{s_k} - u^* \|^2_{W_{2,0}^{1,1/1}(Q)} \leqslant \langle F, e^{-t}(u^*)_t \rangle - \langle \mathscr{S} u^*, e^{-t}(u^*)_t \rangle = 0.$$

Note that the assumption on the existence of a weak accumulation point of the sequence u_s that is different from u^* in H contradicts the uniqueness of the solution to the equation $\mathscr{S} u = F$ (Corollary 4.7). Therefore, there is no necessity to choose the subsequence s_k, i.e., the entire sequence u_s converges to u^* in $W_{2,0}^{1,1/1}(Q)$ as $s \to \infty$.

Since the operator $\mathscr{S}$ is continuous, we have $\mathscr{S} u_s \to F$ in $W_2^{-0,1/1}(Q)$. $\square$

Consider the set of functions $f \in C(\overline{Q}_3)$ satisfying $f(T,\xi) = 0$ and having a continuous derivative with respect to t and denote by $W_{2,T}^1(Q_3)$ its completion with respect to norm (4.56).

Lemma 4.16. *For the arbitrary right-hand sides $f \in L_2(Q)$ and $f_0 \in W_{2,T}^1(Q_3)$, the following inequality is satisfied*

$$\|u_s\|_{W_{2,0}^{1,1/1}(Q)} \leqslant c\|f\|_{L_2(Q)} + c\|f_0\|_{W_{2,T}^1(Q_3)}.$$

Proof. Since $u_s \in H$, Lemma 4.13 implies that

$$c^{-1}\|u_s\|_{W_{2,0}^{1,1/1}(Q)}^2 \leqslant \langle \mathscr{S}u_s, e^{-t}(u_s)_t \rangle. \tag{4.60}$$

On the other hand, if both sides of each equality (4.55) are multiplied by $e^{-t}(g_j^s)_t$, summed over j from 1 to s, and integrated with respect to t from 0 to T, we obtain

$$\langle \mathscr{S}u_s, e^{-t}(u_s)_t \rangle = (f, e^{-t}(u_s)_t)_{L_2(Q)}$$
$$+ \left(\frac{R_3 f_0}{R_1+R_3}, e^{-t}(u_s)_t^- \right)_{L_2(Q_3)} + \left(\frac{R_1 f_0}{R_1+R_3}, e^{-t}(u_s)_t^+ \right)_{L_2(Q_3)}.$$

Taking into account the conditions $f_0(T,\xi) = 0$, $u_s^-(0,\xi) = 0$, and $u_s^+(0,\xi) = 0$ and applying the integration-by-parts formula to the last two terms, we obtain

$$\langle \mathscr{S}u_s, e^{-t}(u_s)_t \rangle = (f, e^{-t}(u_s)_t)_{L_2(Q)} - \left(\frac{R_3(f_0 e^{-t})_t}{R_1+R_3}, u_s^- \right)_{L_2(Q_3)}$$
$$- \left(\frac{R_1(f_0 e^{-t})_t}{R_1+R_3}, u_s^+ \right)_{L_2(Q_3)}$$
$$\leqslant \|f\|_{L_2(Q)} \|(u_s)_t\|_{L_2(Q)} + c\|(f_0 e^{-t})_t\|_{L_2(Q_3)} \|u_s^-\|_{L_2(Q_3)}$$
$$+ c\|(f_0 e^{-t})_t\|_{L_2(Q_3)} \|u_s^+\|_{L_2(Q_3)}$$
$$\leqslant \|f\|_{L_2(Q)} \|u_s\|_{W_{2,0}^{1,1/1}(Q)} + c_1\|(f_0)_t\|_{L_2(Q_3)} \|u_s\|_{W_{2,0}^{1,1/1}(Q)}.$$

Taking into account the last equality and (4.60), we have

$$c^{-1}\|u_s\|_{W_{2,0}^{1,1/1}(Q)} \leqslant \|f\|_{L_2(Q)} + c_1\|(f_0)_t\|_{L_2(Q_3)},$$

which implies the lemma. $\square$

Theorem 4.14. *For the arbitrary right-hand sides $f \in L_2(Q)$ and $f_0 \in W_{2,0}^1(Q_3)$, there exists a unique solution $u^* \in W_{2,0}^{1,1/1}(Q)$ to the equation $\mathscr{S}u = F$, the sequence u_s converges to u^* in the norm of $W_{2,0}^{1,1/1}(Q)$, and $\mathscr{S}u_s \to F$ in $W_2^{-0,1/1}(Q)$.*

Proof. The set $W_{2,0}^{1,0}(Q)$ is dense in $L_2(Q)$. Therefore, there exists a sequence of functions $f^m \in W_{2,0}^{1,0}(Q)$ converging to $f \in L_2(Q)$ in $L_2(Q)$ as $m \to \infty$.

Assume that u_s^m is a sequence of approximate solutions to system (4.55) with the right-hand side given by f^m and f_0. According to Theorem 4.13, u_s^m converges to the solution u^m of the corresponding equation $\mathscr{S} u = F^m$ in $W_{2,0}^{1,1/1}(Q)$ as $s \to \infty$. Let us prove that the sequence of functions u^m is a Cauchy sequence in $W_{2,0}^{1,1/1}(Q)$. Indeed, applying the inequality of Lemma 4.16 to the term $\|u_s^n - u_s^m\|_{W_{2,0}^{1,1/1}(Q)}$, we obtain

$$\|u^n - u^m\|_{W_{2,0}^{1,1/1}(Q)} \leqslant \|u^n - u_s^n\|_{W_{2,0}^{1,1/1}(Q)} + \|u_s^n - u_s^m\|_{W_{2,0}^{1,1/1}(Q)} + \|u_s^m - u^m\|_{W_{2,0}^{1,1/1}(Q)}$$
$$\leqslant c\|f^n - f^m\|_{L_2(Q)} + o(1).$$

Passing to the limit as $s \to \infty$, $m, n \to \infty$, we conclude that u^m is a Cauchy sequence. Since $W_{2,0}^{1,1/1}(Q)$ is complete, there exists an element $u^* \in W_{2,0}^{1,1/1}(Q)$ such that $u^m \to u^*$ in $W_{2,0}^{1,1/1}(Q)$ as $m \to \infty$.

We prove that $u_s \to u^*$ in $W_{2,0}^{1,1/1}(Q)$ as $s \to \infty$. To this end,

$$\|u_s - u^*\|_{W_{2,0}^{1,1/1}(Q)} \leqslant \|u_s - u_s^m\|_{W_{2,0}^{1,1/1}(Q)} + \|u_s^m - u^m\|_{W_{2,0}^{1,1/1}(Q)} + \|u^m - u^*\|_{W_{2,0}^{1,1/1}(Q)}.$$

Theorem 4.13 implies that $\|u_s^m - u^m\|_{W_{2,0}^{1,1/1}(Q)} \to 0$ as $s \to \infty$. Moreover, by Lemma 4.16

$$\|u_s - u_s^m\|_{W_{2,0}^{1,1/1}(Q)} \leqslant c\|f - f^m\|_{L_2(Q)}.$$

Therefore,

$$\varlimsup_{s \to \infty} \|u_s - u^*\|_{W_{2,0}^{1,1/1}(Q)} \leqslant c\|f - f^m\|_{L_2(Q)} + \|u^m - u^*\|_{W_{2,0}^{1,1/1}(Q)}.$$

Passing to the limit as $m \to \infty$, we conclude that $u_s \to u^*$ in $W_{2,0}^{1,1/1}(Q)$ as $s \to \infty$.

Prove that u^* is the solution to the equation $\mathscr{S} u = F$. Indeed, since $\mathscr{S} u^m = F^m$, we obtain

$$\|\mathscr{S} u^* - F\|_{W_2^{-0,1/1}(Q)} \leqslant \|\mathscr{S}(u^* - u^m)\|_{W_2^{-0,1/1}(Q)} + \|F^m - F\|_{W_2^{-0,1/1}(Q)}$$
$$\leqslant c\|u^* - u^m\|_{W_{2,0}^{1,1/1}(Q)} + \|F^m - F\|_{W_2^{-0,1/1}(Q)}.$$

To complete the proof, it is sufficient to pass to the limit as $m \to \infty$. $\square$

If f and f_0 are generalized functions, system (4.55) makes no sense, since the integrals on the right-hand side of (4.55) may not exist. One way to overcome this problem is to replace these integrals by the corresponding bilinear forms, but system (4.55) then has to be considered in terms of the theory of generalized functions.

Another way is to replace the right-hand sides f and f_0 with poor smoothness by similar but smoother functions. Below, we consider this approach and analyze the convergence of the iteration procedures proposed.

Assume that the right-hand side F of the equation $\mathscr{S}_1 u = F$ is an element of the negative space $W_2^{-0,1/1}(Q)$ ($f \in W_2^{-0,1/1}(Q)$, $f_0 \in L_2(Q_3)$). Since $L_2(Q)$ is dense in $W_2^{-0,1/1}(Q)$, we choose a sequence of functions $f^m \in L_2(Q)$ converging to f in $W_2^{-0,1/1}(Q)$. Similarly, assume that f_0^m is a sequence of functions from $W_{2,0}^1(Q_3)$ converging to f_0 in $L_2(Q_3)$. Obviously, with such a choice, the sequence of the right-hand sides F^m converges to F in $W_2^{-0,1/1}(Q)$.

Consider a sequence of approximations $u_{s,m}(t,\xi)$ constructed according to the rule

$$u_{s,m}(t,\xi) = \sum_{i=1}^{s} g_{i,m}^s(t)\phi_i(\xi),$$

where the functions $g_{i,m}^s(t)$ are solutions to system (4.55) whose right-hand side is given by f^m and f_0^m.

Theorem 4.15. *Assume that ε_m is an arbitrary sequence of positive numbers converging to zero. Then, for an arbitrary positive integer $s(m)$ satisfying the condition*

$$\left\| \mathscr{S} u_{s(m),m} - F^m \right\|_{W_2^{-0,1/1}(Q)} < \varepsilon_m$$

(such an $s(m)$ necessarily exists) the sequence $u_{s(m),m}$ converges to the solution of the equation $\mathscr{S}_1 u = F$ in the norm of $W_2^{0,1/1}(Q)$ as $m \to \infty$.

Proof. According to Theorem 4.14, the sequence $u_{s,m}$ converges to the solution $u^m \in W_{2,0}^{1,1/1}(Q)$ of the equation $\mathscr{S} u = F^m$ in the norm of $W_{2,0}^{1,1/1}(Q)$ as $s \to \infty$, and $\left\| \mathscr{S} u_{s,m} - F^m \right\|_{W_2^{-0,1/1}(Q)} \xrightarrow[s \to \infty]{} 0$.

We prove that u^m is a Cauchy sequence in $W_2^{0,1/1}(Q)$. Indeed, applying inequality (4.51) gives

$$\left\| u^m - u^n \right\|_{W_2^{0,1/1}(Q)} \leqslant c \left\| \mathscr{S} u^m - \mathscr{S} u^n \right\|_{W_2^{-0,1/1}(Q)} = c \left\| F^m - F^n \right\|_{W_2^{-0,1/1}(Q)} \xrightarrow[m,n \to \infty]{} 0.$$

Thus, there exists a function $u^* \in W_2^{0,1/1}(Q)$ such that $u^m \to u^*$ in $W_2^{0,1/1}(Q)$ as $m \to \infty$.

Prove that u^* is the solution to the equation $\mathscr{S}_1 u = F$. Indeed, since $\mathscr{S}_1 u^m = \mathscr{S} u^m = F^m$, we obtain

$$\left\| \mathscr{S}_1 u^* - F \right\|_{W_{2,T}^{-1,1/1}(Q)} \leqslant \left\| \mathscr{S}_1 u^* - \mathscr{S}_1 u^m \right\|_{W_{2,T}^{-1,1/1}(Q)} + \left\| F^m - F \right\|_{W_{2,T}^{-1,1/1}(Q)}$$

$$\leqslant c \left\| u^* - u^m \right\|_{W_2^{0,1/1}(Q)} + c \left\| F^m - F \right\|_{W_2^{-0,1/1}(Q)} \xrightarrow[m \to \infty]{} 0.$$

Applying inequality (4.51), we have

$$\|u_{s(m),m} - u^*\|_{W_2^{0,1/1}(Q)} \leqslant \|u_{s(m),m} - u^m\|_{W_2^{0,1/1}(Q)} + \|u^m - u^*\|_{W_2^{0,1/1}(Q)}$$

$$\leqslant c \|\mathscr{S} u_{s(m),m} - \mathscr{S} u^m\|_{W_2^{-0,1/1}(Q)} + \|u^m - u^*\|_{W_2^{0,1/1}(Q)}$$

$$\leqslant c\varepsilon_m + \|u^m - u^*\|_{W_2^{0,1/1}(Q)}.$$

Since the right-hand side of the last inequality tends to zero as $m \to \infty$, the theorem is proved. $\qquad\square$

Now assume that the right-hand side F of the equation $\mathscr{S}_1 u = F$ belongs to $W_{2,T}^{-1,1/1}(Q)$ (e.g. $f \in W_{2,T}^{-1,1/1}(Q)$, $f_0 \in L_2(Q_3)$).

Let $f^m \in L_2(Q)$ be a sequence of functions converging to f in $W_{2,T}^{-1,1/1}(Q)$ and $f_0^m \in W_{2,0}^1(Q_3)$ be a sequence converging to f_0 in $L_2(Q_3)$. The sequence of approximations $u_{s,m}$ satisfies the original conditions.

Theorem 4.16. *Assume that ε_m is an arbitrary sequence of positive numbers converging to zero. Then, for an arbitrary positive integer $s(m)$:*

$$\|\mathscr{S} u_{s(m),m} - F^m\|_{W_{2,T}^{-1,1/1}(Q)} < \varepsilon_m$$

(such an $s(m)$ necessarily exists) the sequence $u_{s(m),m}$ converges in the norm of $L_2(Q)$ as $m \to \infty$ to the solution of the equation $\mathscr{S}_1 u = F$ in the sense of equality (4.54).

Proof. By analogy with Theorem 4.15, we prove that a sequence $u^m \in W_{2,0}^{1,1/1}(Q)$ converges to a function $u^* \in L_2(Q)$ in $L_2(Q)$. Prove that u^* is the solution to the equation $\mathscr{S}_1 u = F$ in the sense of (4.54). Indeed,

$$\langle \mathscr{S} u^m, v \rangle = \langle \mathscr{S} u^m, v \rangle_T = \langle F^m, v \rangle_T, \qquad \forall v \in W_{2,T}^{1,1/1}(Q)$$

or

$$(u^m, \mathscr{S}^* v)_{L_2(Q)} = \langle u^m, \mathscr{S}^* v \rangle_0 = \langle \mathscr{S} u^m, v \rangle = \langle F^m, v \rangle_T$$

for all $v \in W_{2,T}^{1,1/1}(Q)$ such that $\mathscr{S}^* v \in L_2(Q)$.

Passing to the limit as $m \to \infty$, we see that $u^* \in L_2(Q)$ is the solution to the equation $\mathscr{S}_1 u = F$.

The rest of the proof is similar to Theorem 4.15. $\qquad\square$

4.7 On the Unique Solvability of Wave Systems

We study equations of the form

$$\mathscr{L} u \equiv A(u_{tt}) + B^2(u_t) + C(u) = f, \tag{4.61}$$

where A, B, and C are second-order differential operators.

Equations of the form (4.61) arise in applications, for example, when analyzing the dynamics of plane motions of an incompressible viscous fluid, linear waves on a helical flow, small vibrations of an ideal non-rotating stratified liquid moving as a whole at a constant velocity in the direction perpendicular to the stratification direction, etc. [18, 20, 21]. Equation (4.61) generalizes the well-known Sobolev–Gal'perin equation ($B \equiv 0$), which has numerous applications [19, 27, 59, 93, 111]. Initial-boundary value problems for (4.61) were studied from various viewpoints in [20, 21, 60, 62].

The unique solvability was analyzed in [5, 41, 45] for specific types of (4.61) for the case in which the right-hand side is a distribution of a finite-order; approximate methods were also constructed there, and some optimization problems were considered. However, these results deal only with elliptic operators A, B, and C without first-order lower terms and specify the action of the operator only in a single pair of spaces. In the book, we eliminate most of the restrictions imposed on the operators A, B, and C, obtain a "scale" of solvability theorems, and generalize some results obtained earlier.

4.7.1 Basic Notation and Statement of the Operator Equation

In the cylindrical domain $(t,x) \in Q = (0,T) \times \Omega$, we consider (4.61), where $\Omega \subset \mathbb{R}^n$ is a bounded connected domain with regular boundary $\partial \Omega$ and A is a second-order operator in the spatial variables:

$$A(u) \equiv - \sum_{i,j=1}^{n} \frac{\partial}{\partial x_i} \left(a_{ij}(x) \frac{\partial u}{\partial x_j} \right) + \sum_{i=1}^{n} a_i(x) \frac{\partial u}{\partial x_i} + a(x)u.$$

The operators B and C are given by similar differential expressions; moreover, $a_{ij} = a_{ji}$, $b_{ij} = b_{ji}$, and $c_{ij} = c_{ji}$.

We require that the function $u(t,x)$ satisfies the homogeneous initial and boundary conditions (it is known that the case of inhomogeneous conditions can be reduced to the homogeneous one by an appropriate change of the right-hand side of (4.61))

$$u|_{t=0} = u_t|_{t=0} = 0, \qquad u|_{x \in \partial \Omega} = \frac{\partial u}{\partial \mu_B}|_{x \in \partial \Omega} = 0, \tag{4.62}$$

where $\mu_B = \mathbf{Bn}$ is the conormal to the surface $\partial \Omega$, $\mathbf{B} = \{b_{ij}(x)\}_{i,j=1}^{n}$ is the coefficient matrix of the operator B, and $\mathbf{n}$ is the outward normal to the surface $\partial \Omega$.

One can also study other "physical" boundary conditions corresponding to the second and third boundary value problem [21]. To avoid cumbersome expression, we carry out all considerations only for condition (4.62). All considerations can readily be modified for the other cases.

Let H_0^1, W_0^1, and V_0^1 be the completions of the set L_0 of functions infinitely differentiable in $\bar{Q}$ and satisfying the conditions

$$u|_{t=0} = u_t|_{t=0} = \ldots = 0, \qquad u|_{x \in \partial\Omega} = \frac{\partial u}{\partial \mu_B}|_{x \in \partial\Omega} = 0,$$

with respect to the norms

$$\|u\|_{H_0^1}^2 = \sum_{i=1}^n \int_Q u_{tx_i}^2 \, dQ + \sum_{i,j=1}^n \int_Q u_{tx_ix_j}^2 \, dQ,$$

$$\|u\|_{W_0^1}^2 = \sum_{i=1}^n \int_Q u_{tx_i}^2 \, dQ + \sum_{i,j=1}^n \int_Q u_{x_ix_j}^2 \, dQ,$$

$$\|u\|_{V_0^1}^2 = \|u\|_{W_0^1}^2 + \sum_{i,j=1}^n \int_\Omega u_{x_ix_j}^2|_{t=T} \, d\Omega. \qquad (4.63)$$

Let H_T^1, W_T^1, and V_T^1 be the completions of the set L_T of functions infinitely differentiable in $\bar{Q}$ and satisfying the adjoint conditions

$$v|_{t=T} = v_t|_{t=T} = \ldots = 0, \qquad v|_{x \in \partial\Omega} = \frac{\partial v}{\partial \mu_B}|_{x \in \partial\Omega} = 0,$$

with respect to the norms of the spaces $\| \cdot \|_{H_0^1}$, $\| \cdot \|_{W_0^1}$ and $\|v\|_{V_T^1}^2 = \|v\|_{W_T^1}^2 + \sum_{i,j=1}^n \|v_{x_ix_j}^2|_{t=0}\|_{L_2(\Omega)}^2$, respectively.

Further, we need the definition of the spaces $H_0^k, H_T^k, W_0^k, \ldots$ for arbitrary integer k.[1] By H_0^k we denote the completion of the set L_0 in the norm $\| \cdot \|_{H_0^k}$, where the norm of the space H_0^k is defined by induction as:

$$\|u_t\|_{H_0^{k-1}} = \|u\|_{H_0^k}, \qquad \|u\|_{H_0^{k-1}} = \|\bar{u}\|_{H_0^k}, \qquad \bar{u} = \int_0^t u(\tau, x) \, d\tau$$

for all $k \in \mathbb{Z}$. The norm H_0^1 is given by (4.63).

The remaining spaces are defined in a similar way (for the spaces H_T^k, W_T^k, and V_T^k with the subscript T the notation $\bar{v}$ should be understood as $\bar{v} = \int_T^t v(\tau, x) \, d\tau$).

Lemma 4.17. *There are dense continuous embeddings $H_0^k \subset V_0^k \subset W_0^k \subset H_0^{k-1}$ and $H_T^k \subset V_T^k \subset W_T^k \subset H_T^{k-1}$ for arbitrary $k \in \mathbb{Z}$. Moreover, $W_0^0 \subset L_2(Q)$, $W_T^0 \subset L_2(Q)$.*

Proof. Let us prove the lemma, say, for the embedding $V_T^2 \subset W_T^2$. (The remaining cases can be considered in a similar way).

Since $\|v\|_{W_T^2} \leqslant \|v\|_{V_T^2}$ for functions $v \in L_T$, we need to verify the following condition to prove the embedding $V_T^2 \subset W_T^2$: for any sequence $v_n \in L_T$ such that $v_n \to v_0$ in V_T^2 and $v_n \to 0$ in W_T^2 one has $v_0 = 0$ [40].

---

[1] We use the following notation of the spaces $H_0^k, V_0^k, \ldots$: the superscript indicates the number of time derivatives, and the subscript indicates the type of initial conditions.

Let $v_n \in L_T$ be such a sequence (i.e. $v_n \to v_0$ in V_T^2 and $v_n \to 0$ in W_T^2). Since $v_n \in L_T$, we have $(v_n)_t(T,x) = 0$. Therefore, by using the Cauchy–Schwarz integral inequality, we obtain

$$|(v_n)_{tx_i}(0,x)|^2 = \left| \int_0^T (v_n)_{ttx_i}(\tau,x)d\tau \right|^2 \leqslant T \int_0^T (v_n)_{ttx_i}^2 d\tau.$$

By integrating this relation over the domain Ω, we obtain

$$\int_\Omega (v_n)_{tx_i}^2(0,x)d\Omega \leqslant c \int_Q (v_n)_{ttx_i}^2 dQ \leqslant c\|v_n\|_{W_T^2}^2 \to 0 \tag{4.64}$$

as $n \to \infty$, here in throughout the following, c is a positive constant.

The convergence of v_n in the space V_T^2 implies that $(v_n)_{tx_ix_j}(0,x)$ is a Cauchy sequence in $L_2(\Omega)$. Let it converge to $\phi_{ij}(x) \in L_2(\Omega)$. By integrating by parts, we obtain

$$\left((v_n)_{tx_i}|_{t=0}, u_{x_j}\right)_{L_2(\Omega)} = -\left((v_n)_{tx_ix_j}|_{t=0}, u\right)_{L_2(\Omega)},$$

for all $u \in C_0^\infty(\overline{\Omega})$ (the set of compactly supported smooth functions in $\overline{\Omega}$). By taking into account (4.64) and by passing to the limit as $n \to \infty$, we obtain $(\phi_{ij}, u)_{L_2(\Omega)} = 0$. Since the function u is arbitrary, it follows that $\phi_{ij} = 0$, i.e., $\|v_n\|_{V_T^2} \to 0$ as $n \to \infty$. Therefore, $v_0 = 0$ and the embedding operator $V_T^2 \subset W_T^2$ is injective and continuous.

Since the set L_T is dense in both spaces V_T^2 and W_T^2, it follows that the embedding $V_T^2 \subset W_T^2$ is dense. The proof of the lemma is complete. $\qquad\square$

By $(H_0^k)^*, (H_T^k)^*, (W_0^k)^*, \ldots$ we denote the corresponding conjugate spaces. There are natural bilinear forms defined on the pairs of primal and conjugate spaces; for example, $\langle \cdot, \cdot \rangle_{H_0^k}$ stands for the bilinear form on $(H_0^k)^* \times H_0^k$.

We set

$$\langle Au, v \rangle_Q = \sum_{i,j=1}^n (a_{ij}u_{x_i}, v_{x_j})_{L_2(Q)} + \sum_{i=1}^n (a_i u_{x_i}, v)_{L_2(Q)} + (au, v)_{L_2(Q)}, \forall u, v \in W_2^{0,1}(Q),$$

$$\langle Au, v \rangle_\Omega = \sum_{i,j=1}^n (a_{ij}u_{x_i}, v_{x_j})_{L_2(\Omega)} + \sum_{i=1}^n (a_i u_{x_i}, v)_{L_2(\Omega)} + (au, v)_{L_2(\Omega)}, \forall u, v \in W_2^1(\Omega),$$

where $W_2^{0,1}(Q)$ is the completion of the set $C^\infty(\overline{Q})$ in the norm $\|u\|_{W_2^{0,1}(Q)}^2 = \sum_{i=1}^n \|u_{x_i}\|_{L_2(Q)}^2$ and $W_2^1(\Omega)$ is the completion of the set $C^\infty(\overline{\Omega})$ in the norm $\|u\|_{W_2^1(\Omega)}^2 = \sum_{i=1}^n \|u_{x_i}\|_{L_2(\Omega)}^2$.

Consider the relation

$$\langle Au, v_{tt} \rangle_Q - (Bu, B^*v_t)_{L_2(Q)} + \langle Cu, v \rangle_Q = \langle f, v \rangle_{V_T^2}, \tag{4.65}$$

where

$$B^*(v) \equiv -\sum_{i,j=1}^n \frac{\partial}{\partial x_i}\left(b_{ij}(x)\frac{\partial v}{\partial x_j}\right) - \sum_{i=1}^n \frac{\partial(b_i(x)v)}{\partial x_i} + b(x)v. \tag{4.66}$$

Relation (4.65) makes sense for arbitrary functions $u \in H_0^0$, $v \in V_T^2$, and $f \in (V_T^2)^*$. The coefficients of the operator are subjected to the following conditions:

(A) $a_{ij}, c_{ij}, a_i, c_i, a, b, c \in C(\overline{\Omega})$ and $b_{ij}, b_i \in C^1(\overline{\Omega})$.

Let c^* be a positive constant majorizing the corresponding norms of the coefficients of the operator $\mathscr{L}$.

The left-hand side in (4.65) specifies a linear operator $\mathscr{L}_0 : H_0^0 \to (V_T^2)^*$ and the linear adjoint operator $\mathscr{L}_0^* : V_T^2 \to (H_0^0)^*$. Obviously, $\mathscr{L}_0$ is an extension of the classical operator $\mathscr{L}$ (4.61).

4.7.2 A Priori Inequalities: Main Case

By c_f we denote the positive constant in the Friedrichs inequality

$$\|u\|_{L_2(Q)}^2 \leqslant c_f \sum_{i=1}^n \|u_{x_i}\|_{L_2(Q)}^2, \qquad \forall u \in \mathring{W}_2^{0,1}(Q), \tag{4.67}$$

where $\mathring{W}_2^{0,1}(Q)$ is the subspace of $W_2^{0,1}(Q)$ formed by functions vanishing on $(0,T) \times \partial\Omega$.

By using the Cauchy–Schwarz inequality in integral form and the Friedrichs inequality, one can readily prove the estimate

$$\|Bu\|_{L_2(Q)} \leqslant c\|u\|_{\mathring{W}_2^{0,2}(Q)}, \qquad \forall u \in \mathring{W}_2^{0,2}(Q), \tag{4.68}$$

where $\mathring{W}_2^{0,2}(Q)$ is the completion of the set $C^\infty(\overline{Q})$ of functions satisfying the boundary conditions $u = \partial u / \partial \mu_B = 0$ for $x \in \partial\Omega$ with respect to the norm $\|u\|^2 = \sum_{i=1}^n \|u_{x_i x_j}\|_{L_2(Q)}^2$.

Lemma 4.18. *Let condition **(A)** be satisfied. Then there exists a positive constant $c > 0$ such that the inequalities*

$$\|\mathscr{L}_0 u\|_{(V_T^2)^*} \leqslant c\|u\|_{H_0^0}, \quad \|\mathscr{L}_0^* v\|_{(H_0^0)^*} \leqslant c\|v\|_{V_T^2}$$

are valid for arbitrary functions $u \in H_0^0$ and $v \in V_T^2$.

Proof. One should apply the Cauchy–Schwarz inequality in integral form and the Friedrichs inequality (4.67) to the left-hand side of (4.65). $\square$

Therefore, $\mathscr{L}_0$ and $\mathscr{L}_0^*$ are continuous operators.

Remark 4.25. It follows from the definition of the operator $\mathscr{L}_0$ that $R(\mathscr{L}_0) \subset (W_T^2)^*$, where $R(\mathscr{L}_0)$ is the range of the operator $\mathscr{L}_0$, i.e.

$$\mathscr{L}_0 u \notin (V_T^2)^* \setminus (W_T^2)^*$$

for an arbitrary function $u \in H_0^0$. We also have the estimates

$$\|\mathscr{L}_0 u\|_{(V_T^2)^*} \leqslant c_1 \|\mathscr{L}_0 u\|_{(W_T^2)^*} \leqslant c_2 \|u\|_{H_0^0}, \quad \|\mathscr{L}_0^* v\|_{(H_0^0)^*} \leqslant c_1 \|v\|_{W_T^2} \leqslant c_2 \|v\|_{V_T^2}$$

for all $u \in H_0^0$ and $v \in V_T^2$, where $c_1, c_2 > 0$.

Let the operators A and B satisfy the ellipticity condition:

(B1) There exists a number $\alpha > 0$ such that $\langle Au, u \rangle_Q \geqslant \alpha \sum_{i=1}^n \int_Q u_{x_i}^2 dQ$ for any function $u \in W_2^{0,1}(Q)$.

(B2) There exists a number $\alpha_B > 0$ such that $\sum_{i,j=1}^n b_{ij} \xi_i \xi_j \geqslant \alpha_B \sum_{i=1}^n \xi_i^2$ for arbitrary $\xi_i \in \mathbb{R}$, $x \in \overline{\Omega}$.

Lemma 4.19. *(Main Lemma) Let conditions (A), (B1) and (B2) be satisfied. Then there exist positive constants λ, M and c such that for arbitrary functions $u \in H_0^0$ and $v \in V_T^2$ related by the formula $u(t,x) = e^{Mt}(\lambda v_{tt} - v_t + v)$, one has the equality*

$$c \langle \mathscr{L}_0 u, v \rangle_{V_T^2} \geqslant \|v\|_{V_T^2}^2 + \|u\|_{V_0^0}^2.$$

Proof. Since the functions belonging to the space V_T^2 have well-defined traces $v(T,x)$ and $v_t(T,x)$ at the point $t = T$, we have $v|_{t=T} = v_t|_{t=T} = 0$ for $v \in V_T^2$. We take into account the relationship between functions u and v and analyze $\langle \mathscr{L}_0 u, v \rangle_{V_T^2} = I_1 + \ldots + I_{10}$. Let us consider each of the items I_i separately.

By inequality **(B1)**,

$$I_1 = \langle A(e^{Mt} \lambda v_{tt}), v_{tt} \rangle_Q \geqslant \alpha \lambda \sum_{i=1}^n \int_Q e^{Mt} v_{ttx_i}^2 dQ.$$

From the formula $avv_t = (av^2)_t/2 - a_t v^2/2$, the symmetry of the matrix $\{a_{ij}\}_{i,j=1}^n$, the condition $v_t(T,x) = 0$ and inequality **(B1)**, we have

$$I_2 = -\frac{1}{2} \langle A(e^{Mt} v_t), v_{tt} \rangle_Q = \frac{1}{4} \langle Av_t|_{t=0}, v_t|_{t=0} \rangle_\Omega + \frac{M}{4} \langle A(e^{Mt} v_t), v_t \rangle_Q$$

$$\geqslant \frac{\alpha}{4} \sum_{i=1}^n \int_\Omega v_{tx_i}^2|_{t=0} d\Omega + \frac{\alpha M}{4} \sum_{i=1}^n \int_Q e^{Mt} v_{tx_i}^2 dQ.$$

Consider the item I_3. By using the relation $v_{tt} = (e^{-Mt} u + v_t - v)/\lambda$, the formula for integration by parts, and the notation $\int_0^t u \, d\tau = \bar{u}$, we obtain the estimate

$$I_3 = -\frac{1}{2} \langle A(e^{Mt} v_t), v_{tt} \rangle_Q = \frac{1}{2\lambda} \langle Av_{tt}, \bar{u} \rangle_Q - \frac{1}{2\lambda} \langle A(e^{Mt} v_t), v_t - v \rangle_Q$$

$$= \frac{1}{2\lambda^2} \langle A(e^{-Mt} \bar{u}_t), \bar{u} \rangle_Q + \frac{1}{2\lambda^2} \langle A(v_t - v), \bar{u} \rangle_Q - \frac{1}{2\lambda} \langle A(e^{Mt} v_t), v_t - v \rangle_Q.$$

The item $\left\langle A(e^{-Mt}\bar{u}_t),\bar{u}\right\rangle_Q$ can be estimated by analogy with I_2. To estimate two remaining items, we use the inequality

$$|\langle Au,v\rangle_Q| \leqslant \frac{c_0}{\varepsilon}\sum_{i=1}^{n}\int_Q u_{x_i}^2 dQ + \varepsilon\sum_{i=1}^{n}\int_Q v_{x_i}^2 dQ, \qquad \forall u,v \in \overset{\circ}{W}_2^{0,1}(Q), \varepsilon > 0, \quad (4.69)$$

where $c_0 = \left(c^*(n+c_f)\right)^2$. To prove inequality (4.69), one should apply the Cauchy-Buniakovsky inequality and the Friedrichs inequality (4.67) to the definition of $\langle Au,v\rangle_Q$. By using inequality (4.69) for $\varepsilon = 1$, we obtain the estimate

$$I_3 \geqslant \frac{\alpha}{4\lambda^2}\sum_{i=1}^{n}\int_\Omega e^{-Mt}\bar{u}_{x_i}^2|_{t=T}d\Omega + \frac{\alpha M}{4\lambda^2}\sum_{i=1}^{n}\int_Q e^{-Mt}\bar{u}_{x_i}^2 dQ$$
$$- c_1(\lambda)\sum_{i=1}^{n}\int_Q e^{Mt}v_{tx_i}^2 + e^{Mt}v_{x_i}^2 dQ - c_1(\lambda)\sum_{i=1}^{n}\int_Q e^{-Mt}\bar{u}_{x_i}^2 dQ,$$

where $c_1(\lambda)$ is a sufficiently large constant depending on λ.

Let us proceed to the item I_4. By using the integration by parts formula, we obtain

$$I_4 = \left\langle A(e^{Mt}v),v_{tt}\right\rangle_Q = -\left\langle Av|_{t=0},v_t|_{t=0}\right\rangle_\Omega - M\left\langle A(e^{Mt}v),v_t\right\rangle_Q - \left\langle A(e^{Mt}v_t),v_t\right\rangle_Q.$$

To estimate the first term, we use inequality (4.69) with $\varepsilon = \alpha/8$, the second term is estimated by analogy with I_2, and for the third term, we again use (4.69) with $\varepsilon = 1$.

$$I_4 \geqslant -\sum_{i=1}^{n}\int_\Omega \left(\frac{8c_0}{\alpha}v_{x_i}^2 + \frac{\alpha}{8}v_{tx_i}^2\right)|_{t=0}d\Omega + \frac{\alpha M}{2}\sum_{i=1}^{n}\int_\Omega v_{x_i}^2|_{t=0}d\Omega$$
$$+ \frac{\alpha M^2}{2}\sum_{i=1}^{n}\int_Q e^{Mt}v_{x_i}^2 dQ - (c_0+1)\sum_{i=1}^{n}\int_Q e^{Mt}v_{tx_i}^2 dQ.$$

We split the item $-(Bu,B^*v_t)_{L_2(Q)}$ into three parts:

$$-(Bu,B^*v_t)_{L_2(Q)} = -\frac{1}{2}(Bu,Bv_t)_{L_2(Q)} - \frac{1}{2}(Bu,Bv_t)_{L_2(Q)} + (Bu,\Delta Bv_t)_{L_2(Q)},$$

where $\Delta B = B - B^*$.

By analogy with I_2, we have

$$I_5 = -\frac{1}{2}\left(B(e^{Mt}\lambda v_{tt}),Bv_t\right)_{L_2(Q)} = \frac{\lambda}{4}\|Bv_t|_{t=0}\|_{L_2(\Omega)}^2 + \frac{\lambda M}{4}\left\|B\left(e^{Mt/2}v_t\right)\right\|_{L_2(Q)}^2.$$

It is known that the coercivity inequality [45] is valid for an elliptic operator B $(\alpha_B > 0)$. Therefore, there exists a $C > 0$ such that

$$C\|Bu\|_{L_2(Q)}^2 + C\|u\|_{L_2(Q)}^2 \geqslant \|u\|_{W_2^{0,2}(Q)}^2, \qquad (4.70)$$

for all $u \in W_2^{0,2}(Q) \cap \mathring{W}_2^{0,1}(Q)$, where $W_2^{0,2}(Q)$ is the completion of $C^\infty(\overline{Q})$ in the norm

$$\|u\|^2 = \sum_{i,j=1}^{n} \|u_{x_i x_j}\|_{L_2(Q)}^2.$$

By applying the Friedrichs inequality (4.67) to (4.70), we obtain the estimate

$$\|Bu\|_{L_2(Q)}^2 \geq \frac{1}{C}\|u\|_{W_2^{0,2}(Q)}^2 - c_f \sum_{i=1}^{n} \|u_{x_i}\|_{L_2(Q)}^2. \tag{4.71}$$

Therefore,

$$I_5 \geq \frac{\lambda}{4C} \sum_{i,j=1}^{n} \int_\Omega v_{tx_i x_j}^2 |_{t=0} d\Omega + \frac{\lambda M}{4C} \sum_{i,j=1}^{n} \int_Q e^{Mt} v_{tx_i x_j}^2 \, dQ$$

$$-\frac{\lambda c_f}{4} \sum_{i=1}^{n} \int_\Omega v_{tx_i}^2 |_{t=0} d\Omega - \frac{\lambda M c_f}{4} \sum_{i=1}^{n} \int_Q e^{Mt} v_{tx_i}^2 \, dQ.$$

Further, we have

$$I_6 = \frac{1}{2}\left(Be^{Mt} v_t, Bv_t\right)_{L_2(Q)} \geq 0,$$

$$I_7 = -\frac{1}{2}\left(Be^{Mt} v, Bv_t\right)_{L_2(Q)} = \frac{1}{4}\|Bv|_{t=0}\|_{L_2(\Omega)}^2 + \frac{M}{4}\left\|B\left(e^{Mt/2} v\right)\right\|_{L_2(Q)}^2.$$

We again use inequality (4.71) and, arguing by analogy with the estimate I_5, obtain

$$I_7 \geq \frac{1}{4C} \sum_{i,j=1}^{n} \int_\Omega v_{x_i x_j}^2 |_{t=0} d\Omega + \frac{M}{4C} \sum_{i,j=1}^{n} \int_Q e^{Mt} v_{x_i x_j}^2 \, dQ$$

$$-\frac{c_f}{4} \sum_{i=1}^{n} \int_\Omega v_{x_i}^2 |_{t=0} d\Omega - \frac{M c_f}{4} \sum_{i=1}^{n} \int_Q e^{Mt} v_{x_i}^2 \, dQ.$$

To estimate the item I_8, we use the integration by parts formula. Then we have

$$I_8 = -\frac{1}{2}\left(Bu, Bv_t\right)_{L_2(Q)} = -\frac{1}{2}\left(B\bar{u}_t, Bv_t\right)_{L_2(Q)} = \frac{1}{2}\left(B\bar{u}, Bv_{tt}\right)_{L_2(Q)}$$

$$= \frac{1}{2\lambda}\left(B\bar{u}, B\left(e^{-Mt}\bar{u}_t + v_t - v\right)\right)_{L_2(Q)}$$

$$= \frac{1}{4\lambda}\int_\Omega e^{-Mt}(B\bar{u})^2|_{t=T} d\Omega + \frac{M}{4\lambda}\int_Q e^{-Mt}(B\bar{u})^2 dQ + \frac{1}{2\lambda}\left(B\bar{u}, B\left(v_t - v\right)\right)_{L_2(Q)}.$$

This, together with (4.71) and (4.68), implies the estimate

$$
I_8 \geqslant \frac{1}{4\lambda C} \sum_{i,j=1}^{n} \int_{\Omega} e^{-Mt} \bar{u}_{x_i x_j}^2 |_{t=T} d\Omega + \frac{M}{4\lambda C} \sum_{i,j=1}^{n} \int_{Q} e^{-Mt} \bar{u}_{x_i x_j}^2 dQ
$$

$$
- \frac{c_f}{4\lambda} \sum_{i=1}^{n} \int_{\Omega} e^{-Mt} \bar{u}_{x_i}^2 |_{t=T} d\Omega - \frac{Mc_f}{4\lambda} \sum_{i=1}^{n} \int_{Q} e^{-Mt} \bar{u}_{x_i}^2 dQ
$$

$$
- c_1(\lambda) \sum_{i,j=1}^{n} \int_{Q} e^{-Mt} \bar{u}_{x_i x_j}^2 dQ - c_1(\lambda) \sum_{i,j=1}^{n} \int_{Q} e^{Mt} v_{tx_i x_j}^2 + e^{Mt} v_{x_i x_j}^2 dQ.
$$

Consider the item

$$
I_9 = \left(Be^{Mt} \lambda v_{tt}, \Delta B v_t \right)_{L_2(Q)} = -\lambda \left(Bv_t|_{t=0}, \Delta B v_t|_{t=0} \right)_{L_2(\Omega)}
$$

$$
- \lambda M \left(Be^{Mt} v_t, \Delta B v_t \right)_{L_2(Q)} - \lambda \left(Be^{Mt} v_t, \Delta B v_{tt} \right)_{L_2(Q)}.
$$

Since ΔB is a first-order differential operator, we can use the inequality $ab \geqslant -\varepsilon a^2 - \frac{1}{4\varepsilon} b^2$ and the Friedrichs inequality and readily show, by analogy with the preceding, that

$$
-\lambda \left(Bv_t|_{t=0}, \Delta B v_t|_{t=0} \right)_{L_2(\Omega)} \geqslant -\frac{\lambda}{8C} \sum_{i,j=1}^{n} \int_{\Omega} v_{tx_i x_j}^2 |_{t=0} d\Omega - \lambda c_2 \sum_{i=1}^{n} \int_{\Omega} v_{tx_i}^2 |_{t=0} d\Omega,
$$

$$
-\lambda M \left(Be^{Mt} v_t, \Delta B v_t \right)_{L_2(Q)} \geqslant -\frac{\lambda M}{8C} \sum_{i,j=1}^{n} \int_{Q} e^{Mt} v_{tx_i x_j}^2 dQ - \lambda M c_2 \sum_{i=1}^{n} \int_{Q} e^{Mt} v_{tx_i}^2 dQ,
$$

$$
-\lambda \left(Be^{Mt} v_t, \Delta B v_{tt} \right)_{L_2(Q)} \geqslant -\frac{\alpha \lambda}{3} \sum_{i=1}^{n} \int_{Q} e^{Mt} v_{ttx_i}^2 dQ - \lambda c_2 \int_{Q} e^{Mt} \sum_{i,j=1}^{n} v_{tx_i x_j}^2
$$

$$
+ e^{Mt} \sum_{i=1}^{n} v_{tx_i}^2 dQ.
$$

We include all remaining items in I_{10} and estimate it by analogy with the last three inequalities.

$$
I_{10} = \left(Be^{Mt}(v - v_t), \Delta B v_t \right)_{L_2(Q)} + \left\langle C \left(e^{Mt} (\lambda v_{tt} - v_t + v) \right), v \right\rangle_Q \geqslant
$$

$$
- \frac{\alpha \lambda}{3} \sum_{i=1}^{n} \int_{Q} e^{Mt} v_{ttx_i}^2 dQ - c_1(\lambda) \sum_{i=1}^{n} \int_{Q} e^{Mt} \left(v_{tx_i}^2 + v_{x_i}^2 \right) dQ
$$

$$
- c_2 \sum_{i,j=1}^{n} \int_{Q} e^{Mt} \left(v_{tx_i x_j}^2 + v_{x_i x_j}^2 \right) dQ.
$$

We take the sum $I_1 + \ldots + I_{10}$ and set $\lambda = \alpha/(4c_f + 16c_2)$. Now we can readily choose a constant $M(\lambda) > 0$ large enough to ensure that

$$c_3 \langle \mathscr{L}_0 u, v \rangle_{V_T^2} \geq \sum_{i=1}^{n} \int_Q e^{Mt} \left(v_{ttx_i}^2 + v_{tx_i}^2 + v_{x_i}^2 \right) dQ + \sum_{i,j=1}^{n} \int_{\Omega} v_{tx_ix_j}^2 |_{t=0} d\Omega$$

$$+ \sum_{i,j=1}^{n} \int_{\Omega} e^{-Mt} \bar{u}_{x_ix_j}^2 |_{t=T} d\Omega + \sum_{i,j=1}^{n} \int_Q e^{Mt} v_{tx_ix_j}^2 + e^{-Mt} \bar{u}_{x_ix_j}^2 \, dQ$$

for some positive constant $c_3(M, \lambda) > 0$.

In view of the inequality

$$\int_Q e^{-Mt} u_{x_i}^2 \, dQ = \int_Q e^{Mt} \left(\lambda v_{ttx_i} - v_{tx_i} + v_{x_i} \right)^2 \, dQ \leq c_4 \int_Q e^{Mt} \left(v_{ttx_i}^2 + v_{tx_i}^2 + v_{x_i}^2 \right) dQ,$$

we conclude that there exists a constant $c > 0$ such that $c \langle \mathscr{L}_0 u, v \rangle_{V_T^2} \geq \|v\|_{V_T^2}^2 + \|u\|_{V_0^0}^2$. The proof of the lemma is complete. $\qquad\square$

Theorem 4.17. *Let conditions (A), (B1), and (B2) be satisfied. Then there exist constants $c_i > 0$ such that*

$$\|u\|_{V_0^0} \leq c_1 \|\mathscr{L}_0 u\|_{(V_T^2)^*} \leq c_2 \|\mathscr{L}_0 u\|_{(W_T^2)^*} \leq c_3 \|u\|_{H_0^0} \qquad (4.72)$$

for all $u \in H_0^0$.

Proof. It suffices to prove the left inequality. For an arbitrary function $u \in H_0^0$, consider the ordinary differential equation (the Cauchy problem)

$$u(t,x) = e^{Mt} \left(\lambda v_{tt} - v_t + v \right), \qquad v|_{t=T} = 0, \quad v_t|_{t=T} = 0.$$

Obviously, the solution v of this equation exists and belongs to the space $v \in V_T^2$. By Lemma 4.19, $c \langle \mathscr{L}_0 u, v \rangle_{V_T^2} \geq \|v\|_{V_T^2}^2 + \|u\|_{V_0^0}^2$. By using the Schwarz inequality and the inequality $a^2 + b^2 \geq 2ab$, we obtain the estimates

$$c \|\mathscr{L}_0 u\|_{(V_T^2)^*} \|v\|_{V_T^2} \geq c \langle \mathscr{L}_0 u, v \rangle_{V_T^2} \geq \|v\|_{V_T^2}^2 + \|u\|_{V_0^0}^2 \geq 2 \|v\|_{V_T^2} \|u\|_{V_0^0},$$

which imply the assertion of the theorem. $\qquad\square$

Corollary 4.10. The operator $\mathscr{L}_0$ is injective.

4.7.3 Analysis of the System on the Basis of a Single Chain of a Priori Inequalities

Let us show that a meaningful solvability theory of the operator equation $\mathscr{L}u = f$ can already be constructed on the basis of a single chain of a priori inequalities (4.72).

4.7.3.1 Adjoint A Priori Estimate

Our method for proving the a priori inequalities (4.72) can also be used in the proof of a similar chain of the adjoint operator

$$\mathcal{L}^* v \equiv A^*(v_{tt}) - (B^*)^2(v_t) + C^*(v).$$

Let the coefficients of the problem satisfy the following condition:

(C) $a_{ij}, c_{ij}, a, b, c \in C(\overline{\Omega})$ and $b_{ij}, a_i, b_i, c_i \in C^1(\overline{\Omega})$.

Consider the linear operator $\bar{\mathcal{L}}_2 : V_0^2 \to (H_T^0)^*$ given by the relation

$$\langle \bar{\mathcal{L}}_2 u, v \rangle_{H_T^0} = \langle Au_{tt}, v \rangle_Q + (Bu_t, B^* v)_{L_2(Q)} + \langle Cu, v \rangle_Q, \qquad (4.73)$$

and the adjoint operator $\bar{\mathcal{L}}_2^* : H_T^0 \to (V_0^2)^*$ given by the same relation (4.73). Obviously, $\bar{\mathcal{L}}_2$ is the restriction of the operator $\mathcal{L}_0 : H_0^0 \to (V_T^2)^*$ and $\bar{\mathcal{L}}_2^*$ is an extension of the operator $\mathcal{L}_0^* : V_T^2 \to (H_0^0)^*$.

Theorem 4.18. *Let conditions **(B1)**, **(B2)** and **(C)** be satisfied. Then there exist constants $c_i > 0$ such that*

$$\|v\|_{V_T^0} \leqslant c_1 \|\bar{\mathcal{L}}_2^* v\|_{(V_0^2)^*} \leqslant c_2 \|\bar{\mathcal{L}}_2^* v\|_{(W_0^2)^*} \leqslant c_3 \|v\|_{H_T^0} \qquad (4.74)$$

for all functions $v \in H_T^0$.

Proof. Consider the leftmost inequality in (4.74). (The remaining inequalities are easy to prove). For an arbitrary function $v(t,x) \in H_T^0$, consider the function $\bar{u}(\tau, x) = v(T - \tau, x)$. One can readily see that $\bar{u} \in H_0^0$. Let $\bar{v}(\tau, x)$ be the solution of the Cauchy problem

$$\bar{u}(\tau, x) = e^{M\tau}(\lambda \bar{v}_{\tau\tau} - \bar{v}_\tau + \bar{v}), \qquad \bar{v}|_{\tau=T} = \bar{v}_\tau|_{\tau=T} = 0.$$

Obviously, $\bar{v} \in V_T^2$. Therefore, the functions $\bar{u}$ and $\bar{v}$ satisfy the assumptions of Lemma 4.19. By applying this lemma to the operators $A_1 = A^*$, $B_1 = B^*$, and $C_1 = C^*$ (note that $\langle Au, v \rangle_Q = \langle A^* v, u \rangle_Q$, where A^* is defined by analogy with (4.66)) we obtain the inequality

$$\langle A^* \bar{u}, \bar{v}_{\tau\tau} \rangle_Q - (B^* \bar{u}, B \bar{v}_\tau)_{L_2(Q)} + \langle C^* \bar{u}, \bar{v} \rangle_Q \geqslant c^{-1} \|\bar{v}\|_{V_T^2}^2 + c^{-1} \|\bar{u}\|_{V_0^0}^2.$$

By performing the change of variables $t = T - \tau$, we rewrite the last relation in the form

$$\langle Au_{tt}, v \rangle_Q + (Bu_t, B^* v)_{L_2(Q)} + \langle Cu, v \rangle_Q \geqslant c^{-1} \|u\|_{V_0^2}^2 + c^{-1} \|v\|_{V_T^0}^2,$$

where $u(t,x) = \bar{v}(T - t, x) \in V_0^2$.

By using the Schwarz inequality, we obtain

$$\|\bar{\mathcal{L}}_2^* v\|_{(V_0^2)^*} \|u\|_{V_0^2} \geqslant \langle \bar{\mathcal{L}}_2^* v, u \rangle_{V_0^2} \geqslant 2c^{-1} \|u\|_{V_0^2} \|v\|_{V_T^0},$$

which completes the proof. $\square$

Remark 4.26. We could have argued by analogy with Lemma 4.19, directly writing out a relation between functions v and u.

Corollary 4.11. The operator $\bar{\mathscr{L}}_2^*$ is injective.

By analogy with [59–62, 64] one can prove a theorem on existence and uniqueness on the basis of the chain (4.74).

Theorem 4.19. *Let conditions* **(B1)**, **(B2)** *and* **(C)** *be satisfied. Then for an arbitrary right-hand side $f \in (V_T^0)^*$, there exists a unique solution $u \in V_0^2 \subset H_0^0$ of the equation $\mathscr{L}_0 u = f$.*

Corollary 4.12. The range $R(\mathscr{L}_0)$ is dense in $(V_T^2)^*$.

Remark 4.27. By analogy with Theorem 4.19, on the basis of estimate (4.72), one can show that the range $R(\mathscr{L}_0^*)$ of the adjoint operator contains $(V_0^0)^*$, which, in particular, implies that $R(\mathscr{L}_0^*)$ is dense in the space $(H_0^0)^*$.

4.7.3.2 Generalizes Solvability

It was shown in remark 4.25 that $R(\mathscr{L}_0) \subset (W_T^2)^*$; in particular, $R(\mathscr{L}_0) \neq (V_T^2)^*$. Therefore, for functions $f \in (V_T^2)^*$, we face the problem of finding some generalized solution of the equation $\mathscr{L}_0 u = f$.

Definition 4.13. A *generalized solution* of the equation $\mathscr{L}_0 u = f$ with right-hand side $f \in (V_T^2)^*$ is an element $u \in V_0^0$ such that there exists a sequence $u_i \in H_0^0$ such that $\|u_i - u\|_{V_0^0} \to 0$ and $\|\mathscr{L}_0 u_i - f\|_{(V_T^2)^*} \to 0$ as $i \to \infty$.

Remark 4.28. If $\mathscr{L}_0 u = f$ for $u \in H_0^0$, then, obviously, u is a generalized solution in the sense of Definition 4.13.

Definition 4.14. A *generalized solution* of the equation $\mathscr{L}_0 u = f$ with right-hand side $f \in (V_T^2)^*$ is defined as an element $u \in V_0^0$ such that $\langle \mathscr{L}_0^* v, u \rangle_{V_0^0} = \langle f, v \rangle_{V_T^2}$ for all $v \in V_T^2 : \mathscr{L}_0^* v \in (V_0^0)^*$.

The following assertion can be proved on the basis of the chain of inequalities (4.72) by analogy with [62, 64, 75].

Theorem 4.20. *Let conditions* **(B1)**, **(B2)** *and* **(C)** *be satisfied. Then, for an arbitrary right-hand side $f \in (V_T^2)^*$ of the equation $\mathscr{L}_0 u = f$, there exists a unique generalized solution in the sense of Definitions 4.13 and 4.14. Generalized solutions in the sense of Definitions 4.13 and 4.14 are equivalent.*

Corollary 4.13. If a generalized solution u of the equation $\mathscr{L}_0 u = f$ belongs to the space $u \in H_0^0$, then u is an ordinary solution of this equation and $f \in R(\mathscr{L}_0)$. If u is a generalized solution of the equation $\mathscr{L}_0 u = f$ and $f \in R(\mathscr{L}_0)$, then u belongs to the space H_0^0 and is an ordinary solution.

Proof. It suffices to note that the set of functions $v \in V_T^2 : \mathscr{L}_0^* v \in (V_0^0)^*$ is dense in V_T^2 (since $\mathscr{L}_0^*(L_T) \subset (V_0^0)^*$). □

4.7.3.3 Another Type of Adjoint Inequalities

The chain of a priori inequalities (4.72) permits one to prove similar a priori estimates for the adjoint operator. However, there exist other estimates of the operator
$\mathscr{L}_0^*$, which can be derived simply "by passing to the adjoint". Moreover, these estimates permit explicitly indicating an extension of $\mathscr{L}_0$ corresponding to the definition
of generalized solutions of the equation $\mathscr{L}_0 u = f$ in the sense of Definitions 4.13
and 4.14.

Consider the linear operator $\bar{\mathscr{L}}_0 : V_0^0 \to (H_T^2)^*$ (and also the linear operator $\bar{\mathscr{L}}_0^* :$
$H_T^2 \to (V_0^0)^*$)) given by the relation

$$\langle \bar{\mathscr{L}}_0 u, v \rangle_{H_T^2} = \langle Au, v_{tt} \rangle_Q + (B\bar{u}, B^* v_{tt})_{L_2(Q)} + \langle Cu, v \rangle_Q, \qquad (4.75)$$

where $\bar{u} = \int_0^t u \, d\tau$.

The right-hand side of relation (4.75) specifies an extension $\bar{\mathscr{L}}_0 : V_0^0 \to (H_T^2)^*$ of
the operator $\mathscr{L}_0 : H_0^0 \to (V_T^2)^*$ to the entire space V_0^0. It is also obvious that $\bar{\mathscr{L}}_0^*$ is a
restriction of the operator $\mathscr{L}_0^*$.

The following assertion can be proved by analogy with Lemma 4.18.

Lemma 4.20. *Let condition* **(A)** *be satisfied. Then there exist positive constants*
$c_i > 0$ *such that*

$$\|\bar{\mathscr{L}}_0 u\|_{(H_T^2)^*} \leqslant c_1 \|u\|_{W_0^0} \leqslant c_2 \|u\|_{V_0^0}, \quad \|\bar{\mathscr{L}}_0^* v\|_{(V_0^0)^*} \leqslant c_1 \|\bar{\mathscr{L}}_0^* v\|_{(W_0^0)^*} \leqslant c_2 \|u\|_{H_T^2}$$
$$(4.76)$$

for arbitrary $u \in V_0^0$, $v \in H_T^2$.

Therefore, $\bar{\mathscr{L}}_0$ and $\bar{\mathscr{L}}_0^*$ are continuous operators. By analogy with Remark 4.25,
one can describe the range $R(\bar{\mathscr{L}}_0^*)$ of the operator $\bar{\mathscr{L}}_0^*$ in more detail. Indeed, it
follows from (4.76) that the range $R(\bar{\mathscr{L}}_0^*)$ is part of $(W_0^0)^*$.

Theorem 4.21. *Let conditions* **(A)**, **(B1)**, *and* **(B2)** *be satisfied. Then there exist constants* $c_i > 0$ *such that*

$$\|v\|_{V_T^2} \leqslant c_1 \|\bar{\mathscr{L}}_0^* v\|_{(V_0^0)^*} \leqslant c_2 \|\bar{\mathscr{L}}_0^* v\|_{(W_0^0)^*} \leqslant c_3 \|v\|_{H_T^2} \qquad (4.77)$$

for all $v \in H_T^2$.

Proof. It suffices to prove the left inequality. For an arbitrary function $v \in H_T^2$ con-
sider the function $u \in H_0^0 \subset V_0^0$ given by the relation $u(t,x) = e^{Mt}(\lambda v_{tt} - v_t + v)$.
Since $v \in H_T^2$, it follows that the function $v(t,x)$ satisfies the conditions $v(T,x) =$
$v_t(T,x) = 0$. Therefore, one can use Lemma 4.19 for such functions $u(t,x)$ and
$v(t,x)$. We have

$$c\langle u, \bar{\mathscr{L}}_0^* v \rangle_{V_0^0} = c\langle u, \mathscr{L}_0^* v \rangle_{H_0^0} = c\langle \bar{\mathscr{L}}_0 u, v \rangle_{V_T^2} \geqslant \|v\|_{V_T^2}^2 + \|u\|_{V_0^0}^2.$$

By using the Schwarz inequality and the inequality $a^2 + b^2 \geqslant 2ab$, we obtain the estimates

$$c\|\bar{\mathscr{L}}_0^* v\|_{(V_0^0)^*} \|u\|_{V_0^0} \geqslant c\langle u, \bar{\mathscr{L}}_0^* v\rangle_{V_0^0} \geqslant \|v\|_{V_T^2}^2 + \|u\|_{V_0^0}^2 \geqslant 2\|v\|_{V_T^2}\|u\|_{V_0^0},$$

which implies the assertion of the theorem. $\qquad\square$

By analogy with Theorem 4.19, on the basis of he chain of inequalities (4.77), one can prove the solvability theorem.

Theorem 4.22. *Let conditions* **(A)**, **(B1)**, *and* **(B2)** *be satisfied. Then for any* $f \in (V_T^2)^*$, *there exists a unique solution* $u \in V_0^0$ *of the equation* $\bar{\mathscr{L}}_0 u = f$.

We can readily prove a theorem establishing a relationship between the generalized solvability of the equation $\mathscr{L}_0 u = f$ in the sense of Definitions 4.13 and 4.14 and the solvability of the operator equation $\bar{\mathscr{L}}_0 u = f$.

Theorem 4.23. *Let conditions* **(B1)**, **(B2)** *and* **(C)** *be satisfied, and let* $f \in (V_T^2)^*$. *Then the solvability of the equation coincides with the generalized solvability of the equation* $\bar{\mathscr{L}}_0 u = f$ *in the sense of Definitions 4.13 and 4.14.*

Proof. If u is a generalized solution of the equation $\mathscr{L}_0 u = f$ in the sense if Definition 4.13 and $u_i \in H_0^0$ is a sequence specifying this solution, then $\mathscr{L}_0 u_i \to f$ in $(V_T^2)^*$. By virtue of the continuity of the embedding $(V_T^2)^* \subset (H_T^2)^*$, we obtain $\mathscr{L}_0 u_i \to f$ in $(H_T^2)^*$. On the other hand, by virtue of the continuity of the operator $\bar{\mathscr{L}}_0$, we have $\mathscr{L}_0 u_i = \bar{\mathscr{L}}_0 u_i \to \bar{\mathscr{L}}_0 u$ in $(H_T^2)^*$. Hence, it follows that $\bar{\mathscr{L}}_0 u = f$.

But if $u \in V_0^0$ is a solution of the equation $\bar{\mathscr{L}}_0 u = f$, then, by Theorem 4.20, there exists a generalized solution $u^* \in V_0^0$ of the equation $\mathscr{L}_0 u = f$, which, as was shown above, is also a solution of the equation $\bar{\mathscr{L}}_0 u = f$ and coincides with u by virtue of the uniqueness. The proof of the theorem is complete. $\qquad\square$

4.7.4 Construction of a "Scale" of Solvability Theorems

Let us show that the a priori inequalities (4.72) permit one to obtain an entire scale of estimates and prove the corresponding solvability theorems.

4.7.4.1 Shift of the A Priori Inequalities By $\partial^k/\partial t^k$

The a priori inequalities proved above can be "shifted" by the operator of differentiation with respect to the time variable. Let us give related considerations.

We introduce the following notation. Let $u \in L_0$; by $u^{(k)}$ we denote its kth derivative with respect to the variable t for $k \in \mathbb{N} \cup 0$. If $k \in \mathbb{Z}$ and $k < 0$, then $u^{(k)}$ is understood as the function obtained by the $|k|$-fold application of the integration operator $\int_0^t d\tau$ to the function $u(\tau, x)$. The notation $v^{[k]}$ ($v \in L_T$) is introduced in a similar way with the use of the integral operator $\int_T^t d\tau$.

Obviously, the mapping $u \to u^{(k)}$ $(k \in \mathbb{Z})$ is an isometry between the space L_0 with the norm H_0^l (or W_0^l, V_0^l) and the space L_0 with the norm H_0^{l-k} (respectively, W_0^{l-k}, V_0^{l-k}). Consequently, this mapping can be extended by the continuity to the entire space H_0^l (or W_0^l, V_0^l). We preserve the same notation for the extended mapping which is an isometry between H_0^l (respectively, W_0^l or V_0^l) and H_0^{l-k} (respectively, W_0^{l-k} or V_0^{l-k}). One can readily see that $(u^{(k)})^{(l)} = u^{(k+l)}$. A similar property is valid for the mapping $v \to v^{[k]}$.

Note that the above-mentioned mappings should be used with care. For example, the relation $(u^{(1)})^{(-1)} = u$ can be represented in the form

$$\int_0^t u^{(1)}(\tau, x) d\tau = u(t, x), \qquad \forall u \in H_0^0. \tag{4.78}$$

However, the space H_0^0 contains also smooth functions $u(t, x)$ not vanishing for $t = 0$. Nevertheless, formula (4.78) is valid for such functions[2].

Consider the operator $\mathscr{L}_k : H_0^k \to (V_T^{2-k})^*$ $(k \in \mathbb{Z})$ (and the adjoint operator $\mathscr{L}_k^* : V_T^{2-k} \to (H_0^k)^*$) given by the relation

$$\langle \mathscr{L}_k u, v \rangle_{V_T^{2-k}} = (-1)^k \left(\left\langle Au^{(k)}, v^{[2-k]} \right\rangle_Q - \left(Bu^{(k)}, B^* v^{[1-k]} \right)_{L_2(Q)} \right.$$
$$\left. + \left\langle Cu^{(k)}, v^{[-k]} \right\rangle_Q \right).$$

Obviously, if $k = 0$, then the above-presented operator becomes the operator $\mathscr{L}_0$ considered earlier

By using integration by parts, one can readily justify the following assertion.

Lemma 4.21. *The formulas $\mathscr{L}_k \subset \mathscr{L}_{k-1}$, $\mathscr{L}_k^* \subset \mathscr{L}_{k+1}^*$ are valid for the operators $\mathscr{L}_k$, where $A \subset B$ means that A is a restriction of the operator B.*

Theorem 4.24. *Let conditions* (A), (B1), *and* (B2) *be satisfied. Then there exist constants $c_i > 0$ such that*

$$\|u\|_{V_0^k} \leqslant c_1 \|\mathscr{L}_k u\|_{(V_T^{2-k})^*} \leqslant c_2 \|\mathscr{L}_k u\|_{(W_T^{2-k})^*} \leqslant c_3 \|u\|_{H_0^k}. \tag{4.79}$$

for all functions $u \in H_0^k$.

---

[2] Relation (4.78) can be explained from the viewpoint of distribution. One can show that it is meaningful to consider elements of the space H_0^{-1} as functionals on the space $\bar{W}_T^1$ (more precisely $H_0^{-1} \subset (\bar{W}_T^1)^*$), where $\bar{W}_T^1$ is the completion of L_T in the norm $\|v\|_{\bar{W}_T^1}^2 = \|v_t\|_{L_2(Q)}^2$. Then $u^{(1)} \in H_0^{-1}$ should be treated as a functional acting as $\langle u^{(1)}, v \rangle_{\bar{W}_T^1} = -(u, v_t)_{L_2(Q)}$ for all functions $v \in \bar{W}_T^1$, $u \in H_0^0$, including smooth functions $u \in H_0^0$ not satisfying the condition $u(0, x) = 0$. An important difference of $u^{(1)}$ from the Sobolev generalized derivatives the following: the functional $u^{(1)}$ is defined on functions $v \in \bar{W}_T^1$ not necessarily vanishing for $t = 0$.

Proof. Let us prove the leftmost inequality. (The remaining inequalities can be proved in a similar way.) For an arbitrary $u \in H_0^k$, we consider the ordinary differential equation (the Cauchy problem)

$$(-1)^k u^{(k)} = e^{Mt} (\lambda v_{tt} - v_t + v), \quad v|_{t=T} = v_t|_{t=T} = 0.$$

Since $u^{(k)} \in H_0^0$, it follows that the solution v of this equation exists and belongs to the space $v \in V_T^2$. Then $v^{[k]} \in V_T^{2-k}$, and, by Lemma 4.19,

$$\left\langle \mathscr{L}_k u, v^{[k]} \right\rangle_{V_T^{2-k}} = (-1)^k \left(\left\langle Au^{(k)}, v_{tt} \right\rangle_Q - \left(Bu^{(k)}, B^* v_t \right)_{L_2(Q)} + \left\langle Cu^{(k)}, v \right\rangle_Q \right)$$

$$= \left\langle \mathscr{L}_0 \left((-1)^k u^{(k)} \right), v \right\rangle_{V_T^2} \geq c^{-1} \|v\|_{V_T^2}^2 + c^{-1} \left\| (-1)^k u^{(k)} \right\|_{V_0^0}^2$$

$$= c^{-1} \|v^{[k]}\|_{V_T^{2-k}}^2 + c^{-1} \|u\|_{V_0^k}^2.$$

Further, by applying the Schwarz inequality to the left-hand side of the last inequality and the inequality $a^2 + b^2 \geq 2ab$ to the right-hand side, we obtain

$$\|\mathscr{L}_k u\|_{(V_T^{2-k})^*} \|v^{[k]}\|_{V_T^{2-k}} \geq \left\langle \mathscr{L}_k u, v^{[k]} \right\rangle_{V_T^{2-k}} \geq 2c^{-1} \|v^{[k]}\|_{V_T^{2-k}} \|u\|_{V_0^k},$$

which completes the proof. $\qquad\square$

4.7.4.2 Shift of the Adjoint Estimates and the General Solvability Theorem

On the basis of the general a priori estimates (4.79), we prove inequalities for the adjoint operator. Consider the linear operator $\bar{\mathscr{L}}_k : V_0^k \to (H_T^{2-k})^*$ (and also the linear adjoint operator $\bar{\mathscr{L}}_k^* : H_T^{2-k} \to (V_0^k)^*$) given by the relation

$$\left\langle \bar{\mathscr{L}}_k u, v \right\rangle_{H_T^{2-k}} = (-1)^k \left(\left\langle Au^{(k)}, v^{[2-k]} \right\rangle_Q + \left(Bu^{(k-1)}, B^* v^{[2-k]} \right)_{L_2(Q)} \right.$$

$$\left. + \left\langle Cu^{(k-2)}, v^{[2-k]} \right\rangle_Q \right).$$

Obviously, for $k = 0$ and $k = 2$, this definition is in agreement with the definition of the operators $\bar{\mathscr{L}}_0^*$ and $\bar{\mathscr{L}}_2^*$ introduced above. The following assertion can be proved by analogy with Lemma 4.21 with the use of integration by parts.

Lemma 4.22. *One has* $\mathscr{L}_k \subset \bar{\mathscr{L}}_k \subset \mathscr{L}_{k-1}$, $\mathscr{L}_k^* \subset \bar{\mathscr{L}}_{k+1}^* \subset \mathscr{L}_{k+1}^*$.

Theorem 4.25. *Let conditions* **(A)**, **(B1)**, *and* **(B2)** *be satisfied. Then there exist constants* $c_i > 0$ *such that*

$$\|v\|_{V_T^k} \leq c_1 \|\bar{\mathscr{L}}_0^* v\|_{(V_0^{2-k})^*} \leq c_2 \|\bar{\mathscr{L}}_0^* v\|_{(W_0^{2-k})^*} \leq c_3 \|v\|_{H_T^k} \qquad (4.80)$$

for all $v \in H_T^k$.

Proof. We rewrite the desired inequalities in the following form:

$$\|v\|_{V_T^{2-k}} \leqslant c_1 \|\mathscr{L}_0^* v\|_{(V_0^k)^*} \leqslant c_2 \|\mathscr{L}_0^* v\|_{(W_0^k)^*} \leqslant c_3 \|v\|_{H_T^{2-k}}$$

for all $v \in H_T^{2-k}$.

We only prove the leftmost inequality. Consider an arbitrary function $v \in H_T^{2-k}$. Then $v^{[-k]} \in H_T^2$. Let $u(t,x) \in H_0^0 \subset V_0^0$ be the function given by the relation

$$u(t,x) = (-1)^k e^{Mt} \left(\lambda v_{tt}^{[-k]} - v_t^{[-k]} + v^{[-k]} \right).$$

Since $v^{[-k]} \in H_T^2$, it follows that the function $v(t,x)$ satisfies the conditions

$$v^{[-k]}(T,x) = v_t^{[-k]}(T,x) = 0.$$

Therefore, one can apply Lemma 4.19 to the functions $(-1)^k u$ and $v^{[-k]}$. We have

$$c \left\langle \mathscr{L}_0((-1)^k u), v^{[-k]} \right\rangle_{V_T^2} \geqslant \|v^{[-k]}\|_{V_T^2}^2 + \|u\|_{V_0^0}^2 \geqslant 2\|v^{[-k]}\|_{V_T^2}\|u\|_{V_0^0} = 2\|v\|_{V_T^{2-k}}\|u\|_{V_0^0}.$$

$$(4.81)$$

By using the integration by parts formula, we obtain

$$\left\langle \mathscr{L}_0((-1)^k u), v^{[-k]} \right\rangle_{V_T^2} = (-1)^k \left(\left\langle Au, v_{tt}^{[-k]} \right\rangle_Q - \left(Bu, B^* v_t^{[-k]} \right)_{L_2(Q)} \left\langle Cu, v^{[-k]} \right\rangle_Q \right)$$

$$= (-1)^k \left(\left\langle Au, v^{[2-k]} \right\rangle_Q + \left(Bu^{(-1)}, B^* v^{[2-k]} \right)_{L_2(Q)} \left\langle Cu^{(-2)}, v^{[2-k]} \right\rangle_Q \right)$$

$$= \left\langle \mathscr{L}_k(u^{(-k)}), v \right\rangle_{H_T^{2-k}} = \left\langle \mathscr{L}_k^* v, u^{(-k)} \right\rangle_{V_0^k} \leqslant \|\mathscr{L}_k^* v\|_{(V_0^k)^*} \|u^{(-k)}\|_{V_0^k}$$

$$= \|\mathscr{L}_k^* v\|_{(V_0^k)^*} \|u\|_{V_0^0},$$

which implies the desired assertion.   □

Remark 4.29. Just as in the case of Theorem 4.18, one can give another proof of the chain of adjoint a priori inequalities.

The theorem on the unique solvability of the operator $\mathscr{L}$ can be proved by analogy with Theorem 4.19 on the basis of the estimates (4.79) and (4.80).

Theorem 4.26. *Let conditions* **(A)**, **(B1)**, *and* **(B2)** *be satisfied. Then for an arbitrary right-hand side* $f \in (V_T^k)^*$, *there exists a unique solution* $u \in V_0^{2-k}$ *of the equation* $\mathscr{L}_{2-k} u = f$.

Remark 4.30. For $f \in (V_T^k)^*$ one can introduce the notion of generalized solutions of the equation $\mathscr{L}_{2-k} u = f$, prove their existence and uniqueness, and show that they are natural solutions of the equation $\bar{\mathscr{L}}_{2-k} u = f$.

Note that, for the most important special case $B = 0$, the preceding considerations can readily be modified; moreover, the corresponding solvability theorems have the form of criteria, since the spaces V_0^k and H_0^k coincide in this case [76].

4.8 Projection Theorem for Banach and Locally Convex Spaces

As a rule, the existence of generalized solutions of boundary value problems may be reduced to possibility to represent linear continuous functionals using a given bilinear forms.

The classical Vishik–Lax–Milgram Theorem for Hilbert space is well known [48, 113].

Theorem 4.27 (Vishik–Lax–Milgram). *Let H be a Hilbert space, b be a bilinear form bounded on $H \times H$. If there exists a number $c > 0$ such that $c\|x\|_H^2 \leq |b(x,x)|$ $\forall x \in H$, then for any element $f \in H$ there exists a unique element $x \in H$ satisfying the following identity*

$$b(x,y) = (f,y)_H \quad \forall y \in H. \tag{4.82}$$

This theorem is an effective tool for studying elliptic boundary value problems. However, it should be stressed that the natural generalized statements of evolution problems for partial differential equations are not identities (4.82). Moreover, J.-L. Lions proved the following projection theorem for evolution problems.

Theorem 4.28 (J.-L. Lions [50]). *Let F be a Hilbert space and Φ be a linear subspace F with a new inner product $(\cdot,\cdot)_\Phi$. Assume that*

$$\|x\|_F \leq c\|x\|_\Phi \quad \forall x \in \Phi,$$

where $c > 0$. Let $b : F \times \Phi \to \mathbb{R}$ be a bilinear form and the following conditions are satisfied

$$\forall y \in \Phi \quad x \mapsto b(x,y) \in F^*,$$

$$\exists c_1 > 0 : \ |b(y,y)| \geq c_1\|y\|_\Phi^2 \quad \forall y \in \Phi.$$

Then, for every $f \in \Phi^*$, there exists $x \in F$ that satisfies the identity

$$b(x,y) = f(y) \quad \forall y \in \Phi.$$

The following result is well-known for Banach spaces [24].

Theorem 4.29. *Let E be a Banach space, F be a reflexive Banach space, b be a bilinear form bounded on $E \times F$. Then the following statements are equivalent:*

(1)   for any $f \in F^*$ there exists a unique $x \in E$ such that

$$b(x,y) = \langle f,y \rangle_{F^*,F} \quad \forall y \in F;$$

(2)   (i) there exists $c > 0$ such that

$$c\|x\|_E \leq \sup_{y \in B_1(F)} |b(x,y)| \quad \forall x \in E;$$

(ii) if $b(x,y) = 0 \ \forall x \in E$, then $y = 0$.

Proof. Let us introduce an operator $E \xrightarrow{T} F^*$ which acts by the following rule

$$E \ni x \mapsto Tx = b(x, \cdot) \in F^*.$$

The operator T is linear and continuous, and

$$\|Tx\|_{F^*} = \|b(x, \cdot)\|_{F^*} = \sup_{y \in B_1(F)} |b(x,y)| \leq M\|x\|_E.$$

Let Statement 1 holds, i.e. the operator T is bijective. Then by the Banach theorem on inverse operator it is continuously invertible, i.e. $\exists c > 0$:

$$c\|x\|_E \leq \|Tx\|_{F^*}.$$

It is equivalent to Statement (2) (i). Condition (2) (ii) means the totality of $R(T)$ in F^*. It is true when $R(T) = F^*$.

Conversely, let Statement (2) holds. Statement (2) (i) imply that

$$c\|x\|_E \leq \|Tx\|_{F^*},$$

i.e. the operator T is continuously invertible over $R(T) \subseteq F^*$ and the linear manifold $R(T)$ is closed in F^*. Since $R(T)$ is total in F^* (Property 2) (ii), then, taking into account that F is reflexive, we obtain $R(T) = F^*$.   □

Remark 4.31. If E is a reflexive Banach space, and F is a Banach space, then Conditions (1) and (2) of the Theorem 4.29 are reflexive also. If it is possible to represent uniquely all elements of E^* and F^* using the norm b, the the spaces E and F are reflexive [24].

Let us generalized the Theorem 4.29 for locally convex linear topological spaces E and F.

Let E be a Hausdorff locally convex space (l.c.s.) with a conjugate space E^*. Recall that the set $A^* \subseteq E^*$ is called almost closed, if the set $A^* \cap U^o$ is closed in topology $\sigma(E^*, E)$ for every neighborhoods of zero U from E, where $U^o = \left\{ y^* \in E^* : \sup_{x \in U} \langle y^*, x \rangle_{E^*, E} \leq 1 \right\}$ is a polar of the set U [97].

Definition 4.15 ([97]). The space E is called perfectly complete, if every almost closed linear subspace in E^* is $\sigma(E^*, E)$-closed.

Remark 4.32. Every Frechet space is perfectly complete [97]. The space E^* which is conjugate to the Banach space E and is equipped with the topology $\sigma(E^*, E)$ is a perfectly complete space [97]. Strong conjugate space for reflexive Frechet space is perfectly complete [97].

Recall that a barrel in locally convex space is a convex balanced adsorbing and closed subset. Every locally convex space has a fundamental system of zero neighborhoods that consist of barrels. A locally convex space is called barreled of every barrel in it is a zero neighborhood [97].

The following statements hold [97].

Theorem 4.30 (on an open mapping). *A continuous linear mapping of perfectly complete space to a Hausdorff barreled space is open.*

Corollary 4.14. A bijective continuous linear mapping of a perfectly complete to a Hausdorff barreled space is an isomorphism.

Let V be a fundamental system of closed convex and balanced zero neighborhoods of the space E, B be a fundamental system of bounded subset in the space F, $x \mapsto \mu_O(x)$ be a Minkowski functional of the set $O \subseteq E$.

Theorem 4.31. *Let l.c.s. E be perfectly complete and barreled, l.c.s. F be semireflexive, b be a bilinear form which is continuous on $E \times F$. Then the following statements are equivalent:*

(1)   For any $f \in F^*$ there exists a unique $x \in E$ such that

$$b(x,y) = \langle f,y \rangle_{F^*,F} \quad \forall y \in F;$$

(2)   (i) for any neighborhood $O \in V$ there exists a bounded set $P \in B$ such that

$$\mu_O(x) \le \sup_{y \in P}|b(x,y)| \quad \forall x \in E;$$

(ii) if $b(x,y) = 0 \ \forall x \in E$, then $y = 0$.

Proof. Let us argue as in (4.29). Introduce a linear operator $T : E \to F^*$, which acts by the rule

$$E \ni x \mapsto Tx = b(x,\cdot) \in F^*.$$

Let us equip the conjugate space F^* with the strongest topology of uniform convergence $\beta(F^*,F)$. The linear operator $T : E \to F^*$ is continuous. Indeed, $\{P^o : P \in B\}$ is a fundamental system of convex and balanced zero neighborhoods in a strong conjugate space F^*. Let us take $P \in B$, then the space $T^{-1}(P^o)$ is a convex balanced and adsorbing set. Let us show that the set $T^{-1}(P^o)$ is closed. Then the fact that the space E is barreled implies that $T^{-1}(P^o)$ is a zero neighborhood in E. Assume that $Tx = b(x,\cdot) \notin P^o$. Then $\sup_{y \in P}|b(x,y)| > 1$, i.e, $\exists\, y' \in P$: $|b(x,y')| > 1$. The fact that the form b is continuous implies there exists a neighborhood O of the point $x \in E$ such that $|b(x'',y')| > 1 \ \forall x'' \in O$, i.e., $Tx'' \notin P^o \ \forall x'' \in O$.

Assume that Condition (1) is satisfied, i.e., $R(T) = F^*$ and $N(T) = \{0\}$. Since the space E is perfectly complete and the space F^* equipped of the strongest topology of uniform convergence $\beta(F^*,F)$ is barreled [97], then the operator T is an isomorphism (see corollary in 4.14) between E and the space F^* with the topology $\beta(F^*,F)$. Condition (2) (ii) follows from the fact that the operator T is surjective.

Let us prove that Condition (2) (i) is satisfied. Let us select a space $O \in V$. Since the operator T is an isomorphism between E and F^* with the topology $\beta(F^*, F)$, then there exists a set $P \in B$ such that $P^o = T(O)$. The fact that the operator T is injective implies that $T^{-1}(P^o) \subseteq O$. Obviously,

$$\mu_O(x) \leq \mu_{T^{-1}(P^o)}(x)$$

for an arbitrary point $x \in E$. Let us show that the following equity holds:

$$\mu_{T^{-1}(P^o)}(x) = \sup_{y \in P} |b(x,y)| \quad \forall x \in E, \tag{4.83}$$

This completes the proof.

The set $P^o \subseteq F^*$ is a convex and balanced zero neighborhood in a strong conjugate space F^*. The space $T^{-1}(P^o) \subseteq E$ is convex, balanced and adsorbing. Let us take a point $x \in E$. By definition we have

$$\mu_{T^{-1}(P^o)}(x) = \inf\{\lambda : \lambda > 0, x \in \lambda T^{-1}(P^o)\}.$$

If $Tx = b(x, \cdot) \in \lambda P^o$ for some $\lambda > 0$, then $|b(x,y)| \leq \lambda \; \forall y \in P$. Hence, we have

$$\mu_{T^{-1}(P^o)}(x) \geq \sup_{y \in P} |b(x,y)|.$$

Let $\varepsilon > 0$. Then we have $\frac{1}{\sup_{y \in P} |b(x,y)| + \varepsilon} |b(x,y)| < 1 \; \forall y \in P$, i.e.,

$$Tx = b(x, \cdot) \in \left(\sup_{y \in P} |b(x,y)| + \varepsilon\right) P^o.$$

Hence,

$$\mu_{T^{-1}(P^o)}(x) \leq \sup_{y \in P} |b(x,y)| + \varepsilon.$$

Taking into account that $\varepsilon > 0$ is an arbitrary value and the inequality proved above, we obtain equity (4.83).

Conversely, assume that Statements (2) (i) and (2) (ii) holds. Let us show that $N(T) = \{0\}$. Let $Tx = 0$, i.e.

$$b(x,y) = 0 \quad \forall y \in F.$$

Then, $\forall O \in V \; \mu_O(x) = 0$. Since, $\forall O \in V$ we have that $x \in O$. Due to the fact that the space E is Hausdorff, we obtain that $x = 0$, i.e., the operator T is injective.

Statement (2) (ii) implies that the set $R(T)$ is total in F^*. Let us show that $R(T)$ is closed linear subspace in a strong conjugate space F^*. Then the fact that the space F is semi-reflexive implies that $R(T) = F^*$.

Let us take a neighborhoods $O \in V$ and $\tilde{O} \in V$ such that $2\tilde{O} \subseteq O$. Then, since $\exists P \in B$ we have:

$$\mu_{\tilde{O}}(x) \leq \sup_{y \in P}|b(x,y)| = \mu_{T^{-1}(P^o)}(x) \quad \forall x \in E.$$

Hence, $\forall x \in T^{-1}(P^o)$ we have that

$$2\mu_O(x) \leq \mu_{\tilde{O}}(x) \leq \mu_{T^{-1}(P^o)}(x) \leq 1.$$

If $x \in T^{-1}(P^o)$, than $x \in O$, i.e., $T^{-1}(P^o) \subseteq O$. Thus, $R(T) \cap P^o \subseteq T(O)$. Therefore, T is a continuous open (relatively) and bijective linear operator, which acts from E onto $R(T)$. The subspace $R(T)$ with the topology induced by $\beta(F^*, F)$ is perfectly complete [97], hence, it is closed in F^*. □

Remark 4.33. The theorems of Lax–Milgram type were proved in [4, 32, 46, 84, 94, 95, 104] also.

Chapter 5
Computation of Near-Solutions of Operator Equations

In previous chapters, we studied the concept of a near-solution of an operator equation

$$A(x) = y, \qquad x \in E, y \in F, \tag{5.1}$$

where A is a bounded injective linear operator, which acts from a Banach space E into a Banach space F.

Recall that a strong near-solution described in Chap. 2 arises when we introduce a topology $\mathscr{T}_A$ induced in E by the norm $\|x\|_{\bar{E}} = \|A(x)\|_F$, where $\bar{E}$ is a completion of E with respect to the norm $\|x\|_{\bar{E}}$. In this case, the Banach space E is densely embedded into a Banach space $\bar{E}$ and the operator A can be extended by continuity onto the entire space $\bar{E}$. Denoting this extension by $\bar{A}$, we obtain a bounded linear operator $\bar{A}$, which acts from $\bar{E}$ into F. The strong near-solution of (5.1) is such a sequence $x_n \in E$, that $A(x_n)$ converges to y in the space F. The sequence x_n is a Cauchy sequence in the space $\bar{E}$, and hence $x_n \to \widetilde{x}$ as $n \to \infty$, where $\widetilde{x}$ is a generalized solution of (5.1), i.e. $\bar{A}(\widetilde{x}) = y$.

5.1 Construction of Near-Solutions

Let E be a Banach space with a Schauder basis $e_1, e_2, \ldots$, where $e_n \in E$ and $F = H$ be a Hilbert space. Let us introduce the denotation $A(e_k) = \widetilde{e}_k \in F = H$. It is easy to see, that every element $\bar{y} \in R(A)$ can be represented in the following form:

$$\bar{y} = A(x) = \sum_{k=1}^{\infty} \alpha_k \widetilde{e}_k,$$

where $x = \sum_{k=1}^{\infty} \alpha_k e_k$.

Hence, for any sequence ε_n of positive numbers that converges to zero and for any element $y \in F$ there exists such an element

$$y_n = \sum_{k=1}^{n} \beta_k^{(n)} \widetilde{e}_k,$$

D.A. Klyushin et al., *Generalized Solutions of Operator Equations and Extreme Elements*, Springer Optimization and Its Applications 55, DOI 10.1007/978-1-4614-0619-8_5, © Springer Science+Business Media, LLC 2012

for which

$$\|y - y_n\| = \left\| y - \sum_{k=1}^{n} \beta_k^{(n)} \tilde{e}_k \right\| < \varepsilon_n.$$

Consider a finite-dimensional subspace

$$\mathscr{L}_n = \left\{ u : u = \sum_{k=1}^{n} \gamma_k \tilde{e}_k, \ \gamma_k \in \mathbb{R} \right\} \subset F$$

Let $y_n^* = \sum_{k=1}^{n} \beta_k^* \tilde{e}_k$ be the element of the best approximation of y in the subspace $\mathscr{L}_n$:

$$\|y - y_n^*\| = \min_{\beta_k} \left\| y - \sum_{k=1}^{n} \beta_k \tilde{e}_k \right\|,$$

Then,

$$\|y - y_n^*\| = \left\| y - \sum_{k=1}^{n} \beta_k^* \tilde{e}_k \right\| < \varepsilon_n.$$

Put

$$x_n = \sum_{k=1}^{n} \beta_k^* e_k$$

Let us show that x_n is a near-solution of operator equation (5.1). Indeed,

$$A(x_n) = A \left(\sum_{k=1}^{n} \beta_k^* e_k \right) = \sum_{k=1}^{n} \beta_k^* A(e_k) = \sum_{k=1}^{n} \beta_k^* \tilde{e}_k = y_n^* \to y$$

as $n \to \infty$.

Thus, we would obtain near-solutions x_n, if we could define the element of the best approximation of y from the space $\mathscr{L}_n$. This problem is reduced to solving of the system of linear algebraical equations with respect to β_k^* in case of the Hilbert space $F = H$. Since the element $y - y_n^*$ is orthogonal to the subspace $\mathscr{L}_n$, then

$$(y - y_n^*, \tilde{e}_l) = \left(y - \sum_{k=1}^{n} \beta_k^* \tilde{e}_k, \tilde{e}_l \right) = (y, \tilde{e}_l) - \sum_{k=1}^{n} \beta_k^* (\tilde{e}_k, \tilde{e}_l) = 0,$$

where $l = 1, 2, \ldots, n$.

Put $a_{lk} = (\tilde{e}_k, \tilde{e}_l)$, $b_l = (y, \tilde{e}_l)$. Then we obtain the following system of linear algebraical equations

$$\sum_{k=1}^{n} a_{lk} \beta_k^* = b_l, \qquad l = 1, 2, \ldots, n. \tag{5.2}$$

Thus, computation of a near-solution of operator equation (5.1) can be reduced to solving the system of linear algebraical equations (5.2) and this problem is paramount one for such computation.

5.2 Method of Neumann Series

Let us consider a system of linear algebraic equations

$$Ax = b, \tag{5.3}$$

where $A = \left\{ a_{ij} \right\}_{i,j=1}^{n}$ is a non-degenerate matrix of order n (i.e. $det A \neq 0$), $x = (x_1, x_2, \ldots, x_n)$, $b = (b_1, b_2, \ldots, b_n)$ is an element of the space $\mathbb{R}^n$. Using Gauss transformations, we obtain

$$A^* A x = A^* b, \tag{5.4}$$

where A^* is a conjugate matrix. It is clear that system (5.4) is equivalent to original system (5.3), but the matrix $M = A^* A$ is symmetric and positively defined. So, without loss of generality we may consider that the matrix A of original system (5.3) is symmetric and positively defined. This implies that the matrix A has n positive eigenvalues $\mu_1, \mu_2, \ldots, \mu_n$ ($\mu_k > 0$). Let us show that there exists such a positive constant $\mu > 0$ that the norm of the matrix $U = I - \mu A$ (I is a unit matrix) in the space $\mathbb{R}^n$ is less than one: $\|U\| < 1$. Indeed, by the Hirsch–Bendixon Theorem

$$|\mu_k| \leq nq = \delta, \tag{5.5}$$

where

$$q = \max |a_{ij}|, \tag{5.6}$$

a_{ij} is an element of the matrix A.

Put $\mu = 1/\delta$. Then

$$U = I - \mu A = I - \frac{1}{\delta} A. \tag{5.7}$$

Note that inequality (5.5) and, therefore, the parameter μ can be defined more exactly. According to [16], we will call the norm of matrix A any norm $\|A\|$ defined in a vector space of matrices of order n^2, for which the multiplicative inequality $\|AB\| \leq \|A\| \times \|B\|$ holds [16]. As it was stated in [16], modulo of every eigenvalue of a matrix does not exceed any possible norm. Note that the functional $\delta(A) = \delta = nq = n \max |a_{ij}|$ is a so-called M-norm $M(A)$, so Hirsch–Bendixon inequality holds. Taking a N-norm

$$N(A) = \sqrt{\sum_{i,j=1}^{n} |a_{ij}|^2},$$

we have

$$|\mu_k| \leq N(A) \tag{5.8}$$

Since $N(A) \leq M(A)$ [16], inequality (5.8) is more precise than the Hirsch–Bendixon inequality and we can put δ equal to the number $N(A)$ and even an arbitrary norm of the matrix A. Analyzing the proof of the inequality $\|U\| < 1$, we note that the matrix U has the following eigenvalues $\gamma_k = 1 - \mu \mu_k$. Therefore, $|\gamma_k| = \left| 1 - \frac{\mu_k}{\delta} \right| < 1$.

Since $0 < \mu_k/\delta < 1$, then $\|U\| = \max|\gamma_k| < 1$. Thus, the system of linear algebraic equations

$$A_\mu x = \mu A x = \mu b, \qquad \mu = 1/\delta \qquad (5.9)$$

is equivalent to system (5.3), but the matrix $A_\mu = \mu A$ can be represented as $A_\mu = I - U$, where $\|U\| < 1$. Using well-known results of matrix theory (see, for example, [16]) we obtain, the Neumann series

$$A_\mu^{-1} = (I - U)^{-1} = I + U + U^2 + \ldots + U^n + \ldots \qquad (5.10)$$

converges and since $\|U\| < 1$, then A_μ^{-1} can be approximated by a partial sum of series (5.10)

$$A_\mu^{-1} \approx I + U + U^2 + \ldots + U^k = U_k. \qquad (5.11)$$

Thus, putting $\widetilde{x}_k = U_k(\mu b) = \mu(U_k b)$, we obtain that $\widetilde{x}_k \to x$ as $k \to \infty$, and hence $\widetilde{x}_k$ can be considered as approximate solution of system (5.3). We will call this approach to solving system (5.3) the method of Neumann series.

Let us determine the error of kth approximation of this method. Put $\widetilde{q} = \|U\| < 1$, where $\|U\|$ is a norm of U in the space $\mathbb{R}^n$. Then

$$\begin{aligned}
\|A_\mu^{-1}\| &\leq \|I\| + \|U\| + \|U^2\| + \ldots + \|U^k\| + \ldots \\
&\leq 1 + \|U\| + \|U\|^2 + \ldots + \|U\|^k + \ldots \\
&= 1 + \widetilde{q} + \widetilde{q}^2 + \ldots + \widetilde{q}^k + \ldots \\
&= \frac{1}{1 - \widetilde{q}},
\end{aligned} \qquad (5.12)$$

That is why

$$\begin{aligned}
\|x - \widetilde{x}_k\| &= \left\| A_\mu^{-1}(\mu b) - \mu U_k(b) \right\| = |\mu| \times \left\| \left(A_\mu^{-1} - U_k \right)(b) \right\| \\
&\leq |\mu| \times \left(\|U\|^{k+1} + \|U\|^{k+2} + \ldots \right) \times \|b\| \\
&\leq |\mu| \times \|b\| \times \frac{\widetilde{q}^k}{1 - \widetilde{q}} = C\widetilde{q}^k,
\end{aligned}$$

where $\widetilde{q} = \|U\| < 1, C = \frac{|\mu| \times \|b\|}{1 - \widetilde{q}}$.

Thus, we have proved the following inequality for $\widetilde{x}_k$:

$$\|x - \widetilde{x}_k\| \leq C\widetilde{q}^k, \qquad (5.13)$$

so $\widetilde{x}_k$ converges to an exact solution at a rate of geometrical progression with ratio $\widetilde{q} < 1$. This inequality allows to estimate not only the accuracy of the kth approximation, but the norm of the inverse operator A^{-1} also, not computing in itself:

$$\|A^{-1}\| = |\mu| \times \|A_\mu^{-1}\| \leq \frac{\mu}{1 - \widetilde{q}}, \qquad \mu = \frac{1}{\delta}. \qquad (5.14)$$

This estimation is based on the possibility of computing the Euclidian norm of the operator U or, at least, the possibility to majorize this norm $\|U\| = \tilde{q} \le q^* < 1$. This estimation we will call a majorant one.

The process of solving a system of linear algebraic equations can be considered from the other point of view. Indeed, system (5.3) is equivalent to the system

$$(I - U)(x) = \mu b = \tilde{b}, \qquad \mu = \frac{1}{\delta}$$

or

$$\tilde{U}(x) = \tilde{b} + U(x) = x.$$

Thus, solving system (5.3) is equivalent to the fixed point problem for the operator $\tilde{U}$. Let us show that the operator $\tilde{U}$ is a contracting mapping, which acts in the space $\mathbb{R}^n$ with the Euclidean metric. Indeed, for all $u, v \in \mathbb{R}^n$ the following inequality holds

$$\rho(\tilde{U}(u), \tilde{U}(v)) = \left\| \tilde{U}(u) - \tilde{U}(v) \right\| = \|U(u) - U(v)\| \le \|U\| \times \|u - v\| = \tilde{q} \times \rho(u, v),$$

where $\tilde{q} = \|U\| < 1$.

It is known [29] that a fixed point x of the operator can be obtained as a limit of a process of successive approximations (iterative process)

$$x_{n+1} = \tilde{U}(x_n), \qquad n \in \mathbb{N}, \tag{5.15}$$

where x_0 is an arbitrary element from $\mathbb{R}^n$, and the rate of convergence of the sequence x_n to the solution x is determined by the inequality

$$\rho(x_n, x) \le \frac{\tilde{q}^n}{1 - \tilde{q}} \times \rho(x_1, x_0). \tag{5.16}$$

Thus, the solution of an arbitrary system of linear algebraic equations is reduced to the fixed point problem.

Let us consider the relation between the iterative process (5.15) and an approximate solution

$$\tilde{x}_n = \tilde{b} + U\tilde{b} + \ldots + U^n\tilde{b},$$

determined by the Neumann series method. It is easy to see that this approximate solution x_n can be obtained with the help of the process of successive approximations (5.15), if the initial element x_0 of the iterative process is $\tilde{b}$: $x_0 = \tilde{b}$. Indeed,

$$\tilde{x}_n = \tilde{b} + U\tilde{b} + \ldots + U^n\tilde{b} = \tilde{b} + U\tilde{b} + \ldots + U^{n-2}\tilde{b} + U^{n-1}\left(\tilde{b} + U\tilde{b}\right)$$

$$= \tilde{b} + U\tilde{b} + \ldots + U^{n-2}\tilde{b} + U^{n-1}\left[\tilde{U}(\tilde{b})\right]$$

$$= \tilde{b} + U\tilde{b} + \ldots + U^{n-2}\tilde{b} + U^{n-1}x_1.$$

Further,

$$\tilde{x}_n = \tilde{b} + U\tilde{b} + \ldots + U^{n-2}\left(\tilde{b} + Ux_1\right) = \tilde{b} + U\tilde{b} + \ldots + U^{n-2}x_2$$
$$= \ldots = \tilde{b} + U(x_{n-1}) = \tilde{U}x_{n-1} = x_n.$$

Thus, an approximation $\tilde{x}_n$ computed with the help of the Neumann series coincides with the nth iteration of the element $x_0 = \tilde{b}$. This fact allows to formulate a practical recommendation for the approximate solving system (5.3). It is known that one of the most difficult problem of the method of successive approximations is the selection of the initial approximation x_0. If this selection is successful then $x_1 \approx x_0$ and the second factor $\rho(x_1, x_0)$ in (5.16) is small. Thus, we may hope that the right-hand side of (5.16) will take small values. Using the Neumann series, we may take the vector $\tilde{b}$ as an initial approximation x_0 of the iterative process (5.15), which can be far from the exact solution x in the Euclidean metric, so the value $\rho(x_1, x_0)$ is not small. In such cases the method of the Neumann series can be wrong. However, if we will use these "wrong" approximate solutions as an initial approximation x_0 instead of an arbitrary element $x_0 \in \mathbb{R}^n$, then successive refinements of the solution obtained with the help of the process of successive approximations can give satisfactory results.

Let us pass to the solving of the majorant estimation q^* of the norm of the matrix U. It is known that the Euclidean norm of the matrix U, i.e. the norm of the linear self-conjugate of the operator $y = U(x)$ is computed by the formula

$$\tilde{q} = M = \sup_{\|x\|=1} (Ux, x) = \sup_{x \neq \theta} \frac{(Ux, x)}{(x, x)} = (Ux^*, x^*),$$

where x^* is a normed eigenvector corresponding to the largest eigenvalue $\lambda_1 = M$. Let us show that an arbitrary sequence x_n convergent to an eigenvector x^* as $n \to \infty$ after normalization $y_n = \frac{x_n}{\|x_n\|}$ will converge to x^* and $(Uy_n, y_n) \to M$ as $n \to \infty$ also. Indeed, since $\|x_n\| \to \|x^*\| = 1$ as $n \to \infty$, then

$$\|x_n - y_n\| = \left\|x_n - \frac{x_n}{\|x_n\|}\right\| = \left\|x_n\left(1 - \frac{1}{\|x_n\|}\right)\right\| \to 0,$$

hence $y_n = x_n + (y_n - x_n) \to x^*$ as $n \to \infty$.

From the other hand,

$$\|U(x^*) - U(y_n)\| = \|U(x^*) - U(x_n) + U(x_n) - U(y_n)\|$$
$$\leq \|U\| \times \|x^* - x_n\| + \|U\| \times \|x_n - y_n\| \to 0$$

as $n \to \infty$.

Thus, we have proved that $U(y_n) \to U(x^*)$ and $y_n \to x^*$ as $n \to \infty$ and $(Uy_n, y_n) \to (Ux^*, x^*) = M$.

Taking into account that $\|y_n\| = 1$, we have the inequality

$$\|U(y_n)\| \leq \|U\| < 1,$$

so $U(y_n) < 1$ and for large n the value $q^* = \|U(y_n)\|$ can be used as an estimation $\|U\|$, but this estimation is not a majorant (the inequality $\|U\| \leq q^*$ may not be hold). Yet, the inequality $q^* < 1$ which is necessary for computing the condition number of the matrix A holds.

We can take a sequence x_n computed by the method of steepest descent as a sequence convergent to x^* and find the maximum of the functional

$$l(x) = \frac{(Ux, x)}{(x, x)},$$

In this case, the normalizing is not necessary as $\|x_n\| = 1$ and the rate of convergence $(Ux_n, x_n) = l(x_n) \to M$ is geometrical (see [16]).

5.3 The Condition Number of Matrix

The properties of approximate methods for solving a system of linear algebraic equations using Neumann series and successive approximations depend on the condition number of the matrix A. Let us give a non-formal definition of this concept. The inverse matrix A^{-1} is called stable if small changes of elements of the matrix A imply the small changes of elements of the inverse matrix [16]. The matrix A is called ill-posed if the inverse matrix is unstable. There are numerical parameters that describe stability properties of a matrix [16]. These are Turing numbers

$$N = \frac{1}{n} \times N(A) \times N(A^{-1}), \tag{5.17}$$

$$M = \frac{1}{n} \times M(A) \times M(A^{-1}) \tag{5.18}$$

and Todd numbers

$$P = \frac{\max |\mu_i|}{\min |\mu_i|}, \tag{5.19}$$

$$H = \|A\| \times \|A^{-1}\|, \tag{5.20}$$

where $\|A\|$ is the Euclidean norm of the matrix A. As it was pointed out in [16], these condition numbers do not describe the properties of a matrix completely. Thus, we will start investigation of other condition numbers and consider their relations with equations and (5.17)–(5.20). Since all condition numbers depend on the norm of the inverse matrix $\|A^{-1}\|$ it is desirable to propose conditional numbers that do not depend on it. Let us introduce the following condition number:

$$\tau^*(A) = \frac{1}{1 - \|U\|}, \tag{5.21}$$

where the matrix U is defined by the formula (5.7).

Let $\tau_1(A)$ and $\tau_2(A)$ be an arbitrary pair of classical condition numbers (5.17)–(5.20). It was pointed out in [90] that $\tau_1(A)$ and $\tau_2(A)$ are equivalent in the following sense

$$\tau_1(A_m) \to +\infty \ \Leftrightarrow \ \tau_2(A_m) \to +\infty, \quad \text{as } m \to \infty,$$

where A_m is a sequence of non-degenerated matrices of order $n \times n$. The problem of equivalence between $\tau^*(A)$ and an arbitrary classical condition number, for example, $\tau(A) = \|A\| \times \|A^{-1}\|$ arises in a natural way.

Theorem 5.1. *Let A_m be an arbitrary sequence of symmetric positively defined matrix of order $n \times n$. Then $\tau(A_m) \to \infty$ as $m \to \infty$ iff $\tau^*(A_m) \to \infty$ as $m \to \infty$.*

Proof. Necessity. Put $\tau(A) = \|A\| \times \|A^{-1}\|$, where $\|A\|$ is the Euclidean norm. Then we have

$$\tau(A_\mu) = \tau(\mu A) = \tau(A)$$

and

$$\|A_\mu\| = \|I - U\| \le \|I\| + \|U\| \le 2.$$

By virtue of (5.12) the following estimation holds

$$\|A_\mu^{-1}\| \le \frac{1}{1 - \|U\|} = \tau^*(A).$$

Thus,

$$\tau(A) = \tau(A_\mu) = \|A_\mu\| \times \|A_\mu^{-1}\| \le 2\tau^*(A);$$

hence, if $\tau(A_m) \to \infty$ as $m \to \infty$, therefore, $\tau^*(A_m) \to \infty$ as $m \to \infty$.

Sufficiency. Let $\tau^*(A_m) \to \infty$ as $m \to \infty$, then $\|U_m\| \to 1$ as $m \to \infty$, since $U_m = I - \mu^{(m)} A_m$, $\mu^{(m)} = \frac{1}{\delta^{(m)}}$ and $\delta^{(m)} = M(A_m)$, $m \in \mathbb{N}$.

Let us denote by

$$\mu_1^{(m)} \le \mu_2^{(m)} \le \ldots \le \mu_n^{(m)}$$

the eigenvalues of the matrix A_m, then $\mu_1^{(m)}/\delta^{(m)} \to 0$ as $m \to \infty$. Since the norms $\delta^{(m)} = M(A_m)$ and $\|A_m\| = \mu_n^{(m)}$ in $\mathbb{R}^{n^2}$ are equivalent, then there exists such a constant $c > 0$, that $M(A_m) \le c\|A_m\|$. Therefore,

$$\frac{\mu_1^{(m)}}{\delta^{(m)}} = \frac{\mu_1^{(m)}}{M(A_m)} \ge \frac{\mu_1^{(m)}}{c\|A_m\|} = \frac{1}{c} \times \frac{\mu_1^{(m)}}{\mu_n^{(m)}}.$$

This inequality and the relation $\mu_1^{(m)}/\delta^{(m)} \to 0$ as $m \to \infty$ imply that $\mu_1^{(m)}/\mu_n^{(m)} \to 0$ as $m \to \infty$; hence, $\tau(A_m) = \mu_n^{(m)}/\mu_1^{(m)} \to \infty$ as $m \to \infty$. $\qquad\qquad\square$

The theorem implies that the Turing and Todd numbers (5.17)–(5.20) are equivalent to the condition number (5.21). However, the condition number $\tau^*(A)$ is more preferable than Turing and Todd numbers, since in this case it is not required to compute the norm of the inverse matrix and it is sufficient to estimate the norm of the matrix $U = I - \mu A$, $\mu = \frac{1}{\delta}$, $\delta = M(A)$.

5.4 Hotteling Method for Correction Inverse Matrix

A partial sum of the Neumann series U_k (5.11) can be considered as an approximation of the inverse matrix A_μ^{-1} of (5.9), since $U_k \to A_\mu^{-1}$ as $n \to \infty$. In this connection we have to consider the problem of correction elements of the matrix U_k to obtain more precise approximation of A_μ^{-1}. This problem was solved by Hotteling and Schultz [26, 99]. Suppose that we have such an approximation D_0 of the inverse matrix that

$$\|R_0\| \leq q < 1, \qquad (5.22)$$

where $R_0 = I - A_\mu D_0$, I is the unit matrix.

Then, the elements of the inverse matrix A_μ^{-1} can be determined by the following iteration process:

$$D_m = D_{m-1}(I + R_{m-1}), \quad R_m = I - A_\mu D_m, \qquad m \in \mathbb{N}, \qquad (5.23)$$

In addition,

$$\|D_m - A_\mu^{-1}\| \leq \|D_0\| \times \frac{q^{2^m}}{1-q}. \qquad (5.24)$$

Thus, we have a sequence of approximations D_m which quickly converges to A_μ^{-1} (a number of exact decimal place increases with rate of geometrical progression) provided that (5.22) holds. Let us show that we can take any partial sum U_k as an initial approximation D_0 provided that k is equal or greater than 1: $D_0 = U_k$ as $k \geq 1$. Indeed,

$$\|R_0\| = \|I - A_\mu D_0\| = \|I - (I - U)U_k\|$$

$$= \|I - (I - U)\left(I + U + U^2 + \ldots + U^k\right)\|$$

$$= \|U^{k+1}\| \leq \|U\|^{k+1} = \tilde{q}^{k+1} = q < 1.$$

Now, we can formulate the combined method for solving a system of linear algebraic equations $A_\mu x = \mu b$ where $\mu = 1/\delta$:

1. Determine the matrix $U = I - A_\mu$
2. Determine a partial sum of the Neumann series $U_k = I + U + U^2 + \ldots + U^k$ where $(k \geq 1)$
3. Determine the matrix $R_0 = I - A_\mu D_0$ where $D_0 = U_k$
4. Determine the matrix D_m using (5.23)
5. Determine an approximation x_m by the formula

$$x_m = D_m(\mu b). \qquad (5.25)$$

The error of the mth approximation is estimated by the formula

$$\|x - x_m\| \leq \|A_\mu^{-1}(\mu b) - D_m(\mu b)\| = \|(A_\mu^{-1} - D_m)(\mu b)\|$$

$$\leq \|A_\mu^{-1} - D_m\| \times \|\mu b\| \leq \|D_0\| \times |\mu| \times \|b\| \times \frac{q^{2^m}}{1 - q}$$

$$= \|D_0\| \times |\mu| \times \|b\| \times \frac{\tilde{q}^{(k+1) \times 2^m}}{1 - \tilde{q}^{(k+1)}}$$

$$= C\tilde{q}^{(k+1)2^m}, \qquad (5.26)$$

where

$$C = \frac{\|D_0\| \times \|b\| \times |\mu|}{1 - \tilde{q}^{k+1}} \leq \frac{\tilde{q}^{k+1} \times \|b\| \times |\mu|}{1 - \tilde{q}^{k+1}} = C_1, \qquad \tilde{q} = \|U\| < 1.$$

Note that we used the Euclidean norm of the matrix A instead of the norm $N(A)$ which was used by Hotteling and Schultz. It was necessary because the Euclidean norm of the smallest matrix norm and it allows to determine the initial approximation R_0 such that the inequality (5.22) holds; the norm $N(A)$ does not guarantee that we can determine such R_0.

It is easy to see that the Hotteling method is a partial case of the Newton method: if to apply the Newton method to the equation $f(x) = X^{-1} - A = 0$ we obtain the Hotteling method. However, the Hotteling method has high rate of convergence. Indeed, if we put in formula (5.26) for simplicity $c = 1$ and $k = 1$ then $\|x - x_n\| \leq \tilde{q}^{-2^{n+1}}$.

5.5 Exact Solving a System of Linear Algebraic Equations

In many cases the combined method coupled with the orthogonalization allows to obtain an exact solution of a system of linear algebraic equations. Let us consider the orthogonalization and relative geometrical issues. Consider a non-derenerated system of linear algebraic equations $Ax = b$:

$$\begin{cases} a_{11}x_1 + a_{12}x_2 + \ldots + a_{1n}x_n = b_1, \\ a_{21}x_1 + a_{22}x_2 + \ldots + a_{2n}x_n = b_2, \\ \qquad \ldots \\ a_{n1}x_1 + a_{n2}x_2 + \ldots + a_{nn}x_n = b_n. \end{cases} \qquad (5.27)$$

We will call the columns of A

$$a_1 = \begin{pmatrix} a_{11} \\ a_{21} \\ \ldots \\ a_{n1} \end{pmatrix}, \quad a_2 = \begin{pmatrix} a_{12} \\ a_{22} \\ \ldots \\ a_{n2} \end{pmatrix}, \quad \ldots, \quad a_n = \begin{pmatrix} a_{1n} \\ a_{2n} \\ \ldots \\ a_{nn} \end{pmatrix}$$

the basis vectors of the system $Ax = b$. Using the basis vectors $a_1, a_2, \ldots, a_n$ and the vector b, we can rewrite the system in the following way:

$$x_1 a_1 + x_2 a_2 + \ldots + x_n a_n = b, \qquad (5.28)$$

so that the solution of (5.27) is equivalent to the expansion of the vector b by the basic vectors $a_1, a_2, \ldots, a_n$. Using the Hilbert-Schmidt orthogonalization process, we can orthogonalize the system of basis vectors $a_1, a_2, \ldots, a_n$ and obtain the system of vectors $z_1, z_2, \ldots, z_n$. It is easy to see that we can obtain the exact solution of the system $Ax = b$ by the following way:

$$x_n = \frac{(b, z_n)}{(a_n, z_n)}, \quad x_{n-1} = \frac{(b, z_{n-1}) - x_n (a_n, z_n)}{(a_{n-1}, z_{n-1})}, \ldots,$$

$$x_{n-k} = \frac{(b, z_k) - x_{k+1} (a_{k+1}, z_{k+1}) - \ldots - x_n (a_n, z_n)}{(a_{n-k}, z_{n-k})}, \qquad (5.29)$$

where $0 \le k \le (n-1)$.

Unfortunately, the Hilbert-Schmidt orthogonalization process is not stable and can lead to non-exact solution. However, having provided that the basis vectors $a_1, a_2, \ldots, a_n$ are normalized

$$\tilde{a}_1 = \frac{a_1}{\|a_1\|}, \quad \tilde{a}_2 = \frac{a_2}{\|a_2\|}, \ldots, \tilde{a}_n = \frac{a_n}{\|a_n\|},$$

we have the system

$$\tilde{x}_1 \tilde{a}_1 + \tilde{x}_2 \tilde{a}_2 + \ldots + \tilde{x}_n \tilde{a}_n = b, \qquad (5.30)$$

It is not equivalent to (5.28), but its solution is connected with the solutions (5.28) by the formula $\tilde{x}_k = \|a_k\| x_k$, where $k = 1, 2, \ldots, n$.

Definition 5.16. *The system of linear algebraic equations* (5.27) *is called* normed, if all its basic vectors a_i $(i = 1, 2, \ldots, n)$ have unit norm $\|a_i\| = 1$. The operation of transition from an arbitrary system (5.28) to the system (5.30) by normalizing its basic vectors is called *normalization* of a system.

It was mentioned above (see Sect. 5.3), that existing condition numbers do not characterize the matrix completely. The volume of simplex consisting of the basic vectors

$$\text{conv}\{0,\tilde{a}_1,\ldots,\tilde{a}_n\} = S(\tilde{a}_1,\ldots,\tilde{a}_n) = \left\{ u = \sum_{k=1}^{n} \alpha_k \tilde{a}_k, \ \sum_{k=1}^{n} \alpha_k \le 1, \ \alpha_k \ge 0 \right\},$$

is more informative.

The volume V_s can be computed using the Gram determinant of the system $\tilde{a}_1, \tilde{a}_2, \ldots, \tilde{a}_n$. It is maximal if the basic vectors $\tilde{a}_k$, $k = \overline{1,n}$ form an orthogonal system. Therefore, if the volume V_s is small, then then matrix A is ill-posed. The corresponding mathematical tools were developed in [90].

We can reduce the system $Ax = b$ to the equivalent form $A_\mu x = \mu A x = \mu b = \hat{b}$ (see Sect. 5.2) and compute the matrix D_m. Putting $B = D_m$, we obtain the equivalent system $\tilde{A}x = D_m A_\mu x = D_m \hat{b} = \tilde{b}$. Let us show that the system $\tilde{A}x = \tilde{b}$ has flattening basic vectors. Indeed, let ε be an arbitrary small positive number and the matrix D_m be selected in such a way:

$$\left\| A_\mu^{-1} - D_m \right\| \le \frac{\varepsilon}{\left\| A_\mu \right\|}.$$

Denote by $\hat{a}_1, \hat{a}_2, \ldots, \hat{a}_n$ the basic vectors of the system $\tilde{A}x = D_m A_\mu x = D_m \hat{b}$ and let $e_1, e_2, \ldots, e_n$ be orts in the space $\mathbb{R}^n$, i.e.

$$e_1 = (1,0,0,\ldots,0), \ e_2 = (0,1,0,\ldots,0), \ \ldots, e_n = (0,0,\ldots,0,1).$$

Then,

$$\begin{aligned}
\|e_k - \hat{a}_k\| &= \left\| A_\mu^{-1} A_\mu e_k - D_m A_\mu e_k \right\| = \left\| \left(A_\mu^{-1} - D_m \right) A_\mu e_k \right\| \\
&\le \left\| A_\mu^{-1} - D_m \right\| \times \left\| A_\mu \right\| \times \|e_k\| \\
&\le \frac{\varepsilon}{\left\| A_\mu \right\|} \times \left\| A_\mu \right\| = \varepsilon.
\end{aligned}$$

Thus, the basic vectors of the system $\tilde{A}x = D_m A_\mu x = D_m \hat{b} = \tilde{b}$ can be done arbitrary close to the unit vectors e_k and we can use the formulas (5.29) to compute an exact solution of the system $\tilde{A}x = \tilde{b}$ and, therefore, $Ax = b$.

5.6 Solving a System of Linear Algebraic Equations with Guarantee Precision

At first, let us consider the concept of the precision of an approximate solution of a system of linear algebraic equations. Let x be an exact solution of the system $Ax = b$ and $\tilde{x}$ be an approximate solution. Let us introduce the following denotation: $\tilde{b} = A\tilde{x}$ and fix some positive numbers α, β.

Definition 5.17. *The precision of an approximate solution* $\widetilde{x}$ *of a system of linear algebraic equations* $Ax = b$ *is a number*

$$e(\widetilde{x}) = \alpha \|x - \widetilde{x}\|^2 + \beta \|b - \widetilde{b}\|^2,$$

where $\|x - \widetilde{x}\| = \rho(x,\widetilde{x})$ is the Euclidean distance (or distance in some finite-dimensional Banach space) between exact and approximate solutions, $\|b - \widetilde{b}\|$ is a discrepancy, α, β are the positive numbers.

In many cases (but not always!) we may suppose that $\alpha = 1, \beta = 0$, i.e. to determine the precision of a solution as a square of a distance between exact and approximate solutions, so hereinafter we will consider that $e(\widetilde{x}) = \|x - \widetilde{x}\|^2$. As a rule, the precision $e(\widetilde{x})$ is unknown as we do not know an exact solution x, but we would be able to estimate $e(\widetilde{x})$ if we would know the norm of the inverse matrix A^{-1} or majorize it, as far as

$$\|x - \widetilde{x}\| = \left\| A^{-1}b - A^{-1}\widetilde{b} \right\| \le \|A^{-1}\| \times \|b - \widetilde{b}\| \tag{5.31}$$

In Sect. 5.2, we estimated (5.14) for the norm $\|A^{-1}\|$. To do this we used the Euclidean norm of the matrix $U = I - \mu A$ or the majorant estimation q^*. Since these problems are quite difficult, it is helpful to use some easy computable norm for estimation of $\|U\|$. It is easy to see, that the norms $N(U)$ and $M(U)$ are not suitable for this purpose, as $N(I) = \sqrt{n}$ and $M(I) = n$, but we may use the operator norm of the matrix U in the space $l_1^{(n)}$ (the so-called second norm)

$$\|U\|_{II} = \max_k \sum_{i=1}^{n} |u_{ik}|, \tag{5.32}$$

for which $\|I\|_{II} = 1$. In this connection, the following question arises: let A be a positively defined symmetric matrix; can we select such a number $\lambda > 0$ that the norm $\|U\|_{II}$ of the matrix $U = I - \lambda A$ is less then 1: $\|U\|_{II} < 1$?

Definition 5.18. A matrix $A = \{a_{ij}\}_{i,j=\overline{1,n}}$ has *a diagonal domination*, if for all $k = \overline{1,n}$ the following inequality holds

$$a_{kk} > (|a_{1k}| + |a_{2k}| + \ldots + |a_{nk}| - |a_{kk}|) = A_k$$

Theorem 5.2. *If a symmetric matrix A with positive diagonal elements has a diagonal domination then there exists such a number $\lambda > 0$ that the second norm of the matrix $U = I - \lambda A$ is less than 1: $\|U\|_{II} < 1$.*

Proof. Let Λ be a diagonal matrix with diagonal elements $\lambda_1, \lambda_2, \ldots, \lambda_n$ and $U^* = I - A\Lambda$. Then

$$\|U^*\|_{II} = \|I - A\Lambda\|_{II} = \max_{1 \le k \le n} \left\{ \sum_{i=1}^{n} |\lambda_k a_{ik}| + |1 - \lambda_k a_{kk}| \right\}.$$

Let us consider a function $f_k(\lambda_k)$:

$$f_k(\lambda_k) = |\lambda_k a_{1k}| + |\lambda_k a_{2k}| + \ldots + |1 - \lambda_k a_{kk}| + \ldots + |\lambda_k a_{nk}|$$

After transformations we have

$$f_k(\lambda_k) = |\lambda_k| \times \left(|a_{1k}| + |a_{2k}| + \ldots + |a_{(k-1)k}| + |a_{(k+1)k}| + \ldots + |a_{nk}|\right)$$
$$+ |1 - \lambda_k a_{kk}| = |\lambda_k| \times A_k + |1 - \lambda_k a_{kk}|,$$

where

$$A_k = \left(|a_{1k}| + |a_{2k}| + \ldots + |a_{(k-1)k}| + |a_{(k+1)k}| + \ldots + |a_{nk}|\right).$$

If $0 < \lambda_k < \frac{1}{a_{kk}}$, then $\lambda_k a_{kk} < 1$ and therefore $1 - \lambda_k a_{kk} = |1 - \lambda_k a_{kk}|$. Thus,

$$f_k(\lambda_k) = \lambda_k A_k + 1 - \lambda_k a_{kk} = 1 - \lambda_k(a_{kk} - A_k)$$

for all $\lambda_k \in \left(0, \frac{1}{a_{kk}}\right)$.

Since the matrix A has a diagonal domination and its diagonal elements are positive, then $a_{kk} > A_k$. Put

$$\gamma_k = \min\left(\frac{1}{a_{kk}}, \frac{1}{a_{kk} - A_k}\right),$$

Then, we obtain that for all $\lambda_k \in (0, \gamma_k)$ the inequality $0 < f_k(\lambda_k) < 1$ holds. Put $\gamma = \min_{1 \le k \le n} \gamma_k > 0$, the for all $\lambda \in (0, \gamma)$ and for an arbitrary $k = 1, 2, \ldots, n$ the following inequality holds:

$$0 < f_k(\lambda) < 1.$$

Thus, if the matrix Λ has the equal elements $\lambda_k = \lambda$, where $\lambda \in (0, \gamma)$, then for the matrix $U^* = I - A\Lambda = I - \lambda A = U$ we obtain the following estimation:

$$\|U^*\|_{II} = \|U\|_{II} = \max_{1 \le k \le n} f_k(\lambda) < 1.$$

□

This theorem allows to estimate the norm of an inverse matrix for some matrix of an equivalent system of linear system of algebraic equations and estimate the precision of an approximate solution. In addition, using Theorem 5.2 we can determine an approximate solution with a given precision. Indeed, using the combined method for solving a system of linear algebraic equations $Ax = b$ (see Sect. 5.4), we can construct approximations D_m^* for the inverse matrix A^{-1}, which converge quickly to A^{-1} as $m \to \infty$. Let us consider an equivalent system $D_m^* Ax = D_m^* b = b^*$. The matrices $D_m^* A$ converge to the unit matrix, so for some natural m the matrix $B^* = D_m^* A$ will have positive diagonal elements and a diagonal domination, so we may apply Theorem 5.2 to the matrix B^* and construct a matrix $U = I - \lambda B^*$ such that $\|U\|_{II} < 1$. In contrast with the Eulidean norm, the second norm $\|U\|_{II}$ can be easily computed by the formula (5.32). The norm $\|U\|_{II}$ allows to estimate the second norm of the inverse matrix $(B^*)^{-1}$, and therefore the precision of an approximate solution $\tilde{x}$ by the formula (5.31) in the metric of the space $l_1^{(n)}$, and therefore in the metric of the Euclidean space $\mathbb{R}^n$.

Now, let us consider how to determine an approximate solution with a given precision. Note that the solution of this problem is required for the construction of near-solution of an operator equation $Ax = y, x \in E, y \in H$ (see Sect. 5.1). Indeed, we can compute approximations $\tilde{x}_n = \sum_{k=1}^{n} \tilde{\beta}_k e_k$ of the exact solution $x_n = \sum_{k=1}^{n} \beta_k^* e_k$ of the system (5.2) that forms a near-solution of (5.1) with a given precision $\delta_n > 0$, i.e. $\|\tilde{x}_n - x_n\|_E < \delta_n$. If $\delta_n \to 0$ as $n \to \infty$, the elements $\tilde{x}_n$ are the near-solutions of (5.1) also. Indeed,

$$0 \le \|A\tilde{x}_n - y\|_H = \|A\tilde{x}_n - Ax_n + Ax_n - y\|_H$$
$$\le \|A\| \times \|\tilde{x}_n - x_n\|_E + \|Ax_n - y\|_H \le \|A\| \times \delta_n + \|Ax_n - y\|_H \to 0$$

where $n \to \infty$.

5.7 Characterization of a Classic Solution Using Neumann Series

Let us consider an operator equation

$$Ax = y, \tag{5.33}$$

where A is a compact injective linear operator, which acts in a separable Hilbert space H, $x, y \in H$, y is the known element in H, and x is an unknown solution of system (5.33). Using the Gauss transformations, i.e. applying an adjoint operator A^* to the left-hand and right-hand sides of (5.33), we obtain the operator equation $A^*Ax = A^*y = y^*$, where $B = A^*A$ is an injective symmetric positively defined and compact operator (we will say that such an operator satisfies the condition α)). Thus, we may suppose that the operator A of system (5.33) is symmetric and positively defined also, so it satisfies the condition α).

In an investigation of the operator equation (5.33) the following question arises: for which right-hand sides $y \in H$ does the classical solution exist, and for which does the generalized solution exist? To answer this question, we can use the Neumann series (see Sect. 5.2). Consider the operator $U = I - \mu A$, where $\mu = \frac{1}{\|A\|}$ and $\|A\|$ is a norm of the operator A in the Euclidean space R^n. If A satisfies the conditions α), then $\|U\| < 1$. Unfortunately, this fact is not true, if H is a infinite dimensional separable Hilbert space. More precisely, if the operator A in an infinite dimensional separable Hilbert space H satisfies the conditions α), then for any non-negative μ the inequality $\|U\| = \|I - \mu A\| > 1$ holds. Hence, the Neuman series

$$A_\mu^{-1} = (I - U)^{-1} = I + U + U^2 + \ldots + U^n + \ldots \tag{5.34}$$

can diverge.

Therefore the formula (5.34) for the inverse operator A_μ^{-1} is not correct. However, if we use the series

$$y + Uy + U^2y + \ldots + U^n y + \ldots \tag{5.35}$$

instead of (5.34) in the Hilbert space H for a fixed element $y \in H$, then we can determine exactly when the series (5.35) converges or diverges. Let us determine the structure of the operator U and its norm. It is known [40] that a self-adjoint compact operator A which acts in a Hilbert space H can be represented in the form

$$Ax = \sum_{i=1}^{\infty} \lambda_i (x, e_i) e_i, \tag{5.36}$$

where e_i is an orthonormal basis consisting of the eigenvectors of the operator A corresponding to the non-zero eigenvalues $\lambda_i \neq 0$ (if the operator A is definitely defined, then $\lambda_i > 0$). Hence, the operator U has the form

$$Ux = Ix - \mu Ax = x - \sum_{i=1}^{\infty} \mu \lambda_i (x, e_i) e_i$$

$$= \sum_{i=1}^{\infty} \mu (x, e_i) e_i - \sum_{i=1}^{\infty} \mu \lambda_i (x, e_i) e_i$$

$$= \sum_{i=1}^{\infty} (1 - \mu \lambda_i) (x, e_i) e_i,$$

So, the values $\widetilde{\lambda}_i = 1 - \mu \lambda_i$ are eigenvalues of the operator U, and e_i are corresponding eigenvectors of U. Indeed, if x is an eigenvector of the operator A, and λ is its eigenvalue, then x is an eigenvalue of the operator U with the eigenvalue $\widetilde{\lambda} = 1 - \mu \lambda$; if, vice versa, x is an eigenvalue of the operator U with an eigenvalue $\widetilde{\lambda}$, then x is an eigenvalue of the operator A, and the corresponding eigenvalue is equal to $\lambda = \frac{1 - \widetilde{\lambda}}{\mu}$.

Let us estimate the operator U. Since U is a self-adjoint operator, then we have [40]

$$\|U\| = \sup_{\|x\|=1} |(Ux, x)| = \sup_{\|x\|=1} \left| \left(\sum_{i=1}^{\infty} (1 - \mu \lambda_i) (x, e_i) e_i, \sum_{k=1}^{\infty} (x, e_k) e_k \right) \right|$$

$$= \sup_{\|x\|=1} \left| \sum_{i=1}^{\infty} (1 - \mu \lambda_i) (x, e_i)^2 \right| \geq 1, \tag{5.37}$$

since $\lambda_i \to 0$ as $i \to \infty$. From the other hand, if $\mu = \frac{1}{\|A\|}$, then $0 \leq 1 - \mu \lambda_i \leq 1$, as $\lambda_i \geq 0$ and $\lambda_i \leq \|A\|$ (recall that A is a positively defined self-adjoint operator) [40]. Therefore,

$$\|U\| = \sup_{\|x\|=1} \left| \sum_{i=1}^{\infty} (1 - \mu \lambda_i) (x, e_i)^2 \right| \leq \sup_{\|x\|=1} \sum_{i=1}^{\infty} (x, e_i)^2 = 1.$$

Therefore, $\|U\| = 1$.

Theorem 5.3. *Let A be an injective linear positively defined self-adjoint compact operator, which acts in a Hilbert space H and $U = I - \frac{1}{\|A\|}A$, then the operator equation*

$$Ax = y \tag{5.38}$$

has a classical solution $x \in H$ iff the Neumann series

$$y + Uy + U^2y + \ldots + U^ky + \ldots \tag{5.39}$$

converges in the Hilbert space H.

Proof. Sufficiency. Put $y_1 = \mu y$, where $\mu = \frac{1}{\|A\|}$ and suppose that for an element y the series (5.39) converges. Let us show that the sum of the series

$$x = y_1 + Uy_1 + U^2y_1 + \ldots + U^ky_1 + \ldots$$

is a classical solution of the operator equation $Ax = y$. We have

$$A_\mu x = \mu Ax = \lim_{n \to \infty} \mu A \Big(\sum_{k=0}^n U^k y_1 \Big) = \lim_{n \to \infty} \sum_{k=0}^n (\mu A) U^k y_1$$

$$= \lim_{n \to \infty} \sum_{k=0}^n (I - U) U^k y_1 = \lim_{n \to \infty} \big(I + U + \ldots + U^n - U - U^2 - \ldots - U^{n+1} \big) y_1$$

$$= \lim_{n \to \infty} \big(I - U^{n+1} \big) y_1 = \lim_{n \to \infty} y_1 - \lim_{n \to \infty} U^{n+1} y_1 = y_1 = \mu y,$$

as $\lim_{n \to \infty} U^{n+1} y_1 = 0$. Thus, $Ax = y$ and hence the element x is a classical solution of (5.38).

Necessity. Let x be a classical solution of the operator equation (5.38). Denote by $S_n y$ a partial sum of the series (5.39)

$$S_n y = y + Uy + \ldots + U^n y.$$

Then,

$$S_n y = Ax + UAx + \ldots + U^n Ax,$$

and therefore

$$S_n y_1 = \mu S_n y = \mu Ax + U(\mu A)x + U^2(\mu A)x + \ldots + U^n(\mu A)x$$

$$= (I - U)x + U(I - U)x + U^2(I - U)x + \ldots + U^n(I - U)x$$

$$= Ix - Ux + Ux - U^2 x + \ldots + U^n x - U^{n+1} x = x - U^{n+1} x.$$

Thus, $\mu S_n y = x - U^{n+1} x$.

Let us show that $U^{n+1} x \to 0$ as $n \to \infty$ (it would imply that $\mu S_n y \to x$ as $n \to \infty$, i.e. the convergence of the series (5.39) to an element $\frac{1}{\mu}x$). Using the method of mathematical induction, we can prove that the nth order of the operator U has the form

$$U^n(x) = \sum_{i=1}^{\infty} \left(1 - \frac{\lambda_i}{\|A\|}\right)^n (x, e_i)e_i.$$

Put $v_i = 1 - \frac{\lambda_i}{\|A\|}$, then $0 \le v_i < 1$ and

$$U^n(x) = \sum_{i=1}^{\infty} v_i^n(x, e_i)e_i.$$

Let x be an arbitrary element of the Hilbert space H and $\varepsilon > 0$. Since

$$x = \sum_{i=1}^{\infty}(x, e_i)e_i,$$

there exists such a natural number N, that

$$\left\|\sum_{i=N+1}^{\infty}(x, e_i)e_i\right\| = \left(\sum_{i=N+1}^{\infty}|(x, e_i)|^2\right)^{1/2} < \frac{\varepsilon}{2}.$$

Further, since $v_i \in [0, 1)$, we can select such a number $n_0 \in \mathbb{N}$, that for all $n \ge n_0$ the following inequality holds

$$\left\|\sum_{i=1}^{N} v_i^n(x, e_i)e_i\right\| = \left(\sum_{i=1}^{N} v_i^n|(x, e_i)|^2\right)^{1/2} < \frac{\varepsilon}{2}$$

for a fixed N. Thus, we have

$$\|U^n x\| = \left\|\sum_{i=1}^{\infty} v_i^n(x, e_i)e_i\right\| \le \left\|\sum_{i=1}^{N} v_i^n(x, e_i)e_i\right\| + \left\|\sum_{i=N+1}^{\infty} v_i^n(x, e_i)e_i\right\|$$

$$\le \frac{\varepsilon}{2} + \left(\sum_{i=N+1}^{\infty} v_i^{2n}|(x, e_i)|^2\right)^{1/2} \le \frac{\varepsilon}{2} + \left(\sum_{i=N+1}^{\infty} |(x, e_i)|^2\right)^{1/2}$$

$$\le \frac{\varepsilon}{2} + \frac{\varepsilon}{2} = \varepsilon.$$

So, we proved that a sequence of operators U^n converges to zero in the topology of pointwise convergence.   □

Remark 5.34. Theorem 5.3 can be interpreted as a criterium of the belonging of a right-hand side y of an operator equation $Ax = y$ to a range $R(A)$ of the operator A: $y \in R(A)$ iff the Neumann series (5.39) generated by the element y converges in a Hilbert space H. Therefore, if for a given $y \in H$ the Neumann series (5.39) diverges then the operator equation $Ax = y$ has a generalized solution only; the inverse statement is true also.

Remark 5.35. It is easy to see that in Theorem 5.3 we used the linearity and continuity of the operator A only.

Note that using the spectral theory of the linear operators we can simplify the proof of Theorem 5.3 and generalize it for a Banach space.

Let E be a Banach space, $A : E \to E$ be a linear continuous operator, defined of the entire space E. Let us consider an operator equation

$$x - Ax = y, \tag{5.40}$$

where y is a given element in E.

Let us use the method of Neumann series to solve (5.40)

$$x_n = Ax_{n-1} + y, \tag{5.41}$$

where $x_0 \in E$ is a given initial approximation. It is easy to see that a sequence x_n generated by the iterative procedure (5.41) can be represented in the form

$$x_n = y + Ay + A^2 y + \ldots + A^{n-1} y + A^n x_0. \tag{5.42}$$

Let us show that there exists a simple relation between classical solvability of (5.40) and convergence of the sequence x_n generated by the equalities (5.41) for a class of operators described below.

Definition 5.19. An operator A is called *"correct"*, if for any $x \in E$ the sequence $\{A^n x\}$ converges in E.

Let us define the following linear operator B corresponding to the "correct" operator A in E:

$$Bx = \lim_{n \to \infty} A^n x, \qquad x \in E.$$

The Banach–Steinhaus Theorem implies the fact that the operator B is bounded. In addition, for any elements $y^* \in E^*$ and $x \in E$ the following relation holds:

$$\langle (A^*)^n y^*, x \rangle = \langle y^*, A^n x \rangle \to \langle y^*, Bx \rangle = \langle B^* y^*, x \rangle, \quad n \to \infty.$$

Thus, the sequence $\{(A^*)^n y^*\}$ converges to $B^* y^*$ in the topology $\sigma(E^*, E)$. Using this fact and the condition of the strong convergence from the definition of a "correct" operator A, we have

$$\langle y^*, A^{2n} x \rangle = \langle (A^*)^n y^*, A^n x \rangle \to \langle B^* y^*, Bx \rangle = \langle y^*, B^2 x \rangle, \quad n \to \infty.$$

This property implies that the sequence $\{A^{2n} x\}$ converges to $B^2 x$ in the topology $\sigma(E, E^*)$. Thus, we proved that $B^2 = B$. In addition, since $BAx = \lim_{n \to \infty} A^{n+1} x = Bx = ABx$, then $B = AB = BA$. Hence, $R(B) = N(I - A)$.

Let us formulate the main result.

Theorem 5.4. *Let A be a "correct" linear continuous operator, which acts in a Banach space E. Then $y \in R(I - A)$ iff for any $x_0 \in E$ the sequence (5.42) converges to a solution of (5.40) in a Banach space E.*

Proof. Suppose that for any $y \in E$ the sequence (5.42) converges to $x \in E$. Let us show that x is a solution of (5.40) with the right-hand side y, i.e. $y \in R(I-A)$. Indeed, since

$$Bx_n = B\left(y + Ay + A^2y + \ldots + A^{n-1}y + A^n x_0\right) = nBy + Bx_0,$$

then

$$By = 0.$$

Therefore, the following relations hold

$$(I-A)(y + Ay + A^2y + \ldots + A^{n-1}y) = y - A^n y \to y - By = y, \quad n \to \infty.$$

Further, we have

$$y + Ay + A^2y + \ldots + A^{n-1}y \to x - Bx_0$$

as $n \to \infty$. Thus, we proved the equality $(I-A)(x - Bx_0) = y$, but, from the other hand, taking into account the condition $R(B) \subset N(I-A)$, we have that $(I-A)(x - Bx_0) = (I-A)x$. Thus, we have

$$(I-A)x = (I-A)(x - Bx_0) = y.$$

Now, suppose that $y \in R(I-A)$. Then for any $x \in E$ we have that $y = x - Ax$. Let us take an arbitrary element $x_0 \in E$ and consider a sequence

$$x_n = y + Ay + A^2y + \ldots + A^{n-1}y + A^n x_0, \quad n \geq 1.$$

Hence,

$$y + Ay + A^2y + \ldots + A^{n-1}y = x - A^n x.$$

Since $A^n x \to Bx$ and $A^n x_0 \to Bx_0$ as $n \to \infty$, then

$$x_n \to x - Bx + Bx_0, \quad n \to \infty.$$

Moreover, it is easy to see that $x - Bx + Bx_0$ is a solution of (5.40). Indeed,

$$(I-A)(x - Bx + Bx_0) = x - Ax + (I-A)B(x_0 - x) = y,$$

as $R(B) \subset N(I-A)$. $\qquad\square$

Note, that the "correct" self-adjoint operators which act in a Hilbert space, are described simply.

Statement 5.1 ([66]). *Let A be a self-adjoint linear continuous operator, which acts in a Hilbert space H. Then the operator A is "correct" iff $\|A\| \leq 1$ and -1 is not an eigenvalue of the operator A.*

Proof. Let $\|A\| \leq 1$. Then,

$$A = \int_0^1 \lambda\, dE_\lambda.$$

Let us show that for any $x \in H$ the sequence

$$A^n x = \int_0^1 \lambda^n \, dE_\lambda \, x$$

converges in H as $n \to \infty$.

Denote by H_1 a proper subspace of the operator A corresponding to an eigenvalue $\lambda = 1$ (it is possible that $H_1 = \{\theta\}$), and denote by H_2^δ and H_3^δ the subspaces

$$x = \int_0^{1-\delta} dE_\lambda \, x.$$

$$x = \int_{-1}^{-1+\delta} dE_\lambda \, x + \int_{1-\delta}^{1} dE_\lambda \, x,$$

respectively $(0 < \delta < 1)$. Then, $H = H_1 \oplus H_2^\delta \oplus H_3^\delta$. Let P_1, P_2 and P_3 be the orthogonal projectors onto the subspaces H_1, H_2^δ, and H_3^δ, respectively. Then,

$$A^n x = A^n P_1 x + A^n P_2 x + A^n P_3 x = P_1 x + A^n P_2 x + A^n P_3 x \tag{5.43}$$

Let us estimate the second and the third term in (5.43)

$$\|A^n P_2 x\| = \left\| \int_{-1+\delta}^{1-\delta} \lambda^n dE_\lambda P_2 x \right\| \le (1-\delta)^2 \|P_2 x\|,$$

$$\|A^n P_3 x\| \le \left\| A^{n-1} P_3 x \right\| \le \dots \le \|P_3 x\|.$$

Since

$$\|P_3 x\| = \int_{-1}^{-1+\delta} d(E_\lambda x, x) + \int_{1-\delta}^{1} d(E_\lambda x, x)$$

and the number -1 is not a point of discontinuity of the function $\lambda \mapsto (E_\lambda x, x)$, then for each of the fixed elements $x \in H$ we have

$$\lim_{n \to \infty} \|P_3 x\| = 0.$$

Let ε be an arbitrary positive number. Select $\delta > 0$ such that the inequality $\|P_3 x\| \le \frac{\varepsilon}{2}$ holds. Then we have for sufficiently large n

$$\|A^n (P_2 x + P_3 x)\| \le (1-\delta)^n \|P_2 x\| + \frac{\varepsilon}{2} < \varepsilon,$$

i.e. the sequence $A^n x$ converges to $P_1 x$.

Conversely, let the operator A be "correct". Suppose that $\|A\| > 1$. For any $\varepsilon \in (0, \|A\| - 1)$ consider a subspace H_ε consisting of an element

$$x = \int_{1+\frac{\varepsilon}{2}}^{1+\varepsilon} dE_\lambda x.$$

Take an arbitrary element $x \in H_\varepsilon$. We have the following estimation for the sequence $\{A^n x\}$:

$$\|A^n x\| = \left\| \int_{1+\frac{\varepsilon}{2}}^{1+\varepsilon} \lambda^n dE_\lambda x \right\| \geq \left(1 + \frac{\varepsilon}{2}\right)^n \|x\| \to +\infty,$$

This contradicts the "correctness" of the operator A.

The convergence of the method of simple iteration for solving the equation $x - Ax = f$ with a linear self-adjoin operator A, which acts in a Hilbert space H was studied in [38]. It was supposed that $\|A\| = 1$. Let us formulate the main result – Krasnoselskii's Theorem.

Theorem 5.5 (M. A. Krasnoselskii). *Let -1 be not an eigenvalue of the operator A. Let the equation $x - Ax = f$ have a solution for a given $f \in H$ (possibly, non-unique). Then for any initial approximation $x_0 \in H$ the successive approximations*

$$x_{n+1} = Ax_n + f, \quad n = 0, 1, 2, \dots$$

converge to a solution of the operator equation $x - Ax = f$.

It is clear that Theorem 5.4 and Statements 5.1 imply the Krasnoselskii's Theorem stated above.

$\square$

Theorem 5.6. *Let A be a self-adjoint non-negatively defined linear continuous operator, which acts in a Hilbert space H and $U = I - \frac{1}{\|A\|}A$. Then, $y \in R(A)$ iff for any $x_0 \in H$ the sequence*

$$x_n = y + Uy + U^2 y + \dots + U^{n-1} y + U^n x_0$$

converges in the space H.

Proof. The operator U is "correct" and $R(I - U) = R(A)$. $\square$

Chapter 6
General Scheme of the Construction of Generalized Solutions of Operator Equations

In this chapter, we will consider a general approach to the construction of generalized solution of a linear operator equation.

Let E, F be linear topological spaces and $\mathscr{L} : E \to F$ be a linear operator defined on the set $D(\mathscr{L}) = E$. Suppose that the operator $\mathscr{L} : E \to F$ is invertible, i.e. the equation $\mathscr{L}u = f$ have no more than one solution. In theory, we always can achieve this by considering a narrowing of the operator $\mathscr{L}$ onto the factor space $E/\operatorname{Ker}\mathscr{L}$. If the range $R(\mathscr{L})$ of the operator $\mathscr{L}$ does not coincide with F then for the right-hand side $f \in F \setminus R(\mathscr{L})$ there arises a problem of the construction of some generalized solution of the equation $\mathscr{L}u = f$ (if $f \in R(\mathscr{L})$, we will call the element $u \in E$: $\mathscr{L}u = f$ a classical solution as before). The natural approach to this problem is the following one: let us introduce in E and F the topologies $\mathscr{T}_E$, $\mathscr{T}_F$, which are consistent with the structures of the linear spaces E and F, so that in the linear topological spaces $(E, \mathscr{T}_E)$, $(F, \mathscr{T}_F)$ the operator $\mathscr{L}$ acts continuously and the right-hand side $f \in F \setminus R(\mathscr{L})$ of the equation $\mathscr{L}u = f$ belongs to the closure $\overline{R(\mathscr{L})}$ of the set $R(\mathscr{L})$ in the linear topological space $(F, \mathscr{T}_F)$ (it's an ideal case when $\overline{R(\mathscr{L})} = F$). Further, let us extend the operator $\mathscr{L}$ by continuity onto the completion of the space E in "topology $\mathscr{T}_E$", where we will look for a generalized solution of the equation $\mathscr{L}u = f$ with the right-hand side $f \in F \setminus R(\mathscr{L})$.[1]

6.1 Generalized Solution of Linear Operator Equations in Locally Convex Topological Spaces

Let E, F be linear spaces; E', F' be the corresponding algebraic conjugate spaces. Let E^*, F^* be such linear spaces that (E, E^*), (F, F^*) form dual pairs. It is clear, that $E^* \subset E', F^* \subset F'$.

Let us consider a linear injective operator $\mathscr{L} : E \to F$, defined on the entire space E, $\mathscr{L}' : F' \to E'$ is an algebraically adjoint operator.

---

[1] Namely, such schemes were implemented in the previous chapters.

D.A. Klyushin et al., *Generalized Solutions of Operator Equations and Extreme Elements*, Springer Optimization and Its Applications 55, DOI 10.1007/978-1-4614-0619-8_6, © Springer Science+Business Media, LLC 2012

Let us suppose that the operator $\mathscr{L}$ is weakly continuous (i.e. it is continuous in the weak topologies $\sigma(E,E^*)$ and $\sigma(F,F^*)$). Under the assumptions made the narrowing of the operator $\mathscr{L}'$ onto the space F^* specifies the operator $\mathscr{L}^* : F^* \to E^* \subset E'$, where $\varphi \in D(\mathscr{L}^*) = F^*$, i.e. for all $\varphi \in F^*$ the element $\mathscr{L}'\varphi = \mathscr{L}^*\varphi$ belongs to the space E^* [97].

In addition, suppose that the range $R(\mathscr{L}) \subset F$ of the operator $\mathscr{L}$ is a total subset of the space F in duality (F,F^*). Consider the linear equation

$$\mathscr{L}u = f \tag{6.1}$$

and pass to the determination of a generalized solution for this equation.

Let us consider $\mathscr{U} = \{\alpha\}$ which is the system of non-empty centrally symmetric subsets of the space F^* satisfying the following conditions:

(1) The union of two arbitrary sets from $\mathscr{U}$ is contained in some set from $\mathscr{U}$.
(2) The product of an arbitrary set $\alpha \in \mathscr{U}$ by an arbitrary real number $\lambda > 0$ is a set from $\mathscr{U}$.
(3) Every set α and $\mathscr{U}$ is bounded in F^* with respect to the topology $\sigma(F^*,R(\mathscr{L}))$.
(4) The set $\mathbf{N} = \bigcup_{\alpha \in \mathscr{U}} \alpha$ is total in F^* with respect to the duality $(F^*,R(\mathscr{L}))$.

It is easy to prove that every of the sets $\mathscr{L}^*(\alpha)$, where $\alpha \in \mathscr{U}$, is bounded in the space E^* with respect to the topology $\sigma(E^*,E)$. Indeed, if we suppose that some of the sets $\mathscr{L}^*(\alpha_0)$ are unbounded in E^*, then there exists such a sequence $l_n \in \mathscr{L}^*(\alpha_0)$ and an element $u \in E$ that $l_n(u) \geq n$. This implies the existence of a sequence $\varphi_n \in \alpha_0 \subset F^*$ ($\mathscr{L}^*\varphi_n = l_n$) such that

$$l_n(u) = (\mathscr{L}^*\varphi_n)(u) = \varphi_n(\mathscr{L}u) \geq n.$$

The latter inequality contradicts the condition (3).

It is also easy to prove, that $\mathbf{M} = \mathscr{L}^*(\mathbf{N})$ is a total subset E^* in duality (E^*,E). Indeed, if there exists $u \in E$, such that $l(u) = 0$ for all $l \in \mathbf{M} = \mathscr{L}^*(\mathbf{N})$ then $\varphi(\mathscr{L}u) = 0$ for all $\varphi \in \mathbf{N}$. By virtue of the totality of the set $\mathbf{N}$ in duality $(F^*,R(\mathscr{L}))$ we get the fact that $\mathscr{L}u = 0$, and the injectivity of the operator $\mathscr{L}$ implies that $u = 0$.

Let us consider on a linear set E the topology $\mathscr{T}_E$ of the uniform convergence defined by a system of neighborhoods of zero

$$o_\alpha = \{u \in E \,|\, |(\mathscr{L}^*\varphi)(u)| \leq 1, \varphi \in \alpha\}, \qquad \alpha \in \mathscr{U}$$

in other words, by the system of semi-norms

$$p_\alpha(u) = \inf_{\lambda > 0, \frac{1}{\lambda}u \in o_\alpha} \lambda = \sup_{\varphi \in \alpha} |(\mathscr{L}^*\varphi)(u)| \qquad u \in E, \quad \alpha \in \mathscr{U}.$$

The set E with this topology we will denote by $E_{\mathscr{T}}$.

It is easy to see that $E_{\mathscr{T}}$ is a Hausdorff locally convex linear topological space. Let us denote by $\bar{E}_{\mathscr{T}}$ a completion $E_{\mathscr{T}}$ with respect to the topology $\mathscr{T}_E$ (or, to be more precisely, with respect to the corresponding Hausdorff uniform structure

[8, 97]). The semi-norms $p_\alpha(u)$ allow the extension by continuity to $\bar{E}_{\mathscr{I}}$ (we will denote these extensions by $\bar{p}_\alpha$) and the system of semi-norms $\bar{p}_\alpha$ specifies the topology of the space $\bar{E}_{\mathscr{I}}$.

Similarly, let us consider a topology $\mathscr{T}_F$ defined by a system of neighborhoods of zero in the set $R(\mathscr{L})$

$$O_\alpha = \{f \in R(\mathscr{L}) \,|\, |\varphi(f)| \leq 1, \varphi \in \alpha\}, \qquad \alpha \in \mathscr{U}$$

or by a system of semi-norms

$$P_\alpha(f) = \inf_{\lambda > 0, \frac{1}{\lambda} f \in O_\alpha} \lambda = \sup_{\varphi \in \alpha} |\varphi(f)|, \qquad f \in R(\mathscr{L}), \quad \alpha \in \mathscr{U}.$$

The set $R(\mathscr{L})$ with respect to this topology (let us denote it by $R_{\mathscr{I}}$) turns into a Hausdorff locally convex linear topological space. Denote by $\bar{R}_{\mathscr{I}}$ the completion $R_{\mathscr{I}}$.

It is easy to see that the operator $\mathscr{L}$ realizes an isomorphism between $E_{\mathscr{I}}$ and $R_{\mathscr{I}}$ (i.e. an isomorphism between linear locally convex topological spaces). Indeed, for an arbitrary $\alpha \in \mathscr{U}$

$$P_\alpha(\mathscr{L}u) = \sup_{\varphi \in \alpha} |\varphi(\mathscr{L}u)| = \sup_{\varphi \in \alpha} |(\mathscr{L}^*\varphi)(u)| = p_\alpha(u), \qquad u \in E.$$

Thus, the operator $\mathscr{L} : E_{\mathscr{I}} \to R_{\mathscr{I}}$ is continuous and the topology $\mathscr{T}_E$ is the weakest of all topologies on E conserving continuity of the operator $\mathscr{L} : E \to R_{\mathscr{I}}$.

Let us extend the operator $\mathscr{L}$ onto the whole space $\bar{E}_{\mathscr{I}}$. Let $\mathscr{E}$ be the minimal Cauchy filter in the space $E_{\mathscr{I}}$ [8]. Taking into account the isomorphism mentioned above, $\mathscr{F} = \mathscr{L}(\mathscr{E})$ is the minimal Cauchy filter in the space $R_{\mathscr{I}}$. Such extension of the operator $\mathscr{L}$, which we shall denote hereinafter by $\bar{\mathscr{L}}$, is a linear continuous injective operator, which realizes an isomorphism between the spaces $\bar{E}_{\mathscr{I}}$ and $\bar{R}_{\mathscr{I}}$.

If to require the following additional condition:

(5) Every of the sets $\alpha \in \mathscr{U}$ is bounded in the space F^* with respect to the topology $\sigma(F^*, F)$ and $\mathbf{N}$ is a total subset F^* in duality (F^*, F),

then the topology $\mathscr{T}_F$ is naturally extended onto the whole space F, and it makes sense to compare elements from the spaces F and $\bar{R}_{\mathscr{I}}$ with each other. Let us denote this topology by $\mathscr{T}_F$ as before, but now it is defined on the whole space F. The space F with the topology $\mathscr{T}_F$ we will denote by $F_{\mathscr{I}}$, and corresponding completion by $\bar{F}_{\mathscr{I}}$. Obviously, $\bar{R}_{\mathscr{I}}$ is a closed linear subset of $\bar{F}_{\mathscr{I}}$, which can coincide with $\bar{F}_{\mathscr{I}}$.

Definition 6.20. A *generalized solution* of the equation $\mathscr{L}u = f$ is such an element $u \in \bar{E}_{\mathscr{I}}$ that $\bar{\mathscr{L}}u = f$.

It is easy to see that a classic solution of the equation $\mathscr{L}u = f$ is a generalized solution also. If u is a generalized solution and $f \in R(\mathscr{L})$ (or $u \in E$) then u is a classic solution.

Thus, the following theorem holds.

Theorem 6.1. *Let conditions (1)–(5) be true, then for an arbitrary element $f \in F \cap \bar{R}_{\mathscr{T}}$ there exists a unique generalized solution of the equation $\mathscr{L}u = f$.*

Proof. The theorem follows from the fact that the operator $\mathscr{L}$ specifies an isomorphism between $\bar{E}_{\mathscr{T}}$ and $\bar{R}_{\mathscr{T}}$.   □

Note that an arbitrary functional $l \in \mathbf{M}$ allows extension by continuity to the whole space $\bar{E}_{\mathscr{T}}$. Indeed, there exists such $\alpha_0 \in \mathscr{U}$ that $\varphi \in \alpha_0$, $\mathscr{L}^* \varphi = l$, i.e. in the neighborhood

$$o_{\alpha_0} = \{ u \in E \,|\, |(\mathscr{L}^* \varphi)(u)| < 1, \varphi \in \alpha_0 \} \in \mathscr{T}_E$$

the functional l is bounded. The extension of l to $\bar{E}_{\mathscr{T}}$ we will denote by $\bar{l}$, and the set of all extended functionals $\bar{l}$, where $l \in \mathbf{M}$ we will denote by $[\mathbf{M}]$. Thus, $\mathbf{M} \subset (E_{\mathscr{T}})^*$, $[\mathbf{M}] \subset (\bar{E}_{\mathscr{T}})^*$, where $(E_{\mathscr{T}})^*, (\bar{E}_{\mathscr{T}})^*$ are conjugate spaces to $E_{\mathscr{T}}, \bar{E}_{\mathscr{T}}$ respectively. Similarly, the functionals $\varphi \in \mathbf{N}$ allow extension onto $\bar{F}_{\mathscr{T}}$, i.e. $\mathbf{N} \subset (F_{\mathscr{T}})^*$. We will denote extended functionals by $\bar{\varphi}$ and the set of all extended functionals by $[\mathbf{N}]$.

Let us consider one more definition of generalized solution, the analogue of which was studied in [5, 33, 62, 63, 89, 91].

Definition 6.21. *A generalized solution of the equation $\mathscr{L}u = f$ is such an element $u \in \bar{E}_{\mathscr{T}}$ that for all $l \in \mathbf{M}$ the following equality holds*

$$\bar{l}(u) = \bar{\varphi}(f), \qquad \mathscr{L}^* \varphi = l.$$

It is easy to prove that if u is a classic solution, then $u \in \bar{E}_{\mathscr{T}}$ is a generalized solution in the sense if Definition 6.21. Indeed, if $\mathscr{L}u = f$ then for all $\varphi \in F^*$ we have $\varphi(\mathscr{L}u) = \varphi(f)$ or $l(u) = (\mathscr{L}^* \varphi)(u) = \varphi(f)$ for all $l \in R(\mathscr{L}^*)$ in particular for an arbitrary $l \in \mathbf{M}$ also. This implies that $\bar{l}(u) = \bar{\varphi}(f)$ for all $l \in \mathbf{M}$.

Theorem 6.2. *Let the system of sets $\mathscr{U}$ satisfy conditions (1)–(5). Then for an arbitrary right-hand side $f \in F \cap \bar{R}_{\mathscr{T}}$ there exists a generalized solution $u \in \bar{E}_{\mathscr{T}}$ of the equation $\mathscr{L}u = f$ in the sense of Definition 6.21.*

Proof. Let $\bar{\mathscr{O}}(f) = \{\bar{O}_\alpha(f)\}_{\alpha \in \mathscr{U}}$ be a set of neighborhoods of the point $f \in \bar{R}_{\mathscr{T}}$ in topology $\mathscr{T}_F$ of the space $\bar{F}_{\mathscr{T}}$. Since the set $R(\mathscr{L})$ is dense in $\bar{R}_{\mathscr{T}}$ in topology $\mathscr{T}_F$ then the set $\{R(\mathscr{L}) \cap \bar{O}_\alpha(f)\}_{\alpha \in \mathscr{U}}$ forms a basis of some filter $\mathscr{F}$, which majorizes the filter $\bar{\mathscr{O}}(f)$. Thus, $\mathscr{F}$ converges to f in $\bar{F}_{\mathscr{T}}$. In addition, since $R(\mathscr{L}) \cap \bar{O}_\alpha(f) \subset F$ then $\{R(\mathscr{L}) \cap \bar{O}_\alpha(f)\}_{\alpha \in \mathscr{U}}$ is a basis of a Cauchy filter in the space $F_{\mathscr{T}}$. It is clear that the set of pre-images $\{R(\mathscr{L}) \cap \bar{O}_\alpha(f)\}_{\alpha \in \mathscr{U}}$ under the mapping $\mathscr{L} : E \to F$ forms a basis of some filter $\mathscr{E}$ in the space $E_{\mathscr{T}}$.

It is clear that $\mathscr{E}$ is a Cauchy filter. Indeed, it is sufficient to show that for every $\alpha_0 \in \mathscr{U}$ there exists such a neighborhood $\bar{O}^*(f) \in \bar{\mathscr{O}}(f)$ that the set $\{ \mathscr{L}^{-1}(R(\mathscr{L}) \cap \bar{O}^*(f)) \}$ is small of order o_{α_0}. Take as $\bar{O}^*(f)$ the following neighborhood of the point f:

$$\bar{O}^*(f) = \{g \in \bar{F}_{\mathscr{T}} \,|\, |\bar{\varphi}(f - g)| \leq 1, \varphi \in 2\alpha_0\}.$$

Then for arbitrary elements u_1, u_2 from the set $\mathscr{L}^{-1}(R(\mathscr{L}) \cap \bar{O}^*(f))$ we have that $\mathscr{L}u_1, \mathscr{L}u_2 \in \bar{O}^*(f)$, i.e $|\bar{\varphi}(f - \mathscr{L}u_1)| \leq 1$, $|\bar{\varphi}(f - \mathscr{L}u_2)| \leq 1$ for all $\varphi \in 2\alpha_0$. Hence, $|\varphi(\mathscr{L}u_1 - \mathscr{L}u_2)| \leq 2$ for all $\varphi \in 2\alpha_0$ or $|(\mathscr{L}^*\varphi)(u_1 - u_2)| \leq 1$ for all $\varphi \in \alpha_0$. Thus, the difference $(u_1 - u_2)$ belongs to o_{α_0}, whence it follows that $\mathscr{E}$ is a Cauchy filter in $E_{\mathscr{T}}$.

Since $\bar{E}_{\mathscr{T}}$ is a complete space, then a basis $\mathscr{E}$ of some Cauchy filter in $\bar{E}_{\mathscr{T}}$ converges to an element $u_0 \in \bar{E}_{\mathscr{T}}$. In addition, $\lim_{\mathscr{E}} \bar{\mathscr{L}} = f$ in $\bar{F}_{\mathscr{T}}$.

That is why $\lim_{\mathscr{E}} l = \lim_{\mathscr{E}} \bar{l} = \bar{l}(u_0)$ for all $l \in \mathbf{M}$. From the other hand, for every functional $l \in \mathbf{M} \subset R(\mathscr{L}^*)$ there exists $\varphi \in F^*$ such that $l = \mathscr{L}^*\varphi = \varphi \circ \mathscr{L}$. That is why

$$\lim_{\mathscr{E}} l = \lim_{\mathscr{E}}(\varphi \circ \mathscr{L}) = \varphi\left(\lim_{\mathscr{E}} \mathscr{L}\right) = \varphi\left(\lim_{\mathscr{E}} \bar{\mathscr{L}}\right) = \varphi(f).$$

Thus, u_0 is a generalized solution of the equation $\mathscr{L}u = f$ in the sense of Definition 6.21. $\square$

Let us study the uniqueness of a generalized solution in the sense of Definition 6.21.

Lemma 6.1. *A generalized solution of the equation $\mathscr{L}u = f$ in the sense of Definition 6.21 is unique iff the set $[\mathbf{M}]$ is total in the duality $((\bar{E}_{\mathscr{T}})^*, \bar{E}_{\mathscr{T}})$.*

Proof. If $u_1, u_2 \in \bar{E}_{\mathscr{T}}$ are different generalized solutions of the equation $\mathscr{L}u = f$ then $\bar{l}(u_1 - u_2) = 0$ for all $l \in \mathbf{M}$. That is why the condition $u_1 = u_2$ is equivalent to the totality of the set $[\mathbf{M}]$ in the duality $((\bar{E}_{\mathscr{T}})^*, \bar{E}_{\mathscr{T}})$. $\square$

Remark 6.36. It is easy to specify a simple sufficient condition of the totality of the set $[\mathbf{M}]$ (Lema 6.1). Namely, if $\mathscr{L}^*(\alpha)$ are compact sets in $\mathbf{M} = \mathscr{L}^*(\mathbf{N})$ with respect to the topology $\sigma(\mathbf{M}, E)$ for every $\alpha \in \mathscr{U}$, then by the Mackey–Arens Theorem [8] the topology $\mathscr{T}_E$ is matched with the duality $(E, \mathbf{M})$. This means that $(E_{\mathscr{T}})^* = \mathbf{M}$, hence $(\bar{E}_{\mathscr{T}})^* = [\mathbf{M}]$. Thus, $[\mathbf{M}]$ is a total set in the duality $((\bar{E}_{\mathscr{T}})^*, \bar{E}_{\mathscr{T}})$, i.e. a generalized solution in the sense of Definition 6.21 is unique.

Let us consider the extension of this condition. This will allow us to guarantee the totality exactly of the set $\mathbf{M}$. At first, let us remind the statement on embeddings of completions of uniform spaces.

Statement 6.2. *Let us suppose that over the set L two Hausdorff uniform structures $\mathfrak{U}_1$ and $\mathfrak{U}_2$ are set and the uniform structure $\mathfrak{U}_1$ majorizes $\mathfrak{U}_2$[2], $\mathfrak{L}_1, \mathfrak{L}_2$ is a completion of the set L with respect to these uniform structures. The set $\mathfrak{L}_1$ is embedded into the space $\mathfrak{L}_2$ densely and continuously iff the following condition holds.*

π) *Let $\mathscr{E}$ be the minimal Cauchy filter in L with respect to the uniform structure $\mathfrak{U}_1$, then $\mathscr{E}$ is the minimal Cauchy filter in L with respect to the uniform structure $\mathfrak{U}_2$.*

This statement is an analogue of the condition π) from [41]. Let us illustrate this condition by the following commutative diagram.

---

[2] Note that in common case topologies $\mathscr{T}_1$, $\mathscr{T}_2$ of different uniform structures may be the same.

$$L_1 \xrightarrow{\;j_1\;} \mathfrak{L}_1$$

$$\| \qquad\qquad \downarrow j \qquad\qquad (6.2)$$

$$L_2 \xrightarrow{\;j_2\;} \mathfrak{L}_2$$

Let L_1 be the set L with the topology of the uniform structure $\mathfrak{U}_1$. The operator j_1 realizes embedding of the whole space L_1 into the space $\mathfrak{L}_1$. In a similar manner we define the set L_2 and the operator j_2. The operators j_1, j_2 are injective. Diagram (6.2) specifies the operator $j : \mathfrak{L}_1 \to \mathfrak{L}_2$ defined on $j_1(L_1)$. The operator j is injective and since the uniform structure $\mathfrak{U}_1$ majorizes $\mathfrak{U}_2$ then j is a uniformly continuous operator.

The uniformly continuous operator j can be continued from the set $j_1(L_1)$ which is dense in $\mathfrak{L}_1$ to uniformly continuous operator $\bar{j}$, which is defined on the whole space $\mathfrak{L}_1$ [8]. However, this continuation can be non-injective. Indeed, the minimal Cauchy filter $\mathscr{E}$ in L_1 is a Cauchy filter in L_2, but it can by not minimal (the minimal Cauchy filter which corresponds to $\mathscr{E}$ we denote by $\bar{\mathscr{E}}$). In the space $\mathfrak{L}_1$, there can exist several minimal Cauchy filters, and every of them majorizes $\bar{\mathscr{E}}$. Therefore, all these minimal filters are mapped by the operator $\bar{j}$ to one element $\bar{\mathscr{E}}$. When the condition π) holds true the operator $\bar{j}$ is injective, and hence, we can say about embedding $\mathfrak{L}_1 \subset \mathfrak{L}_2$. Otherwise, the embedding $\mathfrak{L}_1 \subset \mathfrak{L}_2$ does not exist. At first, we have to factorize the space $\mathfrak{L}_1$ (with respect to the above-mentioned equivalence relation), and then we may map the equivalence classes of factorized space to the elements $\mathfrak{L}_2$.

In the case of linear locally convex topological spaces, a uniform structure is set by a topology (i.e. a topology in such space is induced by a uniform structure), so the condition (π) can be rewritten in the following form.

(π_1) Let $\mathscr{E}$ be the minimal Cauchy filter in a locally convex linear topological space L_1, which majorizes the filter $\mathscr{O}_2(0)$ of neighbors of the point 0 in L_2. Then $\mathscr{E} = \mathscr{O}_1(0)$ is a filter of neighborhoods of the point 0 in L_1.

Now, we can formulate the condition of the uniqueness of a generalized solution.

Theorem 6.3. *Let the condition π_1) hold for the set E and the topologies $\mathscr{T}_E$ and $\sigma(E, \mathbf{M})$, then the generalized solution is unique.*

Proof. By Lemma 6.1, it is sufficient to prove that the set $[\mathbf{M}]$ is total in the duality $((\bar{E}_{\mathscr{T}})^*, \bar{E}_{\mathscr{T}})$. From the proved above it follows that the linear spaces $(E, \mathbf{M})$ form a dual pair. Let E_M be a set E with the topology $\sigma(E, \mathbf{M})$. It is clear that $(E_M)^* = \mathbf{M}$. Let us denote by $\bar{E}_M$ the completion of E_M. Then $(\bar{E}_M)^* = [\mathbf{M}]$ [3]. It is easy to see that the topology (and the separable uniform structure) $\mathscr{T}_E$ majorizes the topology $\sigma(E, \mathbf{M})$; hence, by the conditions of the theorem, we have that $\bar{E}_{\mathscr{T}} \subset \bar{E}_M$. This embedding is continuous and dense. Thus, there exists the embedding $(\bar{E}_M)^* \subset (\bar{E}_{\mathscr{T}})^*$,

---

[3] More precisely, $(\bar{E}_M)^*$ coincides with the set of linear continuous functionals, every of which is a continuity of some functional from $\mathbf{M}$ to the whole space $\bar{E}_M$.

i.e. $[\mathbf{M}] \subset (\bar{E}_{\mathscr{T}})^*$. Since $(\bar{E}_M, [\mathbf{M}])$ are in duality then for an arbitrary $u \in \bar{E}_M$, in particular, for an arbitrary $u \in \bar{E}_{\mathscr{T}}$, the condition $l(u) = 0$ for all $l \in [\mathbf{M}]$ implies that $u = 0$? which was to be proved. $\qquad\qquad\qquad\qquad\qquad\qquad\qquad\qquad\qquad\qquad\qquad\square$

Remark 6.37. Instead of the topology $\sigma(E, \mathbf{M})$, we can consider any other topology which is conformed with the duality $(E, \mathbf{M})$ and is connected with $\mathscr{T}_E$ via the condition π_1). An example of this situation is described in Remark 6.36.

In [33, 62, 63, 89, 91], the concept of a near-solution was introduced. Let us consider a general analogue of this definition.

Definition 6.22. An element $u \in \bar{E}_{\mathscr{T}}$ for which there exists a filter $\mathscr{E}$ of the space $E_{\mathscr{T}}$ is called *a generalized solution* of the equation $\mathscr{L}u = f$ if the filter of the space $\bar{E}_{\mathscr{T}}$ with the basis $\mathscr{E}$ converges to $u \in \bar{E}_{\mathscr{T}}$ and $\lim_{\mathscr{E}} \bar{\mathscr{L}} = f$ in $\bar{F}_{\mathscr{T}}$.

In addition, in many cases a sequential analogue of this definition play an important role.

Definition 6.23. An element $u \in \bar{E}_{\mathscr{T}}$ for which there exists a sequence $u_n \in E_{\mathscr{T}}$ is called *a generalized solution* of the equation $\mathscr{L}u = f$, if u_n converges to $u \in \bar{E}_{\mathscr{T}}$ in the space $\bar{E}_{\mathscr{T}}$ and $\lim_{n \to \infty} \mathscr{L}u_n = f$ in $\bar{F}_{\mathscr{T}}$.

It is clear that if u is a generalized solution in the sense of Definition 6.23, then u is a generalized solution in the sense of Definition 6.22. If the space $\bar{E}_{\mathscr{T}}$ (or $\bar{F}_{\mathscr{T}}$) satisfies the first axiom of countability (the neighborhood system of every point has a countable base), then the inverse statement is also true.

The definition of the operator $\bar{\mathscr{L}}$ directly implies that generalized solutions in the sense of Definition 6.20 and 6.22 are equivalent, i.e. under the conditions (1)–(5) for any right-hand side $f \in F \cap \bar{R}_{\mathscr{T}}$ there exists a unique generalized solution $u \in \bar{E}_{\mathscr{T}}$ of the equation $\mathscr{L}u = f$ in the sense of Definition 6.22.

By Theorem 6.2 the following statements are true.

Corollary 6.15. If u is a generalized solution in the sense of Definitions 6.20 or 6.22, then u is a solution in the sense of Definitions 6.21 also.

Corollary 6.16. If a solution by Definitions 6.21 is unique, then Definitions 6.20–6.22 are equivalent.

Corollary 6.17. Let $u \in \bar{E}_{\mathscr{T}}$ be a unique generalized solution in the sense of Definition 6.21 and $f \in R(\mathscr{L})$. Then, $u \in E$ and $\mathscr{L}u = f$.

Remark 6.38. Let the linear spaces $\mathbf{N}$ and $\bar{R}_{\mathscr{T}} \cap F$ be in duality and $u \in E$ be a generalized solution in the sense of Definition 6.21, then $\mathscr{L}u = f$. Note that $\mathbf{N}$ and $\bar{R}_{\mathscr{T}}$ are in duality, e.g., when $\alpha \in U$ are compact spaces in $\mathbf{N}$ with respect to the topology $\sigma(\mathbf{N}, F)$.

6.2 Examples of Generalized Solutions

Since the topology $\mathscr{T}_E$ is specified by the structure $\mathscr{U}$, then comparing different structures $\mathscr{U}_1$, $\mathscr{U}_2$ we can study relations between the spaces $\bar{E}_{\mathscr{T}_1}, \bar{E}_{\mathscr{T}_2}$ with different topologies $\mathscr{T}_1, \mathscr{T}_2$, and therefore between generalized solutions in these spaces.

Thus, there is an opportunity to construct a detailed classification of spaces of generalized solutions in terms of structures $\mathscr{U}$. We will not do this since we consider that the applications of the method proposed are far more important then the construction of a complete theory.

In applications, it is important to know the relation between topologies of the spaces $\bar{E}_{\mathscr{T}}$ and $\bar{F}_{\mathscr{T}}$, from the one hand, and natural topologies of the spaces E and F, from the other hand. Let us consider the most common cases and construct examples of specific structures $\mathscr{U}$ which lead to the topological spaces $\bar{E}_{\mathscr{T}}$ and $\bar{F}_{\mathscr{T}}$.

Hereinafter, we will consider that E and F are Banach spaces and $\mathscr{L} : E \to F$ is an injective linear continuous operator (it is well-known that such an operator is weakly continuous [97]), $D(\mathscr{L}) = E$, $R(\mathscr{L})$ is a dense subset of F, E^*, and F^* are conjugate spaces.

It is clear that (E, E^*) and (F, F^*) are dual pairs, and the set $R(\mathscr{L}) \subset F$ is total in the duality (F, F^*). Since $R(\mathscr{L})$ is densely embedded into F, the adjoint operator $\mathscr{L}^* : F^* \to E^*$ is injective and continuous. It is easy to prove that under these conditions the set $R(\mathscr{L}^*)$ is total in E^* with respect to the duality (E^*, E) [33, 35].

Since E, F are Banach spaces, then the following definition of a generalized solution has the great importance.

Definition 6.24. *A generalized solution* of the equation $\mathscr{L}u = f$ is such an element $u \in \bar{E}_{\mathscr{T}}$ for which there exists a sequence $u_n \in E$ which is convergent to $u \in \bar{E}_{\mathscr{T}}$ in the space $\bar{E}_{\mathscr{T}}$ and $\lim_{n \to \infty} \mathscr{L}u_n = f$ in the space F.

It must be stressed that this definition is different from the previous: the convergence of $\mathscr{L}u_n$ is considered in the space F, rather than $\bar{F}_{\mathscr{T}}$. If the topologies of the spaces F and $\mathscr{T}_F$ are comparable, then it is easy to establish the relation between Definitions 6.23 and 6.24.

Now, let us pass to consideration of examples of specific structures $\mathscr{U}$.

6.2.1 Classical Solvability

Let the set $R(\mathscr{L}^*)$ have zero characteristics, for example, when E is a quasireflexive space [86]. Put

$$\mathscr{U} = \{\alpha_\lambda \,|\, \alpha_\lambda = (\mathscr{L}^*)^{-1}(S_\lambda(E^*) \cap R(\mathscr{L}^*)), \lambda \in \mathbb{R}_+\},$$

where $S_\lambda(E^*)$ is a closed ball of radius λ in the space E^* (E^* is the space conjugate to E) with the center in the point of origin. Conditions (1) and (2) are satisfied obviously. In addition, $\mathbf{N} = \cup \alpha_\lambda = F^*$, i.e. Condition (4) is satisfied. Let us test Condition (3). Let some set $\alpha_\lambda \in \mathscr{U}$ not be a bounded set in F^* in topology $\sigma(F^*, R(\mathscr{L}))$, i.e. there exists such a sequence $\varphi_n \in \alpha_\lambda$ and $f \in R(\mathscr{L})$ that $\varphi_n(f) \geq n$. Since $f \in R(\mathscr{L})$, then there exists such an element $u \in E$, that $\varphi_n(\mathscr{L}u) \geq n$ or $(\mathscr{L}^*\varphi_n)(u) \geq n$, but $\mathscr{L}^*\varphi_n \in \mathscr{L}^*(\alpha_\lambda) \subset S_\lambda(E^*)$. This contradicts to boundedness of the ball $S_\lambda(E^*)$.

Thus, the topology $\mathscr{T}_E$ is determined by the system of semi-norms

$$p_\lambda(u) = \sup_{l \in S_\lambda(E^*) \cap R(\mathscr{L}^*)} |l(u)| = \sup_{l \in \overline{S_\lambda(E^*) \cap R(\mathscr{L}^*)}} |l(u)|, \qquad \lambda \in \mathbb{R}_+,$$

where $\overline{S_\lambda(E^*) \cap R(\mathscr{L}^*)}$ is a closure of the set $S_\lambda(E^*) \cap R(\mathscr{L}^*)$ in topology $\sigma(E^*, E)$.

Since the set $R(\mathscr{L}^*)$ is total in the duality (E^*, E), then it is dense in E^* in weak-* topology $\sigma(E^*, E)$, and since $R(\mathscr{L}^*)$ has non-zero characteristics, then $\overline{S_\lambda(E^*) \cap R(\mathscr{L}^*)}$ contains some ball $S_{t\lambda}(E^*)$ of smaller radius $(0 < t < 1)$. Then we have that

$$\sup_{l \in S_{t\lambda}(E^*)} |l(u)| \le p_\lambda(u) \le \sup_{l \in S_\lambda(E^*)} |l(u)|.$$

By the Hahn–Banach Theorem, the topology $\mathscr{T}_E$ is induced by the norm $\|u\|_E$ and the space $\bar{E}_{\mathscr{T}}$ coincides the the Banach space E. Since the operator $\mathscr{L}$ realizes an isomorphism between $E_{\mathscr{T}}$ and $R_{\mathscr{T}}$ and the space $E_{\mathscr{T}} = E$ is complete, then $R_{\mathscr{T}}$ is a complete space also, i.e. $\bar{R}_{\mathscr{T}} = R(\mathscr{L})$.

Thus, under such selection of the structure $\mathscr{U}$ the generalized solvability coincides with the classical solvability. This means that the concept of the classical solvability of linear operator equations is described in terms of the structure $\mathscr{U}$.

6.2.2 Generalized Strong Solvability

Let us suppose that $\mathscr{U} = \{\alpha \mid \alpha = S_\lambda(F^*), \lambda \in \mathbb{R}\}$. The fact that the operator $\mathscr{L}^*$ is bounded implies that the set $\mathscr{L}^*(S_\lambda(F^*))$ is bounded. It is easy to see that Conditions (1)–(5) are satisfied.

By the Hahn–Banach Theorem, the topology $\mathscr{T}_E$ is induced by the norm

$$\|u\|_1 = \sup_{l \in \mathscr{L}^*(S_1(F^*))} |l(u)| = \sup_{\varphi \in S_1(F^*)} |\varphi(\mathscr{L}u)| = \sup_{\varphi \in F^*} \frac{|\varphi(\mathscr{L}u)|}{\|\varphi\|_{F^*}} = \|\mathscr{L}u\|_F,$$

and the topology $\mathscr{T}_F$ is induced by the norm $\|f\|_F$. This means that $\bar{E}_{\mathscr{T}}$ is a completion of E with respect to the norm $\|\mathscr{L}u\|_F$. The space $F = F_{\mathscr{T}}$ is complete and $\bar{R}_{\mathscr{T}} = F$.

Note that under these conditions $\mathbf{N} = F^*$ and $\mathbf{M} = R(\mathscr{L}^*)$, since the vector spaces $\mathbf{N}$ and $\bar{R}_{\mathscr{T}}$ are in duality (see Remark 6.38).

Let us prove that $([\mathbf{M}], \bar{E}_{\mathscr{T}})$ is a dual pair. Suppose that there exists such an element $u \in \bar{E}_{\mathscr{T}}$, that $\bar{l}(u) = 0$ for any $l \in \mathbf{M}$. Let u_n be a sequence of elements from E, which converges to u in the norm of the space $\bar{E}_{\mathscr{T}}$. By definition an extended functional $\bar{l} \in [\mathbf{M}]$ on the element u takes the value

$$\bar{l}(u) = \lim_{n \to \infty} l(u_n) = \lim_{n \to \infty} (\mathscr{L}^*\varphi)(u_n) = \lim_{n \to \infty} \varphi(\mathscr{L}u_n) = \varphi(\lim_{n \to \infty} \mathscr{L}u_n) = \varphi(f),$$

where $l = \mathscr{L}^* \varphi$, $\varphi \in F^*$, $f = \lim \mathscr{L} u_n$. The latter limit exists by virtue of the fact that the sequence u_n converges in the topology $\mathscr{T}_E$. Thus, $\varphi(f) = 0$ for all $\varphi \in \mathbf{N} = F^*$. Whence it follows that, $f = 0$ in F or $u = 0$ in $\bar{E}_{\mathscr{T}}$.

Thus, when we select the structure $\mathscr{U} = \{\alpha \,|\, \alpha = S_\lambda(F^*), \lambda \in \mathbb{R}\}$ a generalized solution in the sense of Definition 6.21 is unique.[4]

Note that in this case the topologies $\mathscr{T}_E$ and $\mathscr{T}_F$ are normed and naturally connected with the original spaces E and F, since a generalized solution specified by the structure $\mathscr{U} = \{\alpha \,|\, \alpha = S_\lambda(F^*), \lambda \in \mathbb{R}\}$ is called a strong generalized solution. Such approach was exposed in Chap. 2. In this important particular case we will denote the space $\bar{E}_{\mathscr{T}}$ by E_1, and the extended operator $\mathscr{L}$ we will denote by $\mathscr{L}_1 : E_1 \to F$. Taking into account the common results, we can formulate the following statement for the structure $\mathscr{U} = \{\alpha \,|\, \alpha = S_\lambda(F^*), \lambda \in \mathbb{R}\}$.

Theorem 6.4. *For an arbitrary right-hand side $f \in F$ there exists a unique generalized solution $u \in E_1$ of the equation $\mathscr{L} u = f$ in one of the following equivalent senses:*

1. *$\mathscr{L}_1 u = f$,*
2. *$\bar{l}(u) = \varphi(f)$ for all $l = \mathscr{L}^* \varphi$, $\varphi \in F^*$,*
3. *$\exists u_n \in E$, so that $u_n \to u$ in E_1 and $\mathscr{L} u_n \to f$ in F as $n \to \infty$.*

6.2.3 Generalized Weak Solvability

Let $\mathscr{U} = \{\alpha\}$ be a collection of sets, consisting of all finite centrally symmetric subsets F^*. It is easy to see that Conditions (1)–(5) are satisfied and the topology $\mathscr{T}_E$ is specified by the system of neighborhoods of zero

$$o_\alpha = \{u \in E \,|\, |(\mathscr{L}^* \varphi_i)(u)| \leq 1\}, \qquad \alpha = \{\varphi_1, \varphi_2, \dots, \varphi_n\} \subset F^*.$$

Thus, in this case the topology $\mathscr{T}_E$ coincides with weak topology $\sigma(E, R(\mathscr{L}^*))$, therefore the space $\bar{E}_{\mathscr{T}}$ coincides with the space studied in Chap. 2. As it was noted in [89], this approach generalizes the known concepts of a generalized solution which were considered, for example, in [5]. The similar cases were considered also in [33, 91]. Let us denote the space $\bar{E}_{\mathscr{T}}$ by E_2, and the space $\bar{R}_{\mathscr{T}}$ by F_2, and the extended operator $\mathscr{L}$ by $\mathscr{L}_2 : E_2 \to F_2$.

Note that $\mathbf{N} = F^*$, $\mathbf{M} = R(\mathscr{L}^*)$. In addition, the sets α are compact in $\mathbf{N}$ with respect to the topology $\sigma(\mathbf{N}, F)$, and the set $\mathscr{L}^*(\alpha)$ is compact in $\mathbf{M}$ with respect to the topology $\sigma(\mathbf{M}, E)$, i.e. the conditions from Remarks 6.36, 6.38 are satisfied.

If $\mathscr{U} = \{(\mathscr{L}^*)^{-1}(\beta)\}$, where β are finite centrally symmetric subsets of some linear set $A \subset R(\mathscr{L}^*)$, which is total in the duality (E^*, E), then $\bar{E}_{\mathscr{T}}$ coincides with $\tilde{M}$ from Chap. 3.

---

[4] The proof of the uniqueness may be conducted using Theorem 6.3. The corresponding reasonings were formulated in Theorem 6.5 under the proving the existence of the embedding $E_1 \subset E_2$.

6.2.4 A Priori Inequalities

Let M be some equable convex set bounded by norm F^*, which is total in the duality (F^*, F). If

$$\mathscr{U} = \{\lambda M \mid \lambda \in \mathbb{R}_+\},$$

then Conditions (1)–(5) are obviously satisfied and the topology $\mathscr{T}_E$ is specified by the norm

$$\|u\|_M = \sup_{l \in \mathscr{L}^*(M)} |l(u)|,$$

for which the following estimation is true

$$\|u\|_M = \sup_{\varphi \in M} |(\mathscr{L}^* \varphi)(u)| \leq \sup_{\varphi \in S_c(F^*)} |\varphi(\mathscr{L}u)| = c \|\mathscr{L}u\|_F.$$

Estimations of such types are called the a priori ones. There are a number of publications, where a priori estimations are used to study properties of operator (see, e.g., [62, 63] and the bibliographies in these monographs) in order to construct the theory of generalized solvability of linear partial differential equations (in the sense of analogous of Definitions 6.21 and 6.22).

6.3 Properties of the Generalized Solvability in the Spaces E_1, E_2

Let us consider the relations between the concepts of generalized solutions in the most important cases – in the spaces E_1, E_2. We remind that if E, F are Banach spaces, $\mathscr{L} : E \to F$ is an injective linear continuous operator, $D(\mathscr{L}) = E$, $R(\mathscr{L})$ is a dense subset of F, then E_1 is a completion of the set E in the norm $\|u\|_{E_1} = \|\mathscr{L}u\|_F$, and E_2 is a completion E in the weak topology $\sigma(E, R(\mathscr{L}^*))$. It should noted that some proofs cited above duplicate the reasonings of Chap. 2. However, now we consider them from the new point of view – in terms of the structures $\mathscr{U}$.

Theorem 6.5. *The space E_1 is densely and continuously embedded into the E_2.*

Proof. Since in any finite subset of the space F^* there exists a ball $S_\lambda(F^*)$, which contains this set, then the structure $\mathscr{U}$ specifying the space E_1, majorizes the corresponding structure for E_2. So, the topology of the space E_1 majorizes the topology E_2. Thus, to prove the existence of the embedding $E_1 \subset E_2$ it is necessary to test the condition π). Since E_1 is a normed space, then we can consider a Cauchy sequence $u_n \in E$ in E_1, which converges to zero in E_2. Then $\mathscr{L}u_n$ is a Caushy sequence in F, which converges to some f (F is a complete space). For every $\varphi \in F^*$ we have

$$(\mathscr{L}^* \varphi)(u_n) = \varphi(\mathscr{L}u_n) \to \varphi(f) \quad n \to \infty.$$

But since u_n converges to zero in $\sigma(E, R(\mathscr{L}^*))$,

$$(\mathscr{L}^* \varphi)(u_n) \to 0.$$

That is why for any $\varphi \in F^*$ we have $\varphi(f) = 0$. Whence it follows that $f = 0$, i.e. $u_n \in E$ converges to zero in E_1, which was to be proved. $\square$

Corollary 6.18. The space F is densely and continuously embedded into the space F_2.

Proof. It is sufficient to equate the element $u \in E$ with $\mathscr{L}(u) \in R(\mathscr{L})$ and to repeat the reasonings of the main theorem. $\square$

  The following statement is true.

Theorem 6.6. *For any right-hand side $f \in F$ there exists a unique generalized solution $u \in E_2$ of the equation $\mathscr{L}u = f$ in one of the following equivalent senses:*

1. $\bar{\mathscr{L}}_2 u = f$,
2. $\bar{l}(u) = \varphi(f)$ for all $l = \mathscr{L}^* \varphi$, $\varphi \in F^*$,
3. $\exists u_n \in E$, such that $u_n \to u$ in E_2 and $\mathscr{L}u_n \to f$ in F.

Proof. It is necessary to test only latter statement of the theorem. Indeed, since $R(\mathscr{L})$ is a dense subset of the space F, then there exists a sequence $\mathscr{L}u_n = f_n \in R(\mathscr{L})$, which converges to f in F. However, the topology of the space F majorizes the topology of the space F_2, so f_n converges to f in the space F_2 also. The operator $\bar{\mathscr{L}}_2$ realizes an isometry between F_2 and E_2, so u_n converges to some $u \in E_2$ in E_2. $\square$

Corollary 6.19. The definitions of generalized solutions in the spaces E_1 and E_2 (see Theorems 6.4 and 6.6) are equivalent.

Chapter 7
Concept of Generalized Solution of Nonlinear Operator Equation

In this chapter, we consider the concept of a generalized solution of a nonlinear operator equation $A(x) = y$ in metric spaces $x \in E, y \in F$, according to which a generalized solution is an element of a completion of an original metric space E in a metric specified by this operator and the metric of the space F [34]. We study the existence and uniqueness of the generalized solution, and its correctness in the case when the operator A is continuous or uniformly continuous. Also, we consider the problems related with embedding of the completion of the metric space E with respect to two comparable metrics.

Suppose E and F are metric spaces with metric ρ_E and ρ_F, respectively, F is a complete space, and $A : E \to F$ is an injective operator whose domain $D(A)$ coincides with the entire space E and the range $R(A) \subset F$ in dense in F. Consider the operator equation

$$A(x) = y, \quad x \in E, y \in F, \tag{7.1}$$

in the metric spaces E and F. If $y \in R(A)$, then there exists a unique solution $x \in E$ of (7.1). We will call it a classic solution. But if $y \notin R(A)$ then classic equation $x \in E$ does not exist. That is why it is necessary to introduce the concept of a generalized solution of a nonlinear operator equation.

To determine a generalized solution of (7.1) let us consider a new metric on E

$$\rho^*(x,y) = \rho_F(A(x), A(y)). \tag{7.2}$$

The fact that the functional ρ^* defined on the Cartesian product $E \times E$ is a metric on E follows immediately from the injectivity of the operator A.

7.1 Generalized Solution of Nonlinear Operator Equation

Let us denote by E^* the completion of E with respect to the metric ρ^*. Let y be an element of F that does not belong to $R(A)$. Since $R(A)$ is everywhere dense in F, there exists a sequence $y_n \in R(A)$ converging to y in F as $n \to \infty$. Let

D.A. Klyushin et al., *Generalized Solutions of Operator Equations and Extreme Elements*, 137
Springer Optimization and Its Applications 55, DOI 10.1007/978-1-4614-0619-8_7,
© Springer Science+Business Media, LLC 2012

$x_n = A^{-1}(y_n) \in E$. It is easy to see that this sequence is Cauchy in the metric ρ^*. Indeed,

$$\rho^*(x_n, x_m) = \rho_F(A(x_n), A(x_m)) = \rho_F(y_n, y_m) \to 0 \text{ as } n, m \to \infty,$$

since y_n is Cauchy. Hence, in the complete metric space E^* the sequence x_n converges to some element

$$\bar{x} \in E^* : \bar{x} = \lim_{n \to \infty} x_n. \tag{7.3}$$

Definition 7.25. The element $\bar{x}$ which was described above is called *a generalized solution* of (7.1).

Let us show that the generalized solution $\bar{x}$ is defined correctly. Indeed, let $\widetilde{y}_n$ be some other sequence in $R(A)$ converging to y in F and $\widetilde{x}_n = A^{-1}(\widetilde{y}_n)$. It is easy to see that $\widetilde{x}_n$ is equivalent to x_n in E^*. Indeed,

$$\rho^*(x_n, \widetilde{x}_n) = \rho_F(A(x_n), A(\widetilde{x}_n)) = \rho_F(y_n, \widetilde{y}_n) \to 0 \text{ as } n \to \infty,$$

since

$$\rho_F(y_n, \widetilde{y}_n) \leq \rho_F(y_n, y) + \rho_F(y, \widetilde{y}_n) \to 0 \text{ as } n \to \infty.$$

Thus, $\bar{x} = \lim_{n \to \infty} x_n = \lim_{n \to \infty} \widetilde{x}_n$ in E^*.

7.2 Near-Solution of Nonlinear Operator Equation

Definition 7.26. A sequence of elements $x_n = A^{-1}(y_n) \in E$, where $y_n \in F$ is an arbitrary sequence converging to $y \in F$ is called *a near-solution* of the operator equation (7.1), and $\bar{x} = \lim_{n \to \infty} x_n$ in E^* is called *a limit element of a near solution*.

It is easy to see that $\bar{x}$ is a generalized solution of the operator equation (7.1) iff it is a limit element of a near-solution. In some cases, the elements of the sequence x_n can be referred to as a near-solutions themselves.

The term "near-solution", is justified by the following reasons. As it was shown before, the sequence x_n converges to the generalized solution $\bar{x}$ in the complete space E^*, so that $\forall \varepsilon > 0$ if $n > N$ we have $\rho^*(\bar{x}, x_n) < \varepsilon$. Since for small $\varepsilon > 0$ two elements $\bar{x}$ and x_n with distance $\rho^*(\bar{x}, x_n) < \varepsilon$ can be considered almost identical (equal) in E^*. Hence, we may suppose that x_n almost coincides with $\bar{x}$, and it is natural to refer to the element x_n as to the near-solution of the operator equation (7.1).

It is interesting to note that generally in the original space E the sequence x_n is not convergent, but there are arguments in support of rationality of the term "near solution" in E not referring to the completion E^* and generalized solution $\bar{x}$. Indeed, in many practically important problems it is difficult or impossible to define the right-hand side y of (7.1) exactly. Therefore, we have to consider its ε-approximation, i.e. an element $y_\varepsilon \in R(A)$ such that $\rho(y, y_\varepsilon) < \varepsilon$. In this case, there exists an element $x_\varepsilon = A^{-1}(y_\varepsilon)$ in $D(A)$, that can be considered as an ε-approximation of a solution of (7.1) in the sense that its image $y_\varepsilon = A(x_\varepsilon)$ deviates slightly from the right-hand side (7.1). Thus, it is naturally to consider x_ε as an ε-solution or near-solution. It must be stressed that in many cases the quality of a solution x_ε depends on the proximity between its image $y_\varepsilon = A(x_\varepsilon)$ and the element y. As it will be shown below, the effect of stabilization of the family of elements x_ε is a derivative one, i.e. it follows from the fact that y_ε converges to y.

7.3 Existence and Uniqueness of a Generalized Solution

From the construction of the generalized solution $\bar{x}$ it follows that this solution exists for every $y \in F$. Let us prove that for an injective operator $A : E \to F$ the generalized solution is unique. Assume the contrary: let there be two generalized solutions $\bar{x}$ and $\widetilde{x}$ corresponding to y. Then there exist sequences $\bar{x}_n = A^{-1}(\bar{y}_n)$ and $\widetilde{x}_n = A^{-1}(\widetilde{y}_n)$, converging yo $\bar{x}$ and $\widetilde{x}$, respectively (it is natural to suppose that $\bar{y}_n \to y$ and $\widetilde{y}_n \to y$ in F as $n \to \infty$). Now, we have

$$\rho^*(\bar{x},\widetilde{x}) = \lim_{n\to\infty} \rho^*(\bar{x}_n,\widetilde{x}_n) = \lim_{n\to\infty} \rho_F(A(\bar{x}_n),A(\widetilde{x}_n)) = \lim_{n\to\infty} \rho_F(\bar{y}_n,\widetilde{y}_n) = 0.$$

Therefore, $\bar{x} = \widetilde{x}$.

7.4 Correctness of Generalized Solution

Usually, correctness of a solution means that it is continuously dependent on the right-hand side:

$$\forall \varepsilon > 0 \, \exists \delta > 0 \, \rho(\bar{y},\widetilde{y}) < \delta \Rightarrow \rho(\bar{x},\widetilde{x}) < \varepsilon,$$

where $\bar{x}$ and $\widetilde{x}$ are generalized solutions corresponding to right-hand sides $\bar{y}$ and $\widetilde{y}$.

Let us denote by E_0^* a set E with the metric ρ^*, then E^* is a completion of E_0^*. The operator $A(x)$ maps E_0^* into $R(A) \subset F$ and is an isometry E_0^* and $R(A)$, since for any $x_1, x_2 \in E_0^*$ the following equality holds.

$$\rho^*(x_1,x_2) = \rho_{E_0^*}(x_1,x_2) = \rho_F(A(x_1),A(x_2)) = \rho_F(y_1,y_2).$$

Here, $A(x_1) = y_1$ and $A(x_2) = y_2$. We can extend the operator A onto the entire space E^* so that it will map E^* into F in the following way. Let $\bar{x}$ be an arbitrary element in E^*. Then there exists a sequence x_n from E_0^* converging to $\bar{x}$. The sequence x_n is Cauchy in E_0^*; hence, the sequence $y_n = A(x_n)$ is Cauchy also

$$\rho_F(y_n,y_m) = \rho_F(A(x_n),A(x_m)) = \rho^*(x_n,x_m) \to 0$$

as $n, m \to \infty$.

Since F is complete, there exists an element $y \in F$ such that $\lim_{n \to \infty} y_n = y$ (with respect to the metric ρ_F). Let us define the extension $\bar{A}$ of the operator A on the completion E^* by the formula

$$\bar{A}(\bar{x}) = y, \quad \bar{x} \in E^*, \ y \in F.$$

We can see that this extension is correct and the element y is determined uniquely. If $y \notin R(A)$, the element $\bar{x}$ is a generalized solution. Both $\bar{A}$ and A are isometries of the spaces E^* and F. Indeed, for all $\bar{x}, \tilde{x} \in E^*$

$$\rho_{E^*}(\bar{x}, \tilde{x}) = \lim_{n \to \infty} \rho_{E^*}(\bar{x}_n, \tilde{x}_n) = \lim_{n \to \infty} \rho_F(A(\bar{x}_n), A(\tilde{x}_n)) = \lim_{n \to \infty} \rho_F(\bar{y}_n, \tilde{y}_n),$$

where $\bar{y} = \bar{A}(\bar{x})$, $\tilde{y} = \bar{A}(\tilde{x})$, $\bar{x}_n, \tilde{x}_n \in E$, $\bar{x}_n \to \bar{x}$, $\tilde{x}_n \to \tilde{x}$ as $n \to \infty$ in E^*. It follows that $\bar{A}^{-1}$ maps F onto E^* and is an isometry between F and E^*. Therefore, the generalized and classic solutions fill up E^*, and the generalized solutions form a set $E^* \setminus E$ (a complement of E up to E^*). Moreover, the generalized solution $\bar{x} = \bar{A}^{-1}(y)$ is correct in E^*, since the isometry $\bar{A}^{-1}$ is a continuous mapping. Note that in E the existence of continuous inverse operator is not guaranteed and classic solution may not be correct.

7.5 Pseudo-Generalized and Essentially Generalized Solutions

The construction of generalized solution described above does not use the properties of the metric space E, so there is no connection between E and E^*. This follows from the fact that we do not impose any restrictions on the original operator A, except its injectivity. But if we assume that it has some additional topological properties (for example, continuity) then this connection arises. Let us study these properties.

Lemma 7.1. *If A is a continuous injective operator mapping a metric space E into a metric space F, then E is densely embedded into E^*, where E^* is the completion of E in the metric (7.2).*

Proof. Indeed, since $A : E \to F$ is a continuous injective operator, $D(A) = E$, and $R(A)$ is a dense subset of F, then the operator A defines a dense and continuous embedding of E into F. From the other hand, as it was established above, the operator $\bar{A}$ is an isometry between the metric spaces E^* and F. Thus, we have the following commutative diagram

$$
\begin{array}{ccc}
E & \xrightarrow{\ A\ } & F \\[2pt]
\| & & \bar{A} \uparrow \\[2pt]
E & \xrightarrow{\ j\ } & E^*,
\end{array}
$$

where the operator $j : E \to E^*$, defined as $j = \bar{A}^{-1} \circ A$, specifies a dense and continuous embedding of the metric space E into the metric space E^*. $\square$

Lemma 7.1 implies that the topology of E_0^* is weaker that the topology of E.

Under the investigation of generalized solutions of the operator equation (7.1) the completion $\bar{E}$ of the original space E with respect to the metric ρ_E plays an important role (in contrast to the completion E^* of the space E with respect to the metric ρ^*).

Definition 7.27. A generalized solution $\bar{x}$ of the operator equation (7.1) is called *a pseudo-generalized solution* if $\bar{x} \in \bar{E}$.

Definition 7.28. A generalized solution $\bar{x}$ of the operator equation (7.1) is called *an essentially generalized solution*, if $\bar{x} \notin \bar{E}$ (and $\bar{x} \in E^*$).

At a first glance, it seems that pseudo-generalized solution differs slightly from the classical one and the completeness of metric space E is not essential. Indeed, each metric space E has a completion $\bar{E}$ and it is obvious that we can extend the operator A by continuity onto the whole space $\bar{E}$. But the latter statement it not true, i.e. by far not every continuous operator $A : E \to F$ may be extended onto the whole $\bar{E}$ (in contrast to linear continuous operators which act in linear topological spaces). In addition, if such an extension yet exists, it can be not an injective operator even if A is an injective itself. Let us consider these issues in details.

Let us recall that the operator $A : E \to F$ is called uniformly continuous on E, if

$$\forall \varepsilon > 0 \exists \delta(\varepsilon) > 0 : \rho_E(x,y) < \delta(\varepsilon) \Rightarrow \rho_F(A(x),A(y)) < \varepsilon. \qquad (7.4)$$

Let us prove that a uniformly continuous operator A can be extended onto the whole space $\bar{E}$ preserving its properties (i.e. uniform continuity).

Theorem 7.1. *Let $A : E \to F$ be a uniformly continuous operator on E. Then it can be extended to uniformly continuous operator $\bar{A}$, which acts from $\bar{E}$ into F.*

Proof. Let $\bar{x}$ be an arbitrary element in $\bar{E}$. Let us take an arbitrary sequence $\bar{x}_n \in E$ converging to $\bar{x}$ in $\bar{E}$ as $n \to \infty$. The sequence $\bar{x}_n \in E$ is Cauchy in E, i.e.

$$\forall \delta > 0 \exists N(\delta) : \forall n,m > N(\delta) \Rightarrow \rho_E(x_n, x_m) < \delta.$$

By virtue of (7.4)

$$\forall \varepsilon > 0 \exists N(\delta(\varepsilon)) : \forall n,m > N(\delta(\varepsilon)) \Rightarrow \rho^*(x_n, x_m) = \rho_F(A(x_n), A(x_m)) < \varepsilon.$$

Therefore, the sequence $\bar{x}_n$ is Cauchy in the space E_0^* also. Since the operator $A : E \to F$ specifies an isometry between E_0^* and $R(A)$ with the metric ρ_F, then the sequence $\bar{y}_n = A(\bar{x}_n)$ is Cauchy in $R(A)$, hence in F also. Since F is a complete metric space, there exists $y = \lim_{n \to \infty} \bar{y}_n$.

Put $\bar{A}(\bar{x}) = y$. Let us justify the correctness of this definition. Let $\tilde{x}_n \in E$ be other arbitrary sequence converging to $\bar{x}$ in $\bar{E}$ as $n \to \infty$. Then by the triangle inequality $\rho_E(\bar{x}_n, \tilde{x}_n) \to 0$ as $n \to \infty$. Therefore,

$$\forall \delta > 0 \exists N : \forall n > N \Rightarrow \rho_E(\bar{x}_n, \tilde{x}_n) < \delta.$$

Whence it follows that, taking into account the uniform continuity of A, we have

$$\forall \varepsilon > 0 \exists N : \forall n > N \Rightarrow \rho^*(\bar{x}_n, \tilde{x}_n) = \rho_F(A(\bar{x}_n), A(\tilde{x}_n)) < \varepsilon.$$

This means that the sequences $\bar{y}_n = A(\bar{x}_n)$ and $\tilde{y}_n = A(\tilde{x}_n)$ have a common limit – the element $y \in F$.

Let us prove that $\bar{A}$ is a uniformly continuous operator. For an arbitrary $\varepsilon > 0$ we select arbitrary elements $\bar{x}, \tilde{x} \in \bar{E}$ satisfying the inequality $\rho_{\bar{E}}(\bar{x}, \tilde{x}) < \delta(\varepsilon)$, where the value $\delta(\varepsilon)$ is determined from the equality (7.4). Let $\bar{x}_n$ and $\tilde{x}_n$ be arbitrary sequences in E converging to $\bar{x}$ and $\tilde{x}$ respectively. Then,

$$\exists N : \forall n > N \Rightarrow \rho_E(\bar{x}_n, \tilde{x}_n) < \delta(\varepsilon),$$

and, taking into account the uniform continuity of A, we have that for all $n > N$ the equality $\rho_F(A(\bar{x}_n), A(\tilde{x}_n)) < \varepsilon$ holds true. Since the metric ρ_F is continuous, we have

$$\rho_F(\bar{A}(\bar{x}), \bar{A}(\tilde{x})) = \rho_F(\lim_{n \to \infty} A(\bar{x}_n), \lim_{n \to \infty} A(\tilde{x}_n)) = \lim_{n \to \infty} \rho_F(A(\bar{x}_n), A(\tilde{x}_n)) \leq \varepsilon.$$

$$\square$$

Note that the continuity of the operator $A : E \to F$ is not sufficient for the existence of a continuous extension $\bar{A}$ of the operator A onto the whole space $\bar{E}$.

Indeed, consider the continuous (but not uniformly continuous) operator $A(x) = \sin \frac{1}{x}$, which acts between the metric spaces $E = (0, 1]$, $F = [-1, 1]$ of the real numbers $\mathbb{R}$ with natural metric. The completion of the space $E = (0, 1]$ is $\bar{E} = [0, 1]$. But since the limit $\lim_{x \to 0} A(x)$ does not exist, then a continuous extension of the operator A onto $\bar{E} = [0, 1]$ does not exist also.

It is obvious that since we have considered the construction of extension of an injective operator $A : E \to F$ onto the whole space E^*, then the cause of the absence of the continuous extension of A onto $\bar{E}$ in the previous example was the fact that the operator A was not injective. However, this is also not the case. Let us consider the example of a continuous injective operator A with a range $R(A)$ that is a dense subset of the complete metric space F, which nevertheless allows the extension onto the the the whole space $\bar{E}$. Consider a continuous (but not uniformly continuous again) injective operator

$$A(x) = \left(x, \sin \frac{1}{x}\right),$$

which acts from the metric space $E = (0, 1] \subset \mathbb{R}$ into the metric space $F = \overline{A(E)} \subset \mathbb{R}^2$ with the usual Euclidean metric. Here, under $F = \overline{A(E)}$ we mean a completion of the metric space $A(E)$ with the usual metric, i.e.

$$F = \left\{ (x, y) \in \mathbb{R}^2 \,|\, x \in (0, 1], y = \sin \frac{1}{x} \right\} \cup \left\{ (0, y) \in \mathbb{R}^2 \,|\, y \in [-1, 1] \right\}.$$

Since the limit $\lim\limits_{x \to 0} A(x)$ in F does not exist (the set of partial limits consists of the point of the segment $\{(0,y) \in \mathbb{R}^2 \,|\, y \in [-1,1]\}$), then a continuous extension does not exist.

Let us study the problem described from the other point of view. Suppose that some injective continuous operator $A : E \to F$ allows the continuation by continuity on the completion $\bar{E}$. Let us establish the criteria of the injectivity of the extended operator $\bar{A} : \bar{E} \to F$. Suppose that the following condition holds.

$\widetilde{\pi}$) if x_n and x'_n are Cauchy sequences from E in metric ρ_E and $\rho^*(x_n, x'_n) = \rho_F(A(x_n), A(x'_n)) \to 0$ as $n \to \infty$, then $\rho_E(x_n, x'_n) \to 0$ as $n \to \infty$ (cf. analogous condition π) in [41]).

Recall that two sequences x_n and x'_n are equivalent in the metric $\rho(x,y)$ if $\rho(x_n, x'_n) \to 0$ as $n \to \infty$. Using this concept, we can reformulate $\widetilde{\pi}$) in the following way: two Cauchy sequences x_n and x'_n from $\mathscr{E}$ which are equivalent in the metric ρ_E^* are equivalent in the metric $\rho_E(x,y)$ also.

Theorem 7.2. *Let A be a uniformly continuous injective operator mapping the metric space E into a complete metric space F and $\bar{A}$ be an extension by continuity of A on the completion $\bar{E}$ of the metric space E. The extension $\bar{A}$ is an injective operator mapping $\bar{E}$ into F iff the condition $\widetilde{\pi}$) holds.*

Proof. Necessity. Let $\bar{A}$ be an injective mapping, x_n and x'_n be two Cauchy sequences in E ($x_n \to \bar{x}$ and $x'_n \to \bar{x}'$ in $\bar{E}$ as $n \to \infty$), which are equivalent with respect to the metric $\rho^*(x,y)$, then

$$0 = \lim_{n \to \infty} \rho^*(x_n, x'_n) = \lim_{n \to \infty} \rho_F(A(x_n), A(x'_n)) = \rho_F(y, y'),$$

where $y = \lim\limits_{n \to \infty} A(x_n)$, $y' = \lim\limits_{n \to \infty} A(x'_n)$. Since $y = \bar{A}(\bar{x})$, $y' = \bar{A}(\bar{x}')$, then the equality $\bar{A}(\bar{x}) = \bar{A}(\bar{x}')$ and injectivity $\bar{A}$ imply that $\bar{x} = \bar{x}'$. This means that $\rho_E(x_n, x'_n) \to \rho_{\bar{E}}(\bar{x}, \bar{x}') = 0$ as $n \to \infty$.

Sufficiency. Suppose that $\bar{A}(\bar{x}) = \bar{A}(\bar{x}')$ and x_n, x'_n are a Cauchy sequences E converging to $\bar{x}$ and $\bar{x}'$ in $\bar{E}$ as $n \to \infty$ respectively. These sequences are equivalent in the metric $\rho^*(x,y)$, since

$$\lim_{n \to \infty} \rho^*(x_n, x'_n) = \lim_{n \to \infty} \rho_F(A(x_n), A(x'_n)) = \rho_F(\bar{A}(\bar{x}), \bar{A}(\bar{x}')) = 0.$$

By virtue of condition $\widetilde{\pi}$) the sequences x_n and x'_n are equivalent in the metric $\rho_E(x,y)$ also. Therefore, $\bar{x} = \bar{x}'$, and the injectivity of the extension $\bar{A}$ of the operator A onto $\bar{E}$ is proved. $\qquad\square$

From the point of view of the previous theorem let us consider the following issue. Suppose that the extension $\bar{A}$ is weakly injective: if $\bar{x} \neq \bar{x}'$ and at least one element of $\bar{x}$ and $\bar{x}'$ belongs to the space E, then $\bar{A}(\bar{x}) \neq \bar{A}(\bar{x}')$. Using the reasons as in the proof Theorem 7.2, we can show that this property is equivalent to the following condition:

π^*)if the sequence x_n is Cauchy in the metric $\rho_E(x,y)$, $x \in E$ and $x_n \to x$ as $n \to \infty$
  in the metric $\rho^*(x,y)$, then $x_n \to x$ in the metric ρ_E.

It is easy to see that condition (π^*) follows from condition $\widetilde{\pi}$), and in addition, if E and F are linear normed spaces and A is a continuous linear operator, then the equivalence of conditions $\widetilde{\pi}$) and π^*) is clear. Let us prove that in a general case these conditions are not equivalent [78].

Indeed, let us consider the sets $E = F = (0,1)$. Let us take as the operator A the identity mapping $A(x) = x$. Let us consider on E, F the following metrics

$$\rho_E(x,y) = |x-y|, \qquad \rho_F(x,y) = \min\{|x-y|, 1-|x-y|\}.$$

It is easy to see that the functionals ρ_E and ρ_F satisfy the metric axioms.

Let us show that condition π^*) holds. Indeed, let $x_n \in (0,1)$ be a Cauchy sequence in E. Since E is the interval $(0,1) \subset \mathbb{R}$ with usual metric, then x_n converges to some $x \in [0,1]$ in $\mathbb{R}$. Since $\rho_F(x,y) \le \rho_E(x,y)$, then x_n converges to x with respect to the metric $\rho^*(x,y) = \rho_F(x,y)$ also. That is why $x \in (0,1) = E$, which was to be proved.

From the other hand, condition $\widetilde{\pi}$) does not hold. Indeed, let $x_n = 1/n$ and $x'_n = (n-1)/n$. Then it is easy to see that $\rho^*(x_n, x'_n) \to 0$ as $n \to \infty$, but $\rho_E(x_n, x'_n)$ does not converge to zero.

Basing on this example, it is easy to construct analogous counterexamples for operators which act in other metric spaces.

7.6 Relation Between Pseudo-Generalized and Generalized Solutions

The investigation of the concept of a pseudo-generalized solution leads to a quite interesting situation, which at first glance contradicts to the uniqueness of a generalized solution. Indeed, suppose that condition $\widetilde{\pi}$) does not hold true. Then the extension $\bar{A}$ of a uniformly continuous operator A onto the whole space $\bar{E}$ is not an injective operator, so there exists a pair $\bar{x}$ and $\bar{x}'$ consisting of different elements of $\bar{E}$, for which $\bar{A}(\bar{x}) = \bar{A}(\bar{x}') = y$. Now, it is obvious that there are at least two pseudo-generalized (and hence generalized) solutions $\bar{x}$ and $\bar{x}'$ corresponding to the element y (the right-hand side of the operator equation (7.1)). This would have been possible if the completion $\bar{E}$ was embedded into the completion E^* of the space E_0^*. However, actually there is no any contradiction, since by far not always $\bar{E} \subset E^*$; therefore, at least one of the elements $\bar{x}$, $\bar{x}'$ does not belong to E^*. So, this element is not a generalized (and pseudo-generalized) solution, and the supposed contradiction is solved. Thus, there is an important theoretical problem of embedding of completions $\bar{E}$ and E^*.

Let us recall the exact definition.

Definition 7.29. Let E and F be metric spaces and j be an injective mapping of E to F. The metric space E is said to be *continuously (uniformly continuously) embedded* into F using the embedding operator j if j is a continuous (uniformly continuous) operator. If in addition the subspace $j(E) \subset F$ is dense in the metric space F, then there exists a dense embedding of E into F using the embedding operator j.

Note, that the embedding operator j in this definition plays an extremely important role, since the space E can be embedded into F by different embedding operators j, so the phrase "the space E is embedded into F" without clear understanding of the nature of the operator j is incorrect.

Let E be an arbitrary metric space and $\bar{E}$ be its completion. Is is well-known that the completion $\bar{E}$ consists of classes $\bar{x}$ of equivalent Cauchy sequences x_n in E. Let us define the canonical embedding operator j_1 in E in the following way: we map every element $x \in E$ to a class $\bar{x}$ containing the stationary sequence $x_n \equiv x$, $n \in \mathbb{N}$. Then the space E is embedded into its completion $\bar{E}$ by canonical embedding operator j_1. In this case, the metric space E is equipped with two metrics: original metric ρ_E and a metric ρ^* (the operator A is supposed to be continuous). The completion $E_0^* = E$ in the metric ρ^* is denoted by E^*, therefore $E = E_0^*$ is embedded into E^* by an analogous canonical embedding operator j_3. If the original operator A is uniformly continuous, we can define a canonical embedding mapping j_2 of $\bar{E}$ into E^* in the following way: every element $\bar{x} \in \bar{E}$ is mapped to an element $x^* \in E^*$ containing the element $\bar{x}$: $j_2(\bar{x}) = x^*$, $\bar{x} \subset x^*$.

Let us work out all the details of j_2 and show that it is correct. Let $\bar{x}$ be an arbitrary element of $\bar{E}$ and x_n^0 be a Cauchy sequence in the metric ρ_E that belongs to the class $\bar{x}$. Since the sequence x_n is an element of $\bar{x}$, then it is equivalent to x_n^0 in the metric ρ_E, but x_n and x_n^0 are equivalent in the metric ρ^* also. Indeed, let $\varepsilon > 0$ be an arbitrary positive number. Since A is a uniformly continuous operator, then there exists such $\delta > 0$, that the condition $\rho_E(x,y) < \delta$ implies that $\rho_F(A(x), A(y)) < \varepsilon$. Let us select a natural number N in such a way that $\rho_E(x_n, x_n^0) < \delta$ for every $n > N$, then $\rho_F(A(x_n), A(x_n^0)) < \varepsilon$ for every $n > N$. Therefore, x_n and x_n^0 are equivalent sequences in metric $\rho_F(x,y)$. Hence, every sequence x_n in $\bar{x}$ belongs to the class x^*; hence, $\bar{x} \subset x^*$ and the mapping j_2 is defined correctly. Canonical embedding operators j_1, j_3 and the operator j_2 are connected by the following commutative diagram.

$$
\begin{array}{ccc}
E & \xrightarrow{\ j_1\ } & \bar{E} \\
\| & & \downarrow{\scriptstyle j_2} \\
E & \xrightarrow{\ j_3\ } & E^*
\end{array}
$$

Recall that due to commutative property this formula can also be expressed in the form $j_3(x) = j_2(j_1(x))$.

In order for the canonical mapping j_2 to be embedding it should be injective. The criterion for injectivity of j_2 can be formulated in terms of condition $\widetilde{\pi}$).

Theorem 7.3. *Let E and F be metric spaces, F be complete and $A : E \to F$ be a uniformly continuous injective operator, which acts from E into F. The canonical mapping $j_2 : \bar{E} \to E^*$ is injective iff condition $\tilde{\pi}$) holds.*

Proof. Necessity. Let j_2 be an injective mapping, x_n and x'_n be Cauchy sequences in metrics ρ_E, $x_n \in \bar{x}$, $x'_n \in \bar{x}'$ and $\rho^*(x_n, x'_n) \to 0$ as $n \to \infty$. By definition of j_2 the element $j_2(\bar{x})$ is such an element $x^*_1 \in E^*$ that $\bar{x} \subset x^*_1$. Similarly, $j_2(\bar{x}') = x^*_2$: $\bar{x}' \subset x^*_2$. This implies that $x_n \in x^*_1$ and $x'_n \in x^*_2$, but from the other hand $\rho^*(x_n, x'_n) \to 0$ as $n \to \infty$. This means that $x^*_1 = x^*_2$, i.e. $j_2(\bar{x}) = j_2(\bar{x}')$, hence by the injectivity of the operator j_2 we have $\bar{x} = \bar{x}'$. Thus, the sequences x_n and x'_n are equivalent in the metric ρ_E also.

Sufficiency. Let $\bar{x}$ and $\bar{x}'$ be two different elements of the completion $\bar{E}$ and let the condition $\tilde{\pi}$) hold. Suppose that $j_2(\bar{x}) = j_2(\bar{x}') = x^*$. Consider two Cauchy sequences $x_n \in \bar{x}$ and $x'_n \in \bar{x}'$, then $x_n \in j_2(\bar{x}) = x^*$ and $x'_n \in j_2(\bar{x}') = x^*$, so these sequences are equivalent in metric ρ^*. By virtue of the condition $\tilde{\pi}$) the sequences x_n and x'_n are equivalent in metric ρ_E also, so $\bar{x} = \bar{x}'$, but this contradicts to our assumption. The theorem in proved.   □

Definition 7.30. An injective canonical mapping j_2 of completion $\bar{E}$ into E^* is called *a canonical embedding* of completion $\bar{E}$ into E^*.

Theorem 7.3 implies that the canonical mapping j_2 is a canonical embedding if the condition $\tilde{\pi}$) holds. The existence of canonical embedding j_2 of $\bar{E}$ into E^* allows us to extend the metric ρ^* from the space E onto its completion $\bar{E}$ by the formula

$$\rho(\bar{x}, \bar{x}') = \rho^*(x^*, x^{**}) = \rho_F\left(\bar{A}(\bar{x}), \bar{A}(\bar{x}')\right),$$

where $x^* = j_2(\bar{x})$, $x^{**} = j_2(\bar{x}')$. If the condition $\tilde{\pi}$) does not hold, then the functional ρ^* is only a quasi-metric. This fact is very useful in the study of generalized solutions of operator equations. Thus, when the condition $\tilde{\pi}$) holds true, the extension $\bar{A}$ of the operator A onto the completion $\bar{E}$ is an injective operator (as well as the operator A), and the space $\bar{E}$ (as well as the space E) is embedded into E^* with the help of canonical embedding. Therefore, this case differs sharply from the original problem.

Now, pass to the classification of solutions of an operator equation according to various types of convergence in the metric space E:

1. If the right-hand side y of (7.1) belongs to the range $R(A)$ of the operator A, then there exists a classical solution of (7.1).
2. If the right-hand side y of (7.1) does not belong to the range $R(A)$ of the operator A, there exist following fundamentally different possibilities:

   (a) Let $x_n = A^{-1}(y_n)$, $y_n \to y$ as $n \to \infty$ be a near-solution of (7.1); if the sequence x_n is Cauchy in metrics ρ_E, then for this element a pseudo-generalized solution exists, and the limit element $\bar{x}$ of this solution belongs to the completion $\bar{E}$.

   (b) If a near-solution $x_n = A^{-1}(y_n)$ is not Cauchy in metric $\rho_E(x, y)$, then $\bar{x} \notin \bar{E}$, and $\bar{x} \in E^*$ is an essentially generalized solution of (7.1).

Let us examine the conditions under which generalized solutions arise and should be investigated. If a metric space E is compact (relatively compact) and the operator $A : E \to F$ is continuous (uniformly continuous), then generalized solutions occur very seldom. Indeed, in this case by virtue of well-known results [8] the inverse mapping $A^{-1} : R(A) \to E$ is continuous and, and since $R(A)$ is everywhere dense in F, the uniformly continuous operator A^{-1} can be extended by continuity to the whole space F. If E is a complete metric space, then for any $y \in F$ the element $x = A^{-1}y$ is a classical solution. If E is an incomplete metric space, then there arise pseudo-generalized solutions which differ only slightly from classical solutions. In both cases, the essentially generalized solutions do not arise. However, generalized solutions can arise when E and F are non-compact infinite-dimensional spaces (topological spaces or differentiable manifolds, spaces of distributions, classical Banach spaces, Hilbert spaces an so on). We need such spaces to investigate linear and nonlinear integral equations and infinite systems of linear and nonlinear algebraic equations. In this case function spaces and spaces of sequences play the role of space E.

The problem of necessity of the investigation of generalized solutions appears for the following reasons. As it was shown earlier, the generalized solution x of the equation $A(x) = y$ is a limit element of a near-solution x_n, where $x_n = A^{-1}(y_n)$, and $y_n \to y$ as $n \to \infty$. If $y \notin R(A)$, then the near-solution x_n does not converge in E to any element (for essentially generalized solution this sequence is not even a Cauchy sequence in E). Therefore, when solving approximately the operator equation (7.1) it is not obvious which metric is expected to stabilize the near-solution x_n and when the element x_n can be considered as an approximate solution of (7.1). Moreover, it is not clear what the generalized solution $\bar{x}$ is and to which classical function space of sequences it belongs. From the reason we have mentioned it can be deduced that stabilization of the near-solutions takes place in the metric ρ^* in E, and in this case the generalized solution belongs to the completion E^* of the space E in metrics ρ^*. However, this result represents only a principal solution of this problem, since it does not answer the question how to define the sense of convergence of a sequence of near solutions x_n in the metric ρ^* and what elements constitute the space E^*. Unfortunately, the space E^* depends on the operator A. This makes the problem more complicated, because investigating the family of operators A depending on parameter λ involves a family of spaces $E^*(\lambda)$ (in some cases we may have a whole scale of spaces). It is very difficult to obtain interesting results for such family of spaces. Fortunately, we can use the theory of embedded spaces [41]. Indeed, if we suppose that for some family of operators A_λ, ($\lambda \in I$) all metric spaces $E^*(\lambda)$ are continuously embedded, with the help of an operator of natural embedding, into some metric or topological space Σ well-defined structure, then each generalized solution $\bar{x}$ of the operator equation $A_\lambda(x) = y$ belongs to the space Σ and we can identify the element $\bar{x}$ (for example, if Σ is a space of measurable functions, then $\bar{x}$ is some measurable function). We will refer to the metric space Σ as to the basic one. Moreover, under certain restrictions imposed on a near-solution $x_n \in E^*(\lambda) \subset \Sigma$ the convergence of x_n in the topology (or metric) of the space Σ (this convergence holds true always) implies convergence of x_n in the space $E^*(\lambda)$. For example, by the classic Lebesgue theorem on majorants, the convergence of a sequence of integrable functions $x_n(t)$ by measure with conditions $|x_n| \leq g(t)$, where $g(t)$ is an integrable function implies convergence of x_n in the space of integrable functions L_1. This fact allows us to investigate the conditions of stabilization of near-solution x_n in the space Σ (and in the space $E^*(\lambda)$) and to obtain an approximate solution of the operator equation $A_\lambda(x) = y$. Metric or topological vector spaces with the weakest topology (convergency) may play the role of the basic space Σ. For many Banach function spaces (e.g., for Banach spaces of measurable functions and, in particular, for ideal spaces [41]) the basic space Σ is a space $S(a,b)$, consisting of measurable functions $x(t)$ defined on a segment $[a,b]$ with the metric

$$\rho(x,y) = \int_a^b \frac{|x(t) - y(t)|}{1 + |x(t) - y(t)|} \, dt, \quad x(t), y(t) \in S(a,b).$$

In this space, convergence is equivalent to convergence by measure. Also, the space of Schwarz distribution may be considered as the basic space [33]). In the space of sequences, we may select as the basic space Σ the space s of all numerical sequences with the metric

$$\rho(x,y) = \sum_{n=1}^{\infty} \frac{1}{2^n} \times \frac{|x_n - y_n|}{1 + |x_n - y_n|}, \quad x = (x_1, x_2, \ldots), \, y = (y_1, y_2, \ldots),$$

where the convergence is coordinate-wise.

7.7 Example of Operators

Let us consider examples of nonlinear integral and differential operators which may induce a concept of generalized solutions (see, e.g., [37]).

(1) Nemytskii operator. It is said that a function $f(s,u)$ of two arguments $-\infty < u < \infty$, $s \in G$ satisfies the Caratheodory conditions if it is continuous with respect to u almost for all $s \in G$ and is measurable by s for all u. Here, G is a subset of a n-dimensional Euclidean space with finite measure. Denote by f an operator on the set of real functions defined on G by the equality

$$f[u(s)] = f(s, u(s)),$$

where $f(s,u)$ satisfies the Caratheodory conditions. The operator f is called the Nemytskii operator. The operator $f[u(s)]$ is continuous and maps the space $S(G)$ of all measurable functions onto G. In addition, the continuity and boundedness of the operator f follows from the fact that f acts from L^{p_1} into L^{p_2}.

(2) Uryson operator. Let $K(x,t,u)$, $s,t \in G$, $-\infty < u < \infty$ be a function of three arguments. The nonlinear integral operator

$$A[\varphi(s)] = \int_G K(s,t,\varphi(t)) \, dt$$

is called the Uryson operator. If $K(s,t,u)$ is continuous with respect to every variable in the aggregate, where $s,t \in G$ and $u \leq a$, then $A[\varphi]$ is defined in the ball of radius a in the space $C(G)$ and is completely continuous. Under quite common conditions, the operator is defined in L^p and is completely continuous.

(3) Hammerstein operator. One class of the Uryson operators is studied more carefully: that is the Hammerstein operators.

$$A[\varphi(s)] = \int_G K(s,t) f[t, \varphi(t)] \, dt.$$

Let us denote by B the linear integral operator induced by the kernel $K(s,t)$

$$B[\varphi(s)] = \int_G K(s,t) \varphi(t) dt,$$

If the operator f mapping Banach space E_2 into E_2 is continuous and bounded, and the operator B mapping from E_1 into E_2 is completely continuous, then the Hammerstein operator acts from E_1 into E_2 and is completely continuous. Note, that the concept of generalized solution is especially important for completely continuous operators mapping infinite-dimensional spaces, since the inverse operator A^{-1} is not continuous and there exists only essentially generalized solution for every $y \in R(A) \subset F$.

(4) Nonlinear parabolic equation. Consider the nonlinear initial-boundary value problem which occurs in the theory of mass transport in porous media:

$$A(u) \equiv \frac{\partial u}{\partial t} - \sum_{\alpha=1}^{2} \frac{\partial}{\partial x_\alpha}\left(k_\alpha(u) \frac{\partial u}{\partial x_\alpha}\right) = f(x,t),$$

$$u|_{x \in \partial\Omega} = u|_{t=0} = 0$$

where $(x,t) \in Q = \Omega \times (0,T]$, $\Omega \subset \mathbb{R}^2$, $u \in W_{2,\text{bd}}^{+l}(Q)$, $f \in V \subset W_{2,\text{bd}}^{-l}(Q)$, $W_{2,\text{bd}}^{+l}(Q)$ is the Sobolev space consisting of the functions of $W_{2,\text{bd}}^{+l}(Q)$, that satisfy the boundary conditions (bd), $W_{2,\text{bd}}^{-l}(Q)$ is a negative space constructed on $L_2(Q)$ and $W_{2,\text{bd}}^{+l}(Q)$, V is a dense subset of $W_{2,\text{bd}}^{-l}(Q)$. Similar problem, e.g., describes the water transport in unsaturated soil during the drip irrigation. In this case the right-hand side has the following form:

$$f(x,t) = \sum_{i=1}^{m} Q_i \delta(x - x_i).$$

In conclusion, let us consider the problem of construction of the basic space Σ for one-parametric family of operators A_λ, $\lambda \in I$. Each operator A_λ induces on E the metric $\rho_\lambda^*(x,y) = \rho_F(A_\lambda(x), A_\lambda(y))$ and the completion E_λ^* in this metric. Denote by Σ the union of all spaces E_λ^*:

$$\Sigma = \bigcup_{\lambda \in I} E_\lambda^*.$$

Operator of natural embedding of E_λ^* into Σ induces mappings $f_\lambda : E_\lambda^* \to \Sigma$. We can define in Σ the strongest topology $\mathcal{T}$, for which all the mappings f_λ are continuous [8]. Then, the space Σ with the topology $\mathcal{T}$ plays the role of the basic space for the family A_λ, $\lambda \in I$.

7.8 Computation of Generalized Solution

Both classical and generalized solutions can be found exactly only in exceptional cases. One of the complications in this process is the difficulty of construction of the basic space Σ for a given class of nonlinear operators. In this connection, the problem consists of finding approximate solutions which are close to generalized solution $\bar{x}$ in the metric ρ^* or in the metric of basic space Σ. It is interesting that these approximate solutions belong to the original space E, since we may consider them as elements of near solution $\bar{x}_n$ with limit element $\bar{x}$. Thus, to compute approximate solutions x_n we may not need to define the basic space Σ. It is sufficiently only to prove that their images $y_n = A(x_n)$ converge to the right-hand side y of (7.1) in the metric of space F as $n \to \infty$. At a first glance, the problem of constructing the approximate solutions x_n appears to be quite simple: we have to select a sequence y_n from the range $R(A)$, which converges to y, and then the elements $x_n = A^{-1}(y_n)$ would be approximate solutions of the operator equation $A(x) = y$. For that purpose, let us introduce the following definition.

Definition 7.31. Let ε be an arbitrary positive number. We will refer to an element x_ε in the space E as ε-approximation of the generalized solution $\bar{x}$ of the operator equation $A(x) = y$, if $\rho_F(A(x_\varepsilon), y) = \rho_F(A(x_\varepsilon), \bar{A}(\bar{x})) < \varepsilon$.

However, careful analysis of computation of ε-approximations shows great deficiencies of such a "direct approach". First, in many cases it is difficult or impossible to describe the range $R(A)$ of an operator A in a space F. Therefore, it is not clear how to select a sequence y_n converging to the element y in F. Second, even if elements $x_n = A^{-1}(y_n)$ are known exactly, computation of the elements $y_n \in R(A)$ frequently is a very hard problem. That is why it is necessary to find a more effective way of constructing approximate solutions x_n of the operator equation $A(x) = y$ when $y \notin R(A)$. Let us describe one method which is realistic (although not necessarily optimal) for constructing of a sequence converging to the solution in the metric $\rho^*(x, y)$. Note, that this sequence of ε-approximations converges only in the metric ρ^*, whereas in original metric ρ_E it will be divergent, since the near-solution x_n does not converge to any element in the metrics ρ_E. Thus, the proposed approximate solution differs from all other approximations of exact solution of the operator equation (7.1).

Denote by E a separable metric space. Let $S = (a_1, a_2, \ldots)$ be a countable everywhere dense set in E. Let us introduce the following notation: $b_n = A(a_n)$, $n \in \mathbb{N}$, $B = (b_1, b_2, \ldots, b_n, \ldots)$. It is easy to see that B is everywhere dense in F if A is a continuous operator and $R(A)$ is everywhere dense in F. Indeed, let $\varepsilon > 0$ and y be a fixed element (the right-hand side of (7.1)), then there exists an element $\widetilde{y} \in R(A)$ such that $\rho_F(y,\widetilde{y}) < \frac{\varepsilon}{2}$. Let $\widetilde{x} = A^{-1}(y)$. Since A is a continuous operator, there is a number $\delta > 0$ such that $\rho_E(\hat{x},\widetilde{x}) < \delta$ implies $\rho_F(A(\hat{x},\widetilde{x})) < \frac{\varepsilon}{2}$. Since S is everywhere dense in E, there exists a subsequence a_{n_k} of S converging to $\widetilde{x}$. Hence, there exists such an element a_{n_k} for that $\rho_E(\widetilde{x}, a_{n_k}) < \delta$, and therefore

$$\rho_F(y,b_{n_k}) \leq \rho_F(y,\widetilde{y}) + \rho_F(y,b_{n_k}) \leq \frac{\varepsilon}{2} + \frac{\varepsilon}{2} = \varepsilon.$$

Denote by a_{n_1} the first element of S, for which

$$\rho_F(y,b_{n_1}) = \rho_F(y,A(a_{n_1})) < 1.$$

Existence of the element a_{n_1} follows from the density of B in F. Let a_{n_2} be the next element in S after a_{n_1} such that

$$\rho_F(y,b_{n_2}) = \rho_F(y,A(a_{n_2})) < \frac{1}{2}.$$

(existence of a_{n_2} follows from the density of the set $\{b_{n_1}, b_{n_1+1}, \ldots\}$ in F), and so on. Let a_{n_k} be the next element in S after a_{n_k-1} such that

$$\rho_F(y,b_{n_k}) = \rho_F(y,A(a_{n_k})) < \frac{1}{k}. \tag{7.5}$$

Thus, we obtain a sequence of the elements a_{n_k}, which is a near-solution whose limit element is the generalized solution $\bar{x}$ corresponding to the right-hand side y of (7.1). Note, that in constructing this near-solution we did not resort to the operator A and did not check whether $y_k = A(x_k) \in R(A)$.

One of the ways to find the near-solution x_n using this procedure is the Monte-Carlo method [109]. Indeed, we can number the elements of the everywhere countable dense subset S in E not only by natural numbers, but also by rational numbers from the segment $[0,1]$, i.e. map a rational number $r \in [0,1] \cap \mathbb{Q} = Q_{[0,1]}$ to an element $a_r \in S$. (In some cases such enumeration is more suitable than the enumeration by natural numbers.) In the set $Q_{[0,1]}$ (more precisely, in the class $S_{Q_{[0,1]}}$ of all subsets of $Q_{[0,1]}$) we can introduce the uniform distribution of probabilities $p(M), M \in S_{Q_{[0,1]}}$ so that

$$p(r \in (\alpha,\beta)) = \beta - \alpha, \quad 0 \leq \alpha, \beta \leq 1$$

(note that this distribution of probabilities p is not a measure!).

Randomly selecting rational numbers from the segment $[0,1]$ according to the distribution p we can select elements a_r from S and construct a near solution as above: the first element x_1 of the near-solution is the first element a_r, which under random sampling r from $Q_{[0,1]}$ satisfies the inequality $\rho_F(y,A(a_r)) < 1$, the second element x_2 is an element a_r satisfying the inequality $\rho_F(y,A(a_r)) < \frac{1}{2}$ and so on, the kth element x_k is an element a_r satisfying the inequality $\rho_F(t,A(a_r)) < \frac{1}{k}$. It is easy to see that the sequence x_k obtained as a result of repetition of this random experiment with probability 1 converges to the generalized solution $\bar{x}$ in the metric ρ^*.

If in the spaces E and F there are additional structures besides the structure of a metric space (for example, structure of Hilbert space) a near-solution may be found with the help of modified Galerkin method.

Remark 7.39. The theory of generalized solutions of nonlinear operator equations can be extended from the metric spaces E and F onto uniform spaces introduced by A.Weil. In this case, the analogue of the metric $\rho^*(x,y) = \rho_E(A(x),A(y))$ is a pre-image of the uniform structure of the space F in the space E with respect to injective uniformly continuous mapping $A : E \to F$.

7.9 Uniform Structures and Generalized Solutions of Operator Equations

We developed the abstract theory of generalized solutions of linear operator equations in previous chapters and obtained the series of results for equations with operators defined in metric spaces. In this section we give a brief description of the theory of generalized solutions in uniform spaces basing on [69, 106].

7.9.1 Definition of a Generalized Solution of Operator Equation

Let $(E, \mathscr{U}_E)$ and $(F, \mathscr{U}_F)$ be Hausdorff uniform spaces, $A : E \to F$ be injective operator, whose domain $D(A)$ coincides with the whole set E and range $R(A)$ is dense in F in topology $\mathscr{T}_F$ induced by uniform structure $\mathscr{U}_F$[1]

Let us consider an operator equation

$$A(u) = h, \tag{7.6}$$

where $h \in F$; if $h \in R(A) \subseteq F$, then there exists a unique solution $u \in E$ of (7.6), which we will call a classical solution. If $h \in F \backslash R(A)$, then equation (7.6) has not classical solution. There is a need to extend the notion of a solution of an operator equation (7.6) and to introduce generalized solutions.

Let us pass to definition of generalized solution of (7.6).

Consider the following sets on a Cartesian square $E \times E$:

$$\{(u,v) \in E \times E : (A(u),A(v)) \in O\}, \quad O \in \mathscr{U}_F. \tag{7.7}$$

---

[1] Hereinafter, all topological notions are meant respectively to a topology induced by uniform structures, and all uniform structures introduced are meant Hausdorff. By completion of a uniform space we mean a Hausdorff completion.

The set of all subsets having the form (7.7) is a base of Hausdorff uniform structure. We denote it by $\mathscr{U}_A$ (original with respect to A and uniform structure $\mathscr{U}_F$). The uniform structure $\mathscr{U}_A$ is the weakest uniform structure in E. The operator $A : E \to F$ is uniformly continuous with respect to this uniform structure (F is equipped with the uniform structure $\mathscr{U}_F$).

It is easy to understand that the operator A realizes an isomorphism $(E, \mathscr{U}_A)$ in $(F, \mathscr{U}_F)$. Denote by $(\bar{E}_A, \bar{\mathscr{U}}_A)$ and $(\bar{F}, \mathscr{U}_{\bar{F}})$ completions $(F, \mathscr{U}_F)$ and $(F, \mathscr{U}_F)$, respectively. Since the set $R(A)$ is dense in F in topology $\mathscr{T}_F$, then the completion of the space $\left(R(A), \widetilde{\mathscr{U}}_F \right)$, where $\widetilde{\mathscr{U}}_F = \mathscr{U}_F \cap (R(A) \times R(A))$ is a uniformity induced on $R(A)$ by the uniformity $\mathscr{U}_F$, can be equate with the space $(\bar{F}, \mathscr{U}_{\bar{F}})$.

Let us consider the operator $\bar{A} : \bar{E}_A \to \bar{F}$ which is a uniformly continuous extension of the operator $A : E \to F$.

Recall that the operator $\bar{A}$ is defined in the following way. Let u be an arbitrary element from $\bar{E}_A$, i.e., it is a class of $\mathscr{U}_A$-equivalent $\mathscr{U}_A$-Cauchy nets of nets consisting of elements of the space E [2]. The net $(A(u_\alpha))$ is a $\mathscr{U}_F$-Cauchy net with respect to $(u_\alpha) \in u$. Consider the class $h \in \bar{F}$ consisting the net $(A(u_\alpha))$. Let us put $\bar{A}(u) = h$.

The operator $\bar{A}$ is a uniform isomorphism of the uniform spaces $(\bar{E}_A, \bar{\mathscr{U}}_A)$ and $(\bar{F}, \mathscr{U}_{\bar{F}})$. Therefore, for any $\bar{h} \in \bar{F}$ the operator equation

$$\bar{A}(u) = \bar{h} \tag{7.8}$$

has a unique solution $u \in \bar{E}_A$. Moreover, (7.8) is correctly solvable. In other words, for any surrounding $O \in \mathscr{U}_{\bar{F}}$ there is a surrounding $V \in \bar{\mathscr{U}}_A$ such that the inclusion $(h', h'') \in O$ implies the inclusion $(u', u'') \in V$, where $\bar{A}(u') = h'$ and $\bar{A}(u'') = h''$.

Definition 7.32. A generalized solution of an operator equation $A(u) = h$ is such an element $u \in \bar{E}_A$ that $\bar{A}(u) = h$.

Obviously, a classical solution of (7.6) is generalized. If u is a generalized solution of (7.6) and $h \in R(A)$ (or $u \in E$), then u is a classical solution.

Theorem 7.4. *For any element $h \in F$ there exists a unique generalized solution of an equation $A(u) = h$.*

Proof. The operator $\bar{A}$ realize a uniform isomorphism between uniform spaces $(\bar{E}_A, \bar{\mathscr{U}}_A)$ and $(\bar{F}, \mathscr{U}_{\bar{F}})$. $\qquad\square$

Remark 7.40. The following equivalent definition of a generalized solution of equation (7.6) has significant importance in applications: an element $u \in \bar{E}_A$ is called a generalized solution of equation $A(u) = h$, if there exists a net (u_α) of elements of space E such that $\lim\limits_{\alpha} A(u_\alpha) = h$ in $\bar{F}$. Note that the net (u_α) is $\mathscr{U}_A$-convergent to $u \in \bar{E}_A$.

---

[2] Two $\mathscr{U}_0$-Cauchy nets (u_α), (v_β) of elements of a uniform space $(E_0, \mathscr{U}_0)$ are called $\mathscr{U}_0$-equivalent, if for any symmetric surrounding $O \in \mathscr{U}_0$ there exist α_0, β_0 such that $(u_\alpha, v_\beta) \in O$ as soon as $\alpha \geq \alpha_0$, $\beta \geq \beta_0$.

Remark 7.41. Let $\mathscr{A} = \{A_\lambda\}_{\lambda \in \Lambda}$ be a one-parameter family of injective operators $A_\lambda : E \to F$. Consider a uniform structure, which is a $\mathscr{U}_{\mathscr{A}}$-initial uniform structure with respect to the family $\mathscr{A}$ in E [8]. The completion E by the uniform structure $\mathscr{U}_{\mathscr{A}}$ is a natural object for construction of the theory of generalized solvability of the family od equations $A_\lambda (u) = h, \lambda \in \Lambda$.

7.9.2 Generalized Solutions and Embeddings of Uniform Spaces

Assume that the operator A is continuous. Then the spaces $(E, \mathscr{U}_E)$ and $(\bar{E}_A, \bar{\mathscr{U}}_A)$ are related in the following way.

Theorem 7.5. *Let $A : E \to F$ be a continuous injective operator. Then a uniform space $(E, \mathscr{U}_E)$ is continuously and densely embedded into a uniform space $(\bar{E}_A, \bar{\mathscr{U}}_A)$.*

Proof. The operator $A : E \to F$ specifies a dense and continuous embedding $(E, \mathscr{U}_E)$ in $(F, \mathscr{U}_F)$. The operator $\bar{A} : \bar{E}_A \to \bar{F}$ is a uniform isomorphism between the spaces $(\bar{E}_A, \bar{\mathscr{U}}_A)$ and $(\bar{F}, \mathscr{U}_{\bar{F}})$. Let $j : F \to \bar{F}$ be a canonical embedding of the space $(F, \mathscr{U}_F)$ into $(\bar{F}, \mathscr{U}_{\bar{F}})$.

$$
\begin{array}{ccc}
F & \xrightarrow{\;j\;} & \bar{F} \\[2pt]
A \uparrow & & \bar{A}^{-1} \downarrow \\[2pt]
E & \xrightarrow{\;i\;} & \bar{E}_A.
\end{array}
$$

Then the operator $i = \bar{A}^{-1} \circ j \circ A$ specifies a dense and continuous embedding of the space $(E, \mathscr{U}_E)$ into the space $(\bar{E}_A, \bar{\mathscr{U}}_A)$ (see the commutative diagram). □

The completion $(\bar{E}, \mathscr{U}_{\bar{E}})$ of the space E with respect to the uniformity $\mathscr{U}_E$ plays a significant role in studying of solvability of (7.6). It is quite natural to try to consider an element $u \in \bar{E}$ as a generalized solution of the equation $A (u) = h, h \in F$, if there exists a net (u_α) of elements in E such that $\lim_\alpha u_\alpha = u$ in $\bar{E}$ and $\lim_\alpha A (u_\alpha) = h$ in $\bar{F}$.

Remark 7.42. In studying equations with operators, which act in metric spaces, we called such solutions pseudo-generalized.

However, to obtain a theorem on correct solvability of equations in the space $\bar{E}$ which is similar to Theorem 7.5 it is needed that the operator $A : E \to F$ allows an extension by continuity up to uniform isomorphism between the spaces $(\bar{E}, \mathscr{U}_{\bar{E}})$ and $(\bar{F}, \mathscr{U}_{\bar{F}})$.

Statement 7.3 ([8]). *Let $A : E \to F$ be a uniformly continuous operator. Then the operator A can be uniquely extended up to uniformly continuous operator A', which acts from $\bar{E}$ to $\bar{F}$.*

As we mentioned above, the continuity of the operator $A : E \to F$ is insufficient for the existence of a continuous extension $A' : \bar{E} \to \bar{F}$. In addition, if nevertheless such extension exists then it cannot be an injective operator, even if the operator $A : E \to F$ is injective [8].

Let us formulate a criterium of injectivity of an extension by continuity $A' : \bar{E} \to \bar{F}$ of an injective uniformly continuous operator $A : E \to F$.

Theorem 7.6. *Let $A : E \to F$ be a uniformly continuous injective operator and $A' : \bar{E} \to \bar{F}$ is a uniformly continuous extension of the operator A onto the completion $\bar{E}$ of the space E. The operator $A' : \bar{E} \to \bar{F}$ is injective iff the following condition holds*

(π) *if nets (u_α), (v_β) of elements E are $\mathscr{U}_E$-Cauchy nets and $\mathscr{U}_A$-equivalent, then they are $\mathscr{U}_E$-equivalent.*

Proof. Let $A' : \bar{E} \to \bar{F}$ be an injective operator and (u_α), (v_β) be two $\mathscr{U}_E$-Cauchy and $\mathscr{U}_A$-equivalent nets of elements in the space E. Let us put

$$\bar{u} = \lim_\alpha u_\alpha \in \bar{E}, \ \bar{v} = \lim_\beta v_\beta \in \bar{E}, \ h' = \lim_\alpha A(u_\alpha) \in \bar{F}, \ k' = \lim_\beta A(v_\beta) \in \bar{F}.$$

Since

$$\forall O_F \in \mathscr{U}_F \ \exists \, \alpha_0, \ \exists \, \beta_0 : \ \forall \, \alpha \ge \alpha_0, \forall \, \beta \ge \beta_0 \ (A(u_\alpha), A(v_\beta)) \in O_F,$$

then $h' = k'$.

The injectivity of the operator $A' : \bar{E} \to \bar{F}$ and equalities $A'(\bar{u}) = h'$, $A'(\bar{v}) = k'$ imply that $\bar{u} = \bar{v}$, whence

$$\forall O_E \in \mathscr{U}_E \ \exists \, \alpha_0', \ \exists \, \beta_0' : \ \forall \, \alpha \ge \alpha_0', \forall \, \beta \ge \beta_0' \ (u_\alpha, v_\beta) \in O_E,$$

i.e., the nets (u_α), (v_β) are $\mathscr{U}_E$-equivalent.

Let the conditions π) and $A'(\bar{u}) = A'(\bar{v})$ hold for $\bar{u} \in \bar{E}$ and $\bar{v} \in \bar{E}$. Consider two $\mathscr{U}_E$-Cauchy nets of elements in the space E (u_α) and (v_β) such that $\bar{u} = \lim_\alpha u_\alpha \in \bar{E}$, $\bar{v} = \lim_\beta v_\beta \in \bar{E}$. The nets (u_α) and (v_β) are $\mathscr{U}_A$-equivalent, since $A'(\bar{u}) = \lim_\alpha A(u_\alpha)$, $A'(\bar{v}) = \lim_\beta A(v_\beta)$ in $\bar{F}$. By virtue of the condition π) the nets (u_α) and (v_β) are $\mathscr{U}_E$-equivalent, therefore, $\bar{u} = \bar{v}$ in $\bar{E}$ and the operator A' is injective.   □

Let us consider the following property of the extension $A' : \bar{E} \to \bar{F}$.

Definition 7.33. The extension $A' : \bar{E} \to \bar{F}$ of the operator $A : E \to F$ is called *weakly injective*, if $A'(u') \ne A'(u'')$ for all u' and u'' such that $u' \ne u''$ and at least one of the elements u' or u'' belong to the space E.

Note that in this definition we equate an element $u \in E$ with a class of $\mathscr{U}_E$-equivalent $\mathscr{U}_E$-Cauchy nets, which contains a stationary $(u, u, \ldots)$.

Theorem 7.7. *Let $A : E \to F$ be a uniformly continuous injective operator and $A' : \bar{E} \to \bar{F}$ be a uniformly continuous extension of the operator A onto the completions*

$\bar{E}$ of the space E. The extension $A' : \bar{E} \to \bar{F}$ is weakly injective iff the following condition holds:

(π^*) if the set (u_α) of elements E is $\mathscr{U}_E$-Cauchy and $\mathscr{U}_A$-convergent to $u \in E$, then it is $\mathscr{U}_E$-convergent to $u \in E$.

Proof. Let the extension $A' : \bar{E} \to \bar{F}$ be weakly injective and $u \in E$, and the net (u_α) of elements in E be $\mathscr{U}_E$-Cauchy and $\mathscr{U}_A$-convergent to u. Let us put

$$\bar{u} = \lim_\alpha u_\alpha \in \bar{E}, \ h' = A'(\bar{u}) = \lim_\alpha A(u_\alpha) \in \bar{F}, \ k' = A'(u) = A(u) \in \bar{F}.$$

Since

$$\forall O_F \in \mathscr{U}_F \ \exists \alpha_0 \ \forall \alpha \geq \alpha_0 \ \ (A(u_\alpha), A(u)) \in O_F,$$

then $h' = k'$. The fact that $A' : \bar{E} \to \bar{F}$ is weakly injective implies that $\bar{u} = u$, i.e., the net (u_α) is $\mathscr{U}_E$-convergent to $u \in E$.

Let the condition π^* holds and $A'(\bar{u}) = A'(u) = A(u)$ for $\bar{u} \in \bar{E}$ and $u \in E$. Let us consider an arbitrary net $(u_\alpha) \in \bar{u}$. The net (u_α) is $\mathscr{U}_E$-Cauchy and $\mathscr{U}_A$-convergent to $u \in E$. The condition π^* implies that $u = \lim_\alpha u_\alpha$ in E. Therefore, $\bar{u} = u$ in $\bar{E}$, and the operator A' is weakly injective. $\square$

Remark 7.43. The condition $\pi)$ implies $\pi^*)$. In general case these conditions are not equivalent.

We saw repeatedly that in the studying of generalized solutions an important problem of embedding of completions E with respect to two uniformities $\mathscr{U}_E$ and $\mathscr{U}_A$ arises. Let us show that the embedding $\bar{E}$ into $\bar{E}_A$ exists iff the operator A' (i.e. the extension of uniformly continuous operator $A : E \to F$) is injective.

Recall the accurate definition of an embedding of a uniform space into a uniform space.

Definition 7.34. A uniform space $(E_0, \mathscr{U}_0)$ is to be said uniformly (continuously) embedded into a uniform space $(E_1, \mathscr{U}_1)$ by an embedding operator $j : E_0 \to E_1$, if j is a uniformly continuous (continuous) injective operator. Moreover, if the set $j(E_0)$ is dense in E_1, then there exists a dense embedding of E_0 into E_1 using the operator j[3].

Let $(E, \mathscr{U}_E)$ be a uniform space and $(\bar{E}, \mathscr{U}_{\bar{E}})$ be its completion by uniformity $\mathscr{U}_E$. Recall that the completion $\bar{E}$ consists of classes of $\bar{u}$ $\mathscr{U}_E$-equivalent and $\mathscr{U}_E$-Cauchy nets (u_α) of elements of the set E, and the base of uniformity $\mathscr{U}_{\bar{E}}$ is given by the sets

$$\bar{O} = \{(\bar{u}, \bar{v}) \in \bar{E} \times \bar{E} : \exists (u_\alpha) \in \bar{u}, \ (v_\beta) \in \bar{v} \text{ such that } (u_\alpha, v_\beta) \in O\},$$

where O run over $\mathscr{U}_E$ [12].

---

[3] It should be stressed that the operator j in this and similar definitions plays extremely important role: if the space E_0 can be embedded into the space E_1 using different operators j, then the phrase "the space E_0 is embedded into the space E_1" is incorrect if there is no clear understanding for what exactly operator j is used.

Let us define the operator j_1 of canonical embedding of E into $\bar{E}$ in the following way: an element $u \in E$ is mapped to a class $j_1(u) \in \bar{E}$, which contains a stationary sequence $(u, u, ..., u, ...)$. Then the space E is densely and uniformly continuous embedded into its completion $\bar{E}$ using the operator of canonical embedding j_1.

Besides $\mathscr{U}_E$, we consider the uniformity $\mathscr{U}_A$ on E. The space $(E, \mathscr{U}_A)$ is embedded into $(\bar{E}_A, \mathscr{U}_A)$ using the similar operator of canonical embedding j_2.

If the operator $A : E \to F$ is uniformly continuous, then we can define a canonical mapping $\bar{E}$ into $\bar{E}_A$ in the following way: every class $\bar{u} \in \bar{E}$ is mapped to an element $\bar{\bar{u}} = j(\bar{u}) \in \bar{E}_A$, which is a class containing $\bar{u}$: $\bar{\bar{u}} \supseteq \bar{u}$. The map $j : \bar{E} \to \bar{E}_A$ is defined correctly. Indeed, let $\bar{u} \in \bar{E}$, (u_α) and (v_β) belong to $\bar{u}$. Then (u_α) and (v_β) are $\mathscr{U}_A$-Cauchy nets (an image of a $\mathscr{U}_E$-Cauchy net under a uniformly continuous mapping $A : E \to F$ is a $\mathscr{U}_F$-Cauchy net). In addition, the nets (u_α) and (v_β) are $\mathscr{U}_A$-equivalent. Indeed, let O_F be an arbitrary symmetrical surrounding from $\mathscr{U}_F$. The uniform continuity of $A : E \to F$ implies that there exists a surrounding $O_E \in \mathscr{U}_E$ such that $(A(u), A(v)) \in O_F$, if $(u,v) \in O_E$. Let us select α_0 and β_0 such that $(u_\alpha, v_\beta) \in O_E$ for $\alpha \geq \alpha_0$, $\beta \geq \beta_0$. Then $(A(u_\alpha), A(v_\beta)) \in O_F$. Therefore, every class $\bar{u}$ belongs to some class $\bar{\bar{u}} \in \bar{E}_A$, and the mapping j is defined correctly. The definition of canonical embedding j implies the density of the set in $j(\bar{E})$ is the space $\bar{E}_A$.

The relations between the canonical embeddings j_1, j_2 and the canonical embedding j are illustrated by the following commutative diagram.

$$
\begin{array}{ccc}
E & \xrightarrow{\ j_1\ } & \bar{E} \\
\| & & \downarrow{\scriptstyle j} \\
E & \xrightarrow{\ j_2\ } & \bar{E}_A.
\end{array}
$$

Definition 7.35. Let us say that the completion $(\bar{E}, \mathscr{U}_{\bar{E}})$ is canonically embedded into $(\bar{E}_A, \mathscr{U}_A)$, if the canonical embedding $j : \bar{E} \to \bar{E}_A$ is injective.

Theorem 7.8. *Let $A : E \to F$ be a uniformly continuous injective operator. The space $(\bar{E}, \mathscr{U}_{\bar{E}})$ is uniformly continuously and densely embedded into $(\bar{E}_A, \mathscr{U}_A)$ iff the following condition is satisfied:*

(π) *if the nets (u_α) and (v_β) of elements E are $\mathscr{U}_E$-Cauchy and $\mathscr{U}_A$-equivalent, then they are $\mathscr{U}_E$-equivalent.*

Proof. Let the condition π) holds and $\bar{u}$ and $\bar{v}$ are two different elements $\bar{E}$. Assume that $j(\bar{u}) = j(\bar{v}) = \bar{\bar{u}} \in \bar{E}_A$. Let us consider two nets $(u_\alpha) \in \bar{u}$ and $(v_\beta) \in \bar{v}$. Then, $(u_\alpha) \in j(\bar{u}) = \bar{\bar{u}}$ and $(v_\beta) \in j(\bar{v}) = \bar{\bar{u}}$. In other words, (u_α) and (v_β) are $\mathscr{U}_A$-Cauchy. The condition π) implies that the nets (u_α) and (v_β) are $\mathscr{U}_E$-equivalent. Therefore, $\bar{u} = \bar{v}$, that contradicts to the assumption.

Let us prove the necessity. Let the canonical mapping $j : \bar{E} \to \bar{E}_A$ be injective, and (u_α) and (v_β) be two $\mathscr{U}_E$-Cauchy and $\mathscr{U}_A$-equivalent nets of elements of the set E. Let $\bar{u}$ and $\bar{v}$ be elements $\bar{E}$ containing (u_α) and (v_β), respectively. By definition,

$j(\bar{u}) \in \bar{E}_A$: $j(\bar{u}) \supseteq \bar{u}$ and $j(\bar{v}) \in \bar{E}_A$: $j(\bar{v}) \supseteq \bar{v}$. Therefore, $(u_\alpha) \in j(\bar{u})$, $(v_\beta) \in j(\bar{v})$. From the other hand, the nets (u_α) and (v_β) are $\mathcal{U}_A$-equivalent. Then, $j(\bar{u}) = j(\bar{v})$; hence, by virtue of the injectivity of the mapping j the equality $\bar{u} = \bar{v}$ holds. Thus, (u_α) and (v_β) are $\mathcal{U}_E$-equivalent nets.   □

In a similar way we can prove the following general theorem on embedding of completions of uniform spaces.

Theorem 7.9. *Let $\mathcal{U}_0$, $\mathcal{U}_1$ be two uniformity on a set E; $\left(\bar{E}_0, \bar{\mathcal{U}}_0\right)$ and $\left(\bar{E}_1, \bar{\mathcal{U}}_1\right)$ be completions of E by uniformities $\mathcal{U}_0$ and $\mathcal{U}_1$, respectively. The space $\left(\bar{E}_0, \bar{\mathcal{U}}_0\right)$ is uniformly continuously and densely canonically embedded into $\left(\bar{E}_1, \bar{\mathcal{U}}_1\right)$ iff the following conditions are satisfied:*

(1) $\forall O_1 \in \mathcal{U}_1 \ \exists O_0 \in \mathcal{U}_0 : O_0 \subseteq O_1$.
(2)   If the nets (u_α) and (v_β) of elements E are $\mathcal{U}_0$-Cauchy and $\mathcal{U}_1$-equivalent, then they are $\mathcal{U}_0$-equivalent.

Remark 7.44. Let the operator $A^{-1} : R(A) \to E$ be uniformly continuous and the following condition holds: the facts that (u_α) and (v_β) are $\mathcal{U}_E$-equivalent and $(A(u_\alpha))$ and $\left(A(v_\beta)\right)$ $\mathcal{U}_F$-Cauchy imply that $(A(u_\alpha))$ and $\left(A(v_\beta)\right)$ $\mathcal{U}_F$-equivalent. Then the space $\bar{E}_A$ is uniformly continuously canonically embedded into $\bar{E}$. This condition in a nonlinear analogous of the known property of linear operators which allow closuring.

7.9.3 Examples of Generalized Solutions

Let us consider two known approaches to generalized solvability of linear operators equations from the stated point of view (see Chap. 2). Assume that E and F are Banach spaces with norms $\|\cdot\|_E$ and $\|\cdot\|_F$, respectively. An operator $A : E \to F$ is injective, linear and continuous, and the set $R(A)$ is dense in F. Then the adjoint operator $A^* : F^* \to E^*$ is injective and continuous. Moreover, the set $R(A^*)$ is dense in E^* in topology $\sigma(E^*, E)$.

Let us consider a Hausdorff uniform structure $\mathcal{U}_A$ in E with a base of surroundings induced by the sets

$$\{(u,v) \in E \times E : \|Au - Av\|_F < \varepsilon\}.$$

The uniformity of $\mathcal{U}_A$ is an pre-image of the uniformity induced by the strong topology of the space F with respect to the operator A.

Let us complete the the space E by the uniformity $\mathcal{U}_A$. Denote the corresponding completion by $\bar{E}_A$. The continuous extension of the operator A onto $\bar{E}_A$ is an isomorphism between the spaces $\bar{E}_A$ and F, more precisely, taking into account the fact that the uniformity above are normed, it is an isometry of $\bar{E}_A$ onto F. The completion of this kind and corresponding notion of a generalized solution of an operator equation

$$Au = h, \ h \in F, \tag{7.9}$$

are considered in Chaps. 2 and 3.

Let us consider a uniform structure $\mathcal{U}_F^\sigma$ induced by the weak topology $\sigma(F, F^*)$ on the subspace $R(A) \subseteq F$. Consider $R(A) \subseteq F$ on the subspace. Denote by $\mathcal{U}_A^\sigma$ a pre-image of $\mathcal{U}_F^\sigma$ with respect to the operator A. The base of uniformity $\mathcal{U}_A^\sigma$ consist of the sets

$$\bigcap_{k=1}^n \left\{ (u,v) \in E \times E : \left| \langle A^* \varphi_k, u - v \rangle_{E^*, E} \right| < \varepsilon_k \right\}, \ n \in \mathbb{N}, \ \varphi_k \in F^*, \varepsilon > 0.$$

Note that the topology induced by $\mathcal{U}_A^\sigma$ on E coincides with the weak topology $\sigma(E, R(A^*))$. Let $\widetilde{F}$ be a completion $R(A)$ with respect to $\mathcal{U}_F^\sigma$, and $\widetilde{E_A}$ is a completion of E with respect to $\mathcal{U}_A^\sigma$, and $\widetilde{A} : \widetilde{E_A} \to \widetilde{F}$ is an isomorphism of $\widetilde{E_A}$ onto $\widetilde{F}$ induced by the extension of A onto $\widetilde{E_A}$.

Let us show that the space F is continuously and densely embedded into $\widetilde{F}$. The spaces F and $\widetilde{F}$ are induced by the completion $R(A)$ using the corresponding uniform structures. To finish the proof it is necessary to show only that the facts that $h_n \in R(A)$ is strongly Cauchy sequence and converges to zero in the topology $\sigma(F, F^*)$ imply that $\|h_n\|_F \to 0$. Since the space F is Banach, then there exists $h \in F$ such that $\|h_n - h\|_F \to 0$ as $n \to \infty$. For every $h^* \in F^*$, we have

$$\langle h^*, h_n \rangle_{F^*, F} \to \langle h^*, h \rangle_{F^*, F} = 0.$$

Therefore, $h = 0$, i.e., $\|h_n\|_F \to 0$.

Note that $\widetilde{F} = F$ if F is a reflexive space.

We proved that the space $\widetilde{E_A}$ is densely and continuously embedded into the space $\bar{E}_A$ in Sect. 6.3.

The problem

$$u \in \widetilde{E_A} : \ \widetilde{A}u = h, \ h \in F. \tag{7.10}$$

is a general statement of (7.9).

For any $h \in F$ there exists a unique solution $u \in \widetilde{E_A}$ of the operator equation (7.10). Note that $u \in \widetilde{E_A}$ may be considered as an element, for which there exists a sequence $u_n \in E$ such that $\|Au_n - h\|_F \to 0$, $u_n \to u$ in $\widetilde{E_A}$. Indeed, if the set $R(A)$ is dense in the space F, then there exists a sequence $u_n \in E$ such that $\|h_n - h\|_F \to 0$ as $n \to \infty$ ($h_n = Au_n$). The strong uniformity of the space F majorizes the uniformity of the space $\widetilde{F}$. Then, $h_n \to h$ in $\widetilde{F}$. The operator $\widetilde{A}$ is an isomorphism of $\widetilde{E_A}$ onto $\widetilde{F}$, then there exists $u \in \widetilde{E_A}$ such that $u_n \to u$ in $\widetilde{E_A}$.

We shown that the definition of generalized solutions from the spaces $\widetilde{E_A}$ and $\bar{E}_A$ are equivalent in Chap. 7.

Thus, we have done the studying of generalized solvability of abstract operator equations.

7.9.4 Generalized Solution of Operator Equation
in Proximity Spaces

In the 1930s, V. Efremovich [13, 14] tried to describe spaces in which it is possible to define geometrical structure that allows to introduce the concept of uniform continuity of a map besides the uniform structures of A. Weil. These attempts were associated with the concept of proximity between two sets in connection with the following considerations. It is well known that one of the main concept of general topology is an adherent point of set; recall that a point x in a topological space E is called an adherent point of a set $A \subset E$ if every neighborhood V_x of the point x has a non-empty intersection with A: $A \cap V_x \neq \emptyset$. Thus, this concept characterizes the proximity between the point x and the set A. The map f between topological spaces E and F is continuous if it preserves this proximity, i.e. the image $y = f(x)$ of any adherent point x of the set A is an adherent point of the set $f(A)$: $y \in \overline{f(A)}$, where $\overline{f(A)}$ is a closure of $f(A)$. V.Efrempovich introduced a space where the concept of proximity between sets is defined and he called it a proximity space; this space like topological one consists of elements of arbitrary nature (points) and in this space it is possible to say whether two any subsets are proximal or not. In this case, there are maps like continuous maps between topological spaces. After the manner of Yu.Smirnov we will call them δ-maps (and the proximity spaces we will call δ-spaces). The map f which acts from δ-space P into δ-space Q is called a δ-map if it preserves the proximity between sets, i.e. any two proximal sets A and B from P are mapping by f into the sets $f(A)$ and $f(B)$ which are proximal in Q. If f is a bijective map of δ-space P onto the δ-space Q wherein the inverse map f^{-1} is a δ-map also then f is called a homeomorphism and the δ-spaces P and Q are called δ-homeomorphous.

Let us pass to the concise definition of the concept of a proximity space. After the manner of V.Efremovich let us call a set P a proximity space (δ-space) if for any two its subset it is defined whether they are proximal or not (in latter case they are said to be remote sets) such that the following conditions are satisfied:

1. If a set A is proximal to a set B then B is proximal to A.
2. The sum of sets A and B is proximal to a set C iff at least one one of the sets A or B is proximal to C.
3. Two points of a set P are proximal iff they are equal.
4. The entire set P is remote from the empty set.
5. For every remote sets A and B there exist sets C and D, such that $C \cup D = P$ and A remote from C and B is remote from D.

Substituting Condition 3 by weaker condition
3'. Every point $x \in P$ is proximal to itself
   we obtain general δ-spaces.

The examples of natural δ-spaces are metric and topological groups: sets A and B of a metric space P are proximal if the distance between them equals to zero:

$$\rho(A,B) = \inf\{\rho(x,y) : x \in A, y \in B\} = 0$$

Sets A and B in a topological group G are proximal if for any neighborhood U of the unit of the group G the intersection $UA \cap B$) $AU \cap B$, respectively) is not empty.

In any δ-space P, it is possible to introduce a topological structure considering that a set $A \subseteq P$ is closed if it contains all its proximal points. In this case all properties of topological space are easily checkable. For any δ-space P, it is possible to introduce the concept of a Cauchy sequence: it is such a sequence that any two its subsequences are proximal sets. As V. Efremovich stated, the it is naturally to apply the completion process with the help of Cauchy sequences to metrizable δ-spaces. In the most general case it is possible to point out the completion process considering generalized Cauchy sequences (nets) $\{x_\alpha\}$ where the index α runs over directed set. Unfortunately, these vain wishes and hopes associated with the process of completion of δ-spaces were not justified. As it was shown by Yu.Smirnov, the results were very unexpected and they do not agree with the classical concept of completion. In Chap. 2 of the paper [108], Yu.Smirnov defined a δ-extension of this space as a δ-space containing this space as a everywhere dense subset. It is naturally to consider a δ-space which have not any δ-extension different from it as complete spaces, as it was made in the report of Yu.Smirnov to Moscow mathematical society (March 11, 1952). However, as we will see hereinafter, this concept of a complete δ-space does not equivalent to the concept of metric completeness since (see Theorem 8 in Chap. 2 [108]) if to define the concept of completeness in a natural way a δ-space P will be complete iff it is bicompact (in own topology). If P is a Hausdorff space, it is just a compact space. This result contradicts to the classical theorems on completeness of almost all metric spaces which occur in topology and functional analysis. In this connection, Yu.Smirnov rejected from the term "completeness" and called completions "absolutely closed" δ-spaces. However, this does not change the matter.

Let us pass to investigation of generalized solution of equations with continuous maps (operators) which act in δ-spaces. Let A is a continuous operator which acts from a complete Hausdorff δ-space P into a Hausdorff δ-space Q. Since P is complete, hence it is a compact topological space, then by Tikhonov's Theorem this map is a homeomorphism, i.e. the inverse map $A^{-1} : Q \leftarrow P$ is continuous, therefore equations with such operators have not generalized solutions. From our point of view, the reason of this quite strange phenomenon consists in the fact that the properties 1–5 in the definitions of proximity spaces are not correct despite that they seem natural and obvious. This incorrectness is absent in the definition of uniform structures of A.Weil where the concept of completeness well conforms with the classical concept of metric completeness.

Chapter 8
Generalized Extreme Elements

In the previous chapters, we introduced and investigated the concept of a generalized solution of a linear operator equation. In this chapter, we will give a definition of generalized extreme elements of functionals, investigate the existence of generalized extreme elements of a convex continuous functional defined in a Banach space, and illustrate this concept by examples. The chapter contains also auxiliary results having independent significance. The presentation is based on the papers [74, 102].

8.1 Examples of Generalized Extreme Elements

Let E be a Banach space, M be a bounded and closed set in E, and φ be a bounded continuous function on E. We will say that the function φ attains a supremum (or infimum) on M if there exists such an element $x^* \in M$ ($x_* \in M$) that

$$\sup_{x \in M} \varphi(x) = \varphi(x^*) \qquad \left(\inf_{x \in M} \varphi(x) = \varphi(x_*) \right).$$

If E is an infinite-dimensional space, then M can be non-compact, so not each of the bounded continuous functions $x \mapsto \varphi(x)$ ($x \in M$) attains a supremum (or infimum) on M. For example, if E is a non-reflexive Banach space and $M = S_1(E) = \{x : \|x\|_E \leq 1\}$ is a unit ball in E, then there exists a linear continuous functional $f \in E^*$, which does not attain a supremum on the unit ball $S_1(E)$ (this fact is a well-known reflexivity criterion for a Banach space [40]).

In this connection, the following problem arises: to construct the extension by continuity for the function φ on $\bar{M}$ ($M \subset \bar{M}, \bar{M} \not\subset E$) such that the extended function $\bar{\varphi}$ attains a supremum (infimum) on generalized extreme elements $\bar{x}^*$ ($\bar{x}_*$), i.e.

$$\sup_{x \in M} \varphi(x) = \bar{\varphi}(\bar{x}^*) \qquad \left(\inf_{x \in M} \varphi(x) = \bar{\varphi}(\bar{x}_*) \right),$$

where $\bar{x}^*, \bar{x}_* \in \bar{M}$, but $\bar{x}^*, \bar{x}_* \notin M$.

D.A. Klyushin et al., *Generalized Solutions of Operator Equations and Extreme Elements*, Springer Optimization and Its Applications 55, DOI 10.1007/978-1-4614-0619-8_8, © Springer Science+Business Media, LLC 2012

Let us formulate a rigorous definition of a generalized extreme element. At first, note that a topology $\mathscr{T}$, matching the structure of the vector space E, induces a uniform structure on E, so hereinafter we will say about a completion $\bar{M}$ of the set M in the topology $\mathscr{T}$, having in mind the completion of M in corresponding uniform structure.

Definition 8.36. *A generalized extreme element* $\bar{x}^*$ $(\bar{x}_*)$ *of a bounded continuous functional* φ *on a bounded closed set* M *in a Banach space* E *is an element* $\bar{x}^*$ $(\bar{x}_*)$ *of a completion* $\bar{M}$ *of the set* M *by some Hausdorff topology* $\mathscr{T}$ *of the space* E *which has the following properties:*

(1) *The topology* $\mathscr{T}$ *agrees with the structure of the vector space* E.
(2) *The topology* $\mathscr{T}$ *is weaker than the original topology of the Banach space* E.
(3) *The functional* φ *is continuous on* M *in the topology* $\mathscr{T}$, *and*

$$\sup_{x \in M} \varphi(x) = \bar{\varphi}(\bar{x}^*) \qquad (\inf_{x \in M} \varphi(x) = \bar{\varphi}(\bar{x}_*)),$$

where $\bar{\varphi}$ *is an extension of* φ *by continuity on the set* $\bar{M}$ *in the topology* $\mathscr{T}$.

Note that the idea of extension of a solution of an extreme problem posed by D. Hilbert[1], was realized completely in 1930th years by L. Young [117] and E. McShane [71] in the case of one-dimensional problems of variational calculus in the form of "generalized curves" (see also multidimensional extensions in [118]). Similar constructions of extensions were proposed and investigated in the optimal control theory by R.V. Gamkrelidze [22] ("sliding regimes"), J. Warga [115, 116] ("generalized curves" and "generalized control functions"), E. McShane [71] ("relaxed controls"), A. Chouila-Houri [10] ("limit controls") and other mathematicians.

Let us study the issue of the existence of generalized extreme elements for convex continuous functionals on a Banach space, formulate some auxiliary results having independent significance also, and give examples of extreme elements.

Let E be a Banach space which is densely embedded into a Banach space F (i.e. there exists a linear continuous injective operator $k : E \to F$ with a dense range in F). Recall that in such case E is called compactly embedded into F, if the operator k is compact, i.e. a closure of an image of the unit ball $S_1(E)$ in E with respect to the metric F is a compact subset of F. In this case, a conjugate space F^* is embedded into E^* (with the help of the operator $k^* : F^* \to E^*$); therefore, for every continuous linear functional $f \in F^*$ the relation $f \in E^*$ makes sense. In addition, there exists such an element $\bar{x}^* \in F$, that the following equality holds:

$$\|f\|_{E^*} = \sup_{x \in S_1(E)} |f(x)| = \sup_{x \in \overline{S_1(E)}} |f(x)| = |f(\bar{x}^*)|, \qquad \bar{x}^* \in \overline{S_1(E)},$$

since the closure $\overline{S_1(E)}$ of the unit ball $S_1(E)$ in F is a compact set, and a narrowing of the functional $f \in F^*$ on $\overline{S_1(E)}$ is a continuous function. Thus, $\bar{x}^*$ is a generalized

---

[1] D.Hilbert posed his twentieth problem: "Do all variational problems with certain boundary conditions have solutions... if to use an extended interpretation of the solution?" in 1900.

maximal element, on which the functional f attains its norm. Note that in the case of a non-reflexive Banach space E there exist functionals f in an conjugate space E^*, which do not attain a supremum of the unit ball $S_1(E)$ by virtue of the James Theorem. However, if such a functional f belongs to the space F^* then there exists the corresponding generalized maximal element in the closure $\overline{S_1(E)}$ of the unit ball $S_1(E)$ (and even sphere) in topology of the space F. Later we will prove that all functionals $f \in E^*$ have a generalized maximal element in some Banach space.

If there exists a compact embedding $E \subset F$, then for $f \in F^*$ it is possible to point out some intermediate Banach space H ($E \subset H \subset F$), containing a generalized maximal element. Namely, let us consider a completion H of the space E with respect to the space F [41]. Recall that H is a set of elements $x \in F$, for which there are exist such a real number $R > 0$ and a sequence $x_n \in E$, $\|x_n\|_E \leq R$, that $x_n \to x$ in the norm of F. Let us define a functional on each element $x \in H$

$$\|x\|_H = \inf\{R \in \mathbb{R} : \exists x_n \in E, \|x_n\|_E \leq R, \|x_n - x\|_F \to 0\}.$$

Such functional $x \to \|x\|_H$ is a norm on the space H, the space H is complete, and the unit ball $S_1(H)$ in the space H is a closure of the unit ball $S_1(E)$ in the space F (in the norm F) [41]. Thus, the generalized maximal element x^* belongs to the unit ball $S_1(H)$ in the space H.

Let us generalize the previous example and find a generalized maximal element investigating a norm of a linear operator A, which acts from E into a Banach space G. Let E be densely embedded into F. Denote by E_F a vector space E with norm of the space F. Suppose that $A \in \mathscr{L}(E_F, G)$, where $\mathscr{L}(E_F, G)$ is a space of all bounded linear operators mapping E_F into G. Such operator A can be extended by continuity on the entire space F (this extension we will denote by $\bar{A} : F \to G$). By definition,

$$\|A\|_{E \to G} = \sup_{x \in S_1(E)} \|Ax\|_G = \sup_{x \in \overline{S_1(E)}} \|\bar{A}x\|_G,$$

where $\overline{S_1(E)}$ is a closure of the unit ball $S_1(E)$ in the norm of the space F.

Let us show that the functional $\varphi(x) = \|Ax\|_G$, where $x \in E$, is continuous in E_F. Indeed,

$$|\varphi(x) - \varphi(y)| = |\|Ax\|_G - \|Ay\|_G| \leq \|Ax - Ay\|_G$$
$$= \|A(x - y)\|_G \leq \|A\|_{E_F \to G}\|x - y\|_{E_F} = C\|x - y\|_{E_F},$$

where $x, y \in E$.

If E is compactly embedded into F, the the set $\overline{S_1(E)}$ would be compact in F; therefore, there would exists such an element $\bar{x}^* \in \overline{S_1(E)}$, that

$$\|A\|_{E \to G} = \sup_{x \in \overline{S_1(E)}} \|\bar{A}x\|_G = \sup_{x \in \overline{S_1(E)}} \bar{\varphi}(x) = \bar{\varphi}(\bar{x}^*), \qquad \bar{x}^* \in \overline{S_1(E)},$$

where $\bar{\varphi}$ is a continuous extension φ from E onto F.

Thus, in this case, the functional $\varphi(x)$ has a generalized extreme element $\bar{x}^* \in \overline{S_1(E)} \subset F$, on which it attains the norm of the operator $A \in \mathscr{L}(E,G)$. Later we will prove that in a reflexive Banach space E every compact operator A has such property.

8.2 Generalized Extreme Elements for Linear and Positively Homogeneous Convex Functional

In the previous section, we gave the examples of generalized extreme elements when the space E was compactly embedded into a Banach space F. In this connection the problem of the constructive construction of the space F arises. Note that we would take the space E^{**} (the second conjugate space) with a weak-* topology $\sigma(E^{**}, E^*)$. Then the space E would be densely and compactly embedded in the space F and the problem of the existence of generalized extremal elements would be solved. However, in this case the space F is not normed, and that would not be preferrable.

Theorem 8.1. *For every separable Banach space E there exists such a separable Banach space F, that E is densely and compactly embedded into F.*

Proof. Let us consider the space E^* that is conjugate to a separable space E and has weak-* topology $\sigma(E^*, E)$. The unit ball $S_1(E^*)$ in the space E^* is a separable in weak-* topology and metrizable space [7]. Let $\{f_n\}_{n=1}^\infty$ be a countable and everywhere dense subset in $S_1(E^*)$ with respect to the topology $\sigma(E^*, E)$. Let us construct an equable convex set $W^* \subset E^*$, which is compact with respect to the norm of the space E^* and absorbs every of the one-point sets f_n. For that, consider the set V^* consisting of the functionals f_n/n and $-f_n/n$, $n \in \mathbb{N}$. Since $f_n \in S_1(E^*)$, then the set $U^* = V^* \cup \{\Theta_{E^*}\}$, where Θ_{E^*} is a zero element of the space E^*, is compact in the strong metric E^*. Since U^* is a compact and symmetric set with respect to Θ_{E^*}, then a closure $W^* = \overline{\text{conv}\, U^*}$ of a convex hull of the set U^* in the norm E^* is equable and compact set in the metric E^* [40]. The topology $\sigma(E^*, E)$ is weaker than the original topology of the space E^*, so W^* is compact in weak-* topology $\sigma(E^*, E)$ also. Since $f_n/n \in U^* \subset W^*$, then the set W^*, clearly, absorbs all elements f_n. Since $U^* \subset S_1(E^*)$, then

$$W^* = \overline{\text{conv}U^*} \subset \overline{\text{conv}S_1(E^*)} = S_1(E^*)$$

and, therefore,

$$S_1(E) = (S_1(E^*))_E^\circ \subset (W^*)_E^\circ, \tag{8.1}$$

where $(A)_E^\circ$ is a polar of the set $A \subset E^*$ in the duality (E^*, E).

The inclusion (8.1) implies that the polar $(W^*)_E^\circ$ of the set W^* in the duality (E^*, E) is a neighborhood of zero Θ_E in the linear normed space E.

Let us show that the polar $(W^*)^\circ_E$ does not contain any linear manifold. Indeed, if to suppose that a linear set $\{\lambda x^* : \lambda \in (-\infty, \infty)\}$ belongs to $(W^*)^\circ_E$ for some $x^* \in E$ $(x^* \neq \Theta_E)$, then

$$\sup_{f \in W^*} |f(\lambda x^*)| \leq 1, \qquad \forall \lambda \in \mathbb{R};$$

hence, $f(x^*) = 0$ for all $f \in W^*$. However, there exists such a functional $f^* \in S_1(E^*)$ in E^* that $f^*(x^*) = \|x^*\|_E > 0$ and by virtue of the density of the subset $\{f_n\}^\infty_{n=1}$ in the unit ball $S_1(E^*)$ in the weak-* topology $\sigma(E^*, E)$ there exists a sequence f_{n_k} converging to f^* in the topology $\sigma(E^*, E)$, hence $f_{n_k}(x^*) \to f^*(x^*) > 0$. However, $f_{n_k}/n_k \in W^*$, so $(f_{n_k}/n_k)(x^*) = 0$ and $f_{n_k}(x^*) = 0$. We have reached a contradiction.

Now, we can see easily that the set $(W^*)^\circ_E$ can be taken as the unit ball in some normed topology, defined on E. Let $\|x\|_F$ be a norm on E, induced by the polar $(W^*)^\circ_E$ (denote the corresponding linear normed space by E_F). The embedding (8.1) implies that the norm $\|x\|_F$ is weaker than the original norm of the space E and the following equality holds:

$$\|x\|_F \leq \|x\|_E \qquad \forall x \in E.$$

Denote by F a Banach space obtained as a result of completion of the vector space E_F (with the norm $\|x\|_F$). Also, denote by $\widehat{(W^*)^\circ_E}$ the unit ball $S_1(F)$ in F. It is easy to see that the unit ball $S_1(E_F) = (W^*)^\circ_E$ in E_F is dense in the unit ball $S_1(F) = \widehat{(W^*)^\circ_E}$ in F, since the ball $S_1(F)$ is obtained by closuring of the set $S_1(E_F)$ in F.

Let us prove that $W^* \subset F^*$, i.e. every functional $f_0 \in W^* \subset E^*$ can be extended by continuity (in the norm of the space F) onto the entire space F. Indeed, since

$$(W^*)^\circ_E = \left\{ x \in E : \sup_{f \in W^*} |f(x)| \leq 1 \right\},$$

then for an arbitrary element $x \in (W^*)^\circ_E \subset E$ we have $|f_0(x)| \leq 1$. Since the polar $(W^*)^\circ_E$ is a unit ball $S_1(E_F)$, then $f_0 \in (E_F)^*$ and $\|f_0\|_{(E_F)^*} \leq 1$. Since E_F is a dense subset of the space F (i.e. f_0 is defined and continuous on a dense subset of the space F), then f_0 allows the extension by continuity onto the entire space F (with conservation of the norm). This proves the embedding $W^* \subset F^*$ (moreover, $W^* \subset S_1(F^*)$).

Let us prove that

$$S_1(F) = \widehat{(W^*)^\circ_E} = (W^*)^\circ_F. \tag{8.2}$$

Let $x \in S_1(F) = \widehat{(W^*)^\circ_E} \subset F$. Since $S_1(E_F)$ is dense in $S_1(F)$, then there exists a sequence $x_n \in S_1(E_F) = (W^*)^\circ_E$, such that $x_n \to x$ in F. Hence,

$$\sup_{f \in W^*} |f(x_n)| \leq 1.$$

If

$$\sup_{f \in W^*} |f(x)| > 1,$$

then there exists $f_0 \in W^*$, such that $|f_0(x)| > 1$. The latter inequality gives a contradiction, since $1 \geq |f_0(x_n)|$ and $f_0 \in F^*$, i.e. $f_0(x_n) \to f_0(x)$.

Hence,

$$\sup_{f \in W^*} |f(x)| \leq 1,$$

i.e. $x \in (W^*)_F^\circ$.

Now, let us prove that $(W^*)_F^\circ \subset S_1(F) = \widehat{(W^*)_E^\circ}$. Suppose the contrary. Let there exists $x \in (W^*)_F^\circ \subset F$, such that $\|x\|_F > 1$. Then $\|x\|_F > 1 + 2\varepsilon$ for some $\varepsilon > 0$ and $|f(x)| \leq 1$ for any $f \in W^*$. Let us consider the element $x' = x/(1+\varepsilon)$. Then

$$\|x'\|_F = \frac{\|x\|_F}{1+\varepsilon} > \frac{1+2\varepsilon}{1+\varepsilon} > 1 + \varepsilon_1$$

and

$$\sup_{f \in W^*} |f(x')| = \sup_{f \in W^*} \frac{|f(x)|}{1+\varepsilon} \leq \frac{1}{1+\varepsilon} < 1 - \varepsilon_2,$$

where $\varepsilon_1 > 0, \varepsilon_2 > 0$.

Since E_F is densely embedded into F, then there exists such a sequence $x_n \in E_F$, that $x_n \to x'$ in F and $\|x_n - x'\|_F < \varepsilon_2$ for any $n \in \mathbb{N}$. Then, since $W^* \subset S_1(F^*)$, then

$$\sup_{f \in W^*} |f(x_n)| = \sup_{f \in W^*} |f(x') + f(x_n - x')| \leq \sup_{f \in W^*} |f(x')| + \sup_{f \in W^*} |f(x_n - x')| \leq$$

$$\leq 1 - \varepsilon_2 + \sup_{f \in W^*} \|f\|_{F^*} \|x_n - x'\|_F \leq 1 - \varepsilon_2 + \varepsilon_2 = 1.$$

Thus, $x_n \in (W^*)_E^\circ = S_1(E_F)$ and since $x_n \to x'$ in F, then x' belongs to the closure of the set $(W^*)_E^\circ$ in F, i.e. $x' \in S_1(F)$. So, we have the contradiction: $\|x'\|_F > 1$. This way, we prove the inclusion $(W^*)_F^\circ \subset S_1(F) = \widehat{(W^*)_E^\circ}$. This finishes the proof of (8.2).

The unit ball $S_1(F^*)$ in the space F^* is a polar of the set $S_1(F) = \widehat{(W^*)_E^\circ} = (W^*)_F^\circ$ in the duality (F^*, F), since $S_1(F^*)$ is a bipolar $((W^*)_F^\circ)_{F^*}^\circ$ of the set W^* in the duality (F^*, F), where $(A)_{F^*}^\circ$ – is a polar of the set $A \subset F$ in the duality (F^*, F).

Since W^* is a convex equated set, then the bipolar $((W^*)_F^\circ)_{F^*}^\circ$ is a closure of W^* in the weak-* topology $\sigma(F^*, F)$. However, by construction $W^* \subset E^*$ is a compact set in the topology $\sigma(E^*, E)$, and since $W^* \subset F^*$, then W^* is a compact set in the topology $\sigma(F^*, E)$ also, hence W^* is closed in this topology also. Since the topology $\sigma(F^*, F)$ is stronger than $\sigma(F^*, E)$, then W^* is closed in the topology $\sigma(F^*, F)$, hence $((W^*)_F^\circ)_{F^*}^\circ = W^* = S_1(F^*)$.

By construction, the set $W^* = S_1(F^*)$ is compact with respect to the norm of the conjugate space E^*. Therefore, the set F^* is compactly embedded into the space E^*. By the Schauder theorem [119] the embedding $E \subset F$ is compact also.   □

Remark 8.45. Theorem 8.1 implies that there exists such a space F, that for any set of functionals $\{g_1, g_2, \ldots\} \subset E^*$, the space E is compactly embedded into F and the functionals $\{g_1, g_2, \ldots, \} \subset E^*$ are continuous in the norm of the space F.

Indeed, to construct the set W^* we can use the following subset which is countable and dense in $S_1(E^*)$ with respect to the weak-* topology $\sigma(E^*, E)$:

$$\left\{ f_1, \frac{g_1}{\|g_1\|_{E^*}}, f_2, \frac{g_2}{\|g_2\|_{E^*}}, \ldots \right\} \subset S_1(E^*).$$

In this case, the set W^* absorbs every of the functional $\{g_1, g_2, \ldots\} \subset E^*$; therefore, they are continuous in the norm of the space F.

Theorem 8.2. *The narrowing of any linear continuous functional defined of a unit ball of a separable Banach space has a generalized extreme (maximal) element, on which this functional attains its norm.*

Proof. Let E be a separable Banach space and $f \in E^*$. By Theorem 8.1 the set E is compactly embedded into some Banach space F using an operator $k : E \to F$. Let us introduce new norm in E:

$$\|x\|_f = |f(x)| + \|k(x)\|_F, \quad x \in E.$$

Let us denote by E_f and E_F the linear set E with norm $\|x\|_f$ and with norm $\|k(x)\|_F$, respectively. Then, for an arbitrary element $x \in E$ the following inequality holds:

$$\|x\|_f = |f(x)| + \|k(x)\|_F \leq \|f\|_{E^*} \|x\|_E + c\|x\|_E = M\|x\|_E,$$

where M does not depend on $x \in E$. Thus, the space E can be considered as naturally embedded into a completion $\bar{E}_f$ of the space E_f.

It is easy to see, that the functional f is linear and continuous in E_f, since $|f(x)| \leq \|x\|_f$, and hence f allows the extension by continuity onto the entire space $\bar{E}_f$.

There exist two alternatives: (1) the functional $f \in E^*$ is continuous in the norm $\|k(x)\|_F$ (i.e. $f \in (E_F)^*$), (2) the functional $f \in E^*$ is not continuous in the norm $\|k(x)\|_F$ ($f \notin (E_F)^*$). Note, that Remark 8.45 allows to construct such a space F, that E is compactly embedded into F and $f \in F^*$, i.e. to guarantee the first alternative. However, in some cases it is necessary to find a generalized extreme element in some a priori specified space F (only if E is compactly embedded into F). This fact make us study the second alternative also.

Under the first alternative we have that $|f(x)| \leq C\|k(x)\|_F$ and it is easy to see that the norm $\|k(x)\|_F$ and $\|x\|_f$ are equivalent and hence the spaces $\bar{E}_f$ and F are isomorphic. Indeed,

$$\|k(x)\|_F \leq \|x\|_f \leq C\|k(x)\|_F + \|k(x)\|_F = (C+1)\|k(x)\|_F, \quad \forall x \in E.$$

Denote by $\bar{f}(x)$ the extension by continuity of a functional $f(x)$ onto a Banach space F, and denote by $\overline{S_1(E)}$ the closure of the unit ball $S_1(E)$ in E with respect to the norm F. Then, by the Weierstrass Theorem there exists such an element x^* in a compact set $\overline{S_1(E)}$, on which the continuous functional $\bar{f}(x)$ attains its maximal value. This element is a generalized extreme element of the functional f.

Let us consider the second alternative: the functional $f \in E^*$ is unbounded with respect to the norm $\|k(x)\|_F$. In this case, in spite of the inequality $\|k(x)\|_F \leq \|x\|_f$ for all $x \in E$, the space $\bar{E}_f$ is not embedded (using the extension of the operator $l : E_f \to E_F$ of the natural embedding $l(x) = x, x \in E = E_f = E_F$) into a completion of the space E with respect to the norm $\|k(x)\|_F$, i.e. into the space F. Indeed, the condition π) does not hold for the norms $\|x\|_f$ and $\|k(x)\|_F$. Recall that the condition π) means the following implication: if a sequence $x_n \in E$ is Cauchy in the norm $\|\cdot\|_f$ and $k(x_n) \to \Theta_F$ as $n \to \infty$ in the norm F, then $x_n \to \Theta_f, n \to \infty$ in the norm $\|\cdot\|_f$. Since the functional $f(x)$ is unbounded in the norm $\|k(x)\|_F$, then there exists such a sequence $x_n \in E$, that $\|k(x_n)\|_F \leq 1$ and $f(x_n) \geq n$. Put

$$y_n = \frac{x_n}{f(x_n)} \in E.$$

Then,

$$\|k(y_n)\|_F = \frac{\|k(x_n)\|_F}{f(x_n)} \leq \frac{1}{n} \to 0, \qquad n \to \infty;$$

i.e. $k(y_n) \to \Theta_F, n \to \infty$.

Further,

$$\|y_n - y_m\|_f = \left| f\left(\frac{x_n}{f(x_n)} - \frac{x_m}{f(x_m)} \right) \right| + \left\| k\left(\frac{x_n}{f(x_n)} - \frac{x_m}{f(x_m)} \right) \right\|_F$$

$$= \|k(y_n - y_m)\|_F \to 0,$$

as $n, m \to \infty$. Thus, the sequence y_n is a Cauchy sequence in the norm $\|\cdot\|_f$. However,

$$\|y_n\|_f = |f(y_n)| + \|k(y_n)\|_F = \left| f\left(\frac{x_n}{f(x_n)} \right) \right| + \|k(y_n)\|_F = 1 + \|k(y_n)\|_F \to 1,$$

as $n \to \infty$.

Thus, y_n does not converge to Θ_f, i.e. the condition π) does not hold.

Nevertheless, the completion $\bar{E}_f$ of the set E in the norm $\|\cdot\|_f$ can be considered as a space $F \oplus \mathbb{R}$. More precisely, $\bar{E}_f$ is isometrically isomorphous to the space $F \oplus \mathbb{R}$. Indeed, let us consider a mapping j, which acts from E into $F \oplus \mathbb{R}$ by formula

$$j(x) = (k(x), f(x)), \qquad x \in E.$$

Since

$$\|x\|_f = \|j(x)\|_{F \oplus \mathbb{R}},$$

the operator j, mapping the linear space E with the norm $\|x\|_f$ onto $R(j) \subset F \oplus \mathbb{R}$ is an isometry. Let us show that the range $R(j)$ of the operator $j : E_f \to F \oplus \mathbb{R}$ is everywhere dense in the space $F \oplus \mathbb{R}$. Let (y, c) be an arbitrary element in $F \oplus \mathbb{R}$.

Since there exists a dense embedding $E \subset F$, then there exists a sequence $y_n \in E$, such that $k(y_n) \to y$, $n \to \infty$ in F. Since the functional f is unbounded on E_F by hypothesis, then there exists a sequence $x_n \in E$, such that

$$k(x_n) \to \Theta_F, \qquad f(x_n) \geq |f(y_n)| + n.$$

Let us consider elements

$$z_n = y_n + \alpha_n x_n \in E,$$

where the numerical sequence α_n satisfies the condition

$$f(z_n) = f(y_n) + \alpha_n f(x_n) = c,$$

i.e.

$$\alpha_n = \frac{c - f(y_n)}{f(x_n)}.$$

It is easy to see, that

$$\|k(z_n) - y\|_F = \left\| k(y_n) + \frac{c - f(y_n)}{f(x_n)} k(x_n) - y \right\|_F$$

$$\leq \|k(y_n) - y\|_F + \left| \frac{c - f(y_n)}{f(x_n)} \right| \|k(x_n)\|_F$$

$$\leq \|k(y_n) - y\|_F + \frac{|c - f(y_n)|}{f(y_n) + n} \|k(x_n)\|_F \to 0, \qquad n \to \infty.$$

Thus, the sequence z_n satisfies the condition $j(z_n) = (k(z_n), c) \to (y, c)$ in $F \oplus \mathbb{R}$. This implies that the range $R(j)$ of the operator j in $F \oplus \mathbb{R}$ is dense.

Extending the operator $j : E_f \leftrightarrow R(j)$ onto the entire space $\bar{E}_f$ by continuity, we conclude that the extended operator is an isometry between the Banach spaces $\bar{E}_f$ and $F \oplus \mathbb{R}$.

Since the operator $k : E \to F$ is compact, then the image of the unit ball $S_1(E)$ in the space E under the mapping

$$j(S_1(E)) \subset k(S_1(E)) \times [-\|f\|_{E^*}, \|f\|_{E^*}]$$

is a relatively compact set. Hence, the mapping j is compact.

Thus, the space E is compactly embedded into $\bar{E}_f$ and $f \in (\bar{E}_f)^*$. So, the functional f has a generalized maximal element. $\qquad \square$

Let E and F be Banach spaces and A be a bounded linear operator, which acts from E into F. We will say that the operator A attains its norm, if there exists such an element $x^* \in E$, that $\|x^*\|_E = 1$ and $\|Ax^*\|_F = \|A\|$.

Theorem 8.3. *Let E be a reflexive Banach space, and F be an arbitrary Banach space, and A be a compact linear operator, which acts from E into F. Then, A attains its norm.*

Proof. Consider the functional $f(x) = \|Ax\|_F$ defined on the set $x \in E$. Let us show that

$$\sup_{x \in S_1(E)} f(x) = \sup_{x \in S_1(E)} \|Ax\|_F = \|A\| = C.$$

Let $\{x_n\}$ be a sequence of the elements from the unit ball $S_1(E)$, such that $f(x_n) = \|y_n\|_F \to C$ as $n \to \infty$, where $y_n = A(x_n)$. Since the unit ball $S_1(E)$ in the reflexive Banach space E is a compact set in the weak topology $\sigma(E, E^*)$ [40], then by the Eberlein–Schmulian theorem [40] we can derive from the sequence $\{x_n\}$ a subsequence (denote it by $\{x_n\}$ again), which is weakly convergent to some element $x^* \in S_1(E)$. Since A is a compact operator, then $y_n = Ax_n$ converges to some element $y \in F$ in the norm of the space F, therefore, for any functional $f \in F^*$ we have

$$f(y) = \lim_{n \to \infty} f(Ax_n) = \lim_{n \to \infty} (A^*f)x_n = (A^*f)x^* = f(Ax^*),$$

i.e. $y = Ax^*$. Thus,

$$\|Ax^*\|_F = \lim_{n \to \infty} \|Ax_n\|_F = \lim_{n \to \infty} \|y_n\|_F = C.$$

$\square$

Remark 8.46. The compactness of the operator A is a very significant condition, since for any continuous operator A and reflexive Banach space E (and even F) Theorem 8.3 does not hold. Indeed, let $E = F = \ell_2$, and an operator $A : \ell_2 \to \ell_2$ is defined by the formula

$$y = Ax = \left(\frac{1}{2}x_1, \frac{2}{3}x_2, \ldots, \frac{n}{n+1}x_n, \ldots\right) \in \ell_2,$$

where $x = (x_1, x_2, \ldots) \in \ell_2$. It is clear that $\|A\| \leq 1$. From the other hand,

$$\|Ae_n\|_{\ell_2} = \left\|\left(0, \ldots, \frac{n}{n+1}, 0, \ldots\right)\right\|_{\ell_2} = \frac{n}{n+1} \to 1, \quad n \to \infty,$$

where $e_1 = (1, 0, 0, \ldots)$, $e_2 = (0, 1, 0, 0, \ldots)$ and so on. Thus, $\|A\| = 1$, but for any $x \in S_1(\ell_2)$ we have $\|Ax\| < 1$, i.e. the operator A does not attain its norm on the unit sphere.

Yet, if the Banach space E is reflexive, then, as it was proven by J. Lindenstrauss [49], the set of linear continuous operators, which act from E into a Banach space F and attain their norm, is strongly dense in the space $\mathcal{L}(E, F)$ of all linear continuous operators, which act from E into F. In [1, 98, 101, 105, 110, 121], some refinements of this statement were get.

Remark 8.47. If a bounded linear operator A maps a reflexive Banach space E into a reflexive Banach space F and it does not attain its norm on the unit sphere in the Banach space E, then it does not have a general maximal element in any locally convex topology $\mathcal{T}$. Indeed, suppose the contrary: let $\mathcal{T}$ be a locally convex topology (see Definition 8.36) and x^* be an element from $M = \overline{S_1(E)}$, such that

$\|\bar{A}x^*\|_F = \|A\|$, where $\bar{A}$ is an extension of the operator A onto the completion $\bar{E}$ of the space E with respect to the topology $\mathscr{T}$ by continuity. Since $S_1(E)$ is a compact set in the weak topology $\sigma(E, E^*)$, and the topology $\sigma(E, (E_{\mathscr{T}})^*)$ is weaker that the topology $\sigma(E, E^*)$ (since $(E_{\mathscr{T}})^* \subset E^*$), then $S_1(E)$ is compact in the topology $\sigma(E, (E_{\mathscr{T}})^*)$ also. Therefore, the ball $S_1(E)$ is compact in the space $\bar{E}_{\mathscr{T}}$ in topology $\sigma(\bar{E}_{\mathscr{T}}, (E_{\mathscr{T}})^*)$, where $\bar{E}_{\mathscr{T}}$ is a completion of E with respect to the topology $\mathscr{T}$. Whence, it follows that $S_1(E)$ is a closed set in $\bar{E}_{\mathscr{T}}$ in the topology $\sigma(\bar{E}_{\mathscr{T}}, (E_{\mathscr{T}})^*)$, therefore, $S_1(E)$ is a closed set in the topology $\mathscr{T}$, so that the closure $\overline{S_1(E)}$ of the unit ball $S_1(E)$ in the topology $\mathscr{T}$ coincides with $S_1(E)$. Thus, the extremal element $x^* \in \overline{S_1(E)}$ belongs to the set $S_1(E)$. This contradicts to the assumption that the operator A does not attain its norm on the unit ball $S_1(E)$.

8.3 On Compact Embedding into a Banach Space

In connection with results obtained in the previous section, let us consider the conditions under which a linear normed space can be densely and compactly embedded in a Banach space.

Recall the concept of embedding of a linear normed space E into a Banach space F.

Definition 8.37. A linear normed space E is said to be embedded into a Banach space F, if there exists a bounded injective linear operator $j : E \to F$ (embedding operator).

Among all embeddings dense and compact embeddings are the most important ones.

Definition 8.38. If the set $j(E)$ is dense in F, then the space E is said to be densely embedded into the space F.

Definition 8.39. If the operator $j : E \to F$ from Definition 8.37 is compact [2], the space E is said to be compactly embedded into F.

Let us consider spaces E and F. The investigated problem is the following:

When the space E is densely and compactly embedded into the Banach space F?

I.e., we must ascertain which conditions provide the existence of a bounded linear operator $j : E \to F$ with the following properties:

$$Ker(j) = \{0\}, \quad \overline{Im(j)} = F, \quad j \in K(E, F),$$

where $K(E, F)$ is a set of linear compact operators $E \to F$.

---

[2] Recall that a linear continuous operator $T : E \to F$ is called compact if it maps a closed unit ball $S_1(E)$ into the set $T(S_1(E)) \subseteq F$ with a compact closure.

Since the range of a linear compact operator is a separable linear subspace, then none of linear normed spaces E can be embedded densely and compactly into a non-separable Banach space F.

The following theorem demonstrates when a linear normed space E can be densely and compactly embed into a Banach space F.

Theorem 8.4. *Let E and F be infinite dimensional linear normed spaces, and a space F be a Banach space. Then the folowing statements are equivalent:*

(1) The space F is separable, the space E^* is separable in the topology $\sigma(E^*,E)$.
(2) The space E can be densely and compactly embedded into F.

Remark 8.48. The separability of the conjugate space E^* in a weak-* topology $\sigma(E^*,E)$ is equivalent to the existence of a countable total set $M \subseteq E^*$.

The proof of Theorem 8.4 is based on the following statement.

Lemma 8.1. *In a infinite-dimensional linear normed space E there exists a sequence of elements (x_n) such that:*

(1) $\sum\limits_{n=1}^{\infty} \|x_n\|_E < +\infty.$

(2) for $\alpha = (\alpha_n) \in \ell_\infty$ with $\sum\limits_{n=1}^{\infty} \alpha_n x_n = 0$ implies $\alpha_n = 0$ for all $n \in \mathbb{N}$.

Remark 8.49. Obviously, the elements x_n from Lemma 8.1 form a linearly independent system.

Proof (Lemma 8.1). Let us consider a linearly independent system of elements $y_n \in E$ such that $\|y_n\|_E = 1$ for all $n \in \mathbb{N}$. Put

$$m_n = \min\left\{ \left\| \sum_{k=1}^{n} c_k y_k \right\|_E : \tfrac{1}{2} \le \sum_{k=1}^{n} |c_k| \le n \right\}.$$

Obviously, $1 > m_n \ge m_{n+1} > 0$.

Let us define a sequence of elements $x_n \in E$ in the following way:

$$x_n = \lambda_n y_n,$$

where

$$\lambda_{n+1} = \frac{\lambda_n m_n}{2^3} \quad \forall n \in \mathbb{N}, \ \lambda_1 = 1.$$

Let us show that the sequence (x_n) has the desirable properties.
We have:

$$0 < \lambda_{n+k} = \frac{\lambda_{n+k-1} m_{n+k-1}}{2^3} = \frac{\frac{\lambda_{n+k-2} m_{n+k-2}}{2^3} m_{n+k-1}}{2^3}$$

$$< \frac{\lambda_{n+k-2} m_{n+k-2}}{2^{3+3}} < \frac{\lambda_n m_n}{2^{3k}}$$

$$< \frac{1}{2^{3k}} \quad \forall n, k \in \mathbb{N}.$$

Therefore,

$$\sum_{n=1}^{\infty} \|x_n\|_E = \sum_{n=1}^{\infty} \lambda_n \leq \sum_{n=1}^{\infty} \frac{1}{2^{3n}} < +\infty.$$

Statement 1 is proved.

Let us prove Statement 2. Let us consider an arbitrary sequence $\alpha = (\alpha_n) \in \ell_\infty \backslash \{0\}$. Let us show that

$$\sum_{n=1}^{\infty} \alpha_n x_n \neq 0.$$

Let $\bar{\alpha} = \sup_{n \in \mathbb{N}} |\alpha_n| > 0$ and $\bar{n} \in \mathbb{N}$ such that $|\alpha_{\bar{n}}| > \bar{\alpha}/2$. We have

$$\left\| \sum_{n=1}^{\infty} \alpha_n x_n \right\|_E \geq \left\| \sum_{n=1}^{\bar{n}} \alpha_n x_n \right\|_E - \left\| \sum_{n=\bar{n}+1}^{\infty} \alpha_n x_n \right\|_E. \tag{8.3}$$

Let us estimate the sums in the right-hand side of (8.3). We have

$$\left\| \sum_{n=1}^{\bar{n}} \alpha_n x_n \right\|_E = \left\| \sum_{n=1}^{\bar{n}} \alpha_n \lambda_n y_n \right\|_E$$

$$= \max_{1 \leq n \leq \bar{n}} |\alpha_n \lambda_n| \left\| \sum_{n=1}^{\bar{n}} \frac{\alpha_n \lambda_n}{\max_{1 \leq n \leq \bar{n}} |\alpha_n \lambda_n|} y_n \right\|_E$$

$$\geq \max_{1 \leq n \leq \bar{n}} |\alpha_n \lambda_n| \times m_{\bar{n}} \geq |\alpha_{\bar{n}} \lambda_{\bar{n}}| \times m_{\bar{n}}$$

$$> \frac{\bar{\alpha} \times \lambda_{\bar{n}} \times m_{\bar{n}}}{2}. \tag{8.4}$$

Further,

$$\left\| \sum_{n=\bar{n}+1}^{\infty} \alpha_n x_n \right\|_E \leq \bar{\alpha} \sum_{n=\bar{n}+1}^{\infty} \lambda_n \leq \bar{\alpha} \times \lambda_{\bar{n}} \times m_{\bar{n}} \sum_{k=1}^{\infty} \frac{1}{2^{3k}} < \frac{\bar{\alpha} \times \lambda_{\bar{n}} \times m_{\bar{n}}}{4}. \tag{8.5}$$

The estimations (8.4), (8.5), and (8.3) imply that

$$\left\| \sum_{n=1}^{\infty} \alpha_n x_n \right\|_E > 0.$$

Proof (Theorem 8.4). Let a space F be separable, a space E^* be separable in the topology $\sigma(E^*, E)$. Then there exists countable linearly independent sets $\{\phi_n\} \subseteq S_1(F)$, $\{\psi_n\} \subseteq S_1(E^*)$ with the following properties:

(1) The linear span of the set $\{\phi_n\}$ is dense in F.
(2) The set $\{\psi_n\}$ is total in E^*.

Let us construct sequences of elements f_n, e_n^*, respectively, that satisfy conditions of Lemma 8.1, using $\{\phi_n\}$ and $\{\psi_n\}$. Obviously, the linear span $\{f_n\}$ is dense in F also, and $\{e_n^*\}$ is total in E^*.

Let us consider the operator

$$j(x) = \sum_{n=1}^{\infty} \langle e_n^*, x \rangle f_n \quad \forall x \in E.$$

Let us show that $j \in K(E,F)$, $Ker(j) = \{0\}$ and $\overline{Im(j)} = F$, i.e. j is the operator of dense and compact embedding of E into F. Let us consider the following sequence of operators

$$j_m(x) = \sum_{n=1}^{m} \langle e_n^*, x \rangle f_n.$$

It is clear that j_m are bounded and linear operators and $\dim Im(j_m) < +\infty$. Therefore, $j_m \in K(E,F)$. Since

$$\|j - j_m\|_{E \to F} = \sup_{x \in S_1(E)} \|j(x) - j_m(x)\|_F$$

$$= \sup_{x \in S_1(E)} \left\| \sum_{n=m+1}^{\infty} \langle e_n^*, x \rangle f_n \right\|_F$$

$$\leq \sum_{n=m+1}^{\infty} \|e_n^*\|_{E^*} \|f_n\|_F \to 0 \quad \text{as} \quad m \to \infty,$$

then $j \in K(E,F)$.

Let $x \in E$ be such that

$$j(x) = \sum_{n=1}^{\infty} \langle e_n^*, x \rangle f_n = 0.$$

Then $\langle e_n^*, x \rangle = 0$ for all $n \in \mathbb{N}$. The totality of $\{e_n^*\}$ implies that $x = 0$. Therefore, $Ker(j) = \{0\}$.

Let us show that $\overline{Im(j)} = F$. Arguing by contradiction, assume that there exist $f^* \in F^* \backslash \{0\}$ such that $\langle f^*, j(x) \rangle = 0$ for all $x \in E$, i.e.,

$$\left\langle f^*, \sum_{n=1}^{\infty} \langle e_n^*, x \rangle f_n \right\rangle = \sum_{n=1}^{\infty} \langle e_n^*, x \rangle \langle f^*, f_n \rangle = \left\langle \sum_{n=1}^{\infty} \langle f^*, f_n \rangle e_n^*, x \right\rangle = 0 \quad \forall x \in E.$$

Thus, we have $\sum_{n=1}^{\infty} \langle f^*, f_n \rangle e_n^* = 0$. Hence, $\langle f^*, f_n \rangle = 0$ for all $n \in \mathbb{N}$. Since a linear span of the set $\{f_n\}$ is dense in F, then $f^* = 0$. This contradicts to the assumption $f^* \neq 0$.

Let there exist an operator $j \in K(E,F)$ such that $Ker(j) = \{0\}$ and $\overline{Im(j)} = F$. Since the subspace $Im(j) \subseteq F$ is separable and dense in F, then the space F is separable. Further, the conjugate space $Im(j)^*$ has a countable total subset. Since

$j \in L(E, Im(j))$, $Ker(j) = \{0\}$, then E^* has a countable total subset also. It is equivalent to the separability of the space E^* in the topology $\sigma(E^*, E)$.

Theorem 8.5. *Let E and F be infinite dimensional separable spaces, and the space F is a Banach space. Then the spaces E and E^* can be densely and compactly embedded into the space F.*

Proof. The assumption implies that the space E^* is separable in the topology $\sigma(E^*, E)$, and the space E^{**} be separable in the topology $\sigma(E^{**}, E^*)$. Next, it is necessary to apply Theorem 8.4. □

8.4 Generalized Extreme Elements for General Convex Functionals

Let F^* be a space conjugate to a Banach space F. By Banach–Alaoglu Theorem the ball $S_1(F^*)$ is compact in the topology $\sigma(F^*, F)$. This statement is true for any convex, closed, and bounded set $X \subset F^*$. Therefore, the minimization problem

$$f(x) \to \inf_{x \in X},$$

where the functional $f : F^* \to \mathbb{R}$ is lower semi-continuous in the topology $\sigma(F^*, F)$, has solutions. But sometimes it is necessary to consider extremal problems on convex and bounded subsets of Banach spaces, which are not isomorphous to conjugate spaces (for example, in the space $L_1(0, 1)$). As a rule, these problems have not solutions.

In this section we will show that any convex continuous functional defined on a convex, closed, and bounded subset of a Banach space has a generalized extreme element. In the previous section we studied generalized solutions compactly embedding a Banach space into another Banach space. Now, we will construct an isometrical and dense in a weak topology embedding of of the original Banach space into an conjugate Banach space. This space will depend on elements of extreme problems – functional and feasible set.

Let us cite some auxiliary results which provide a basis for the construction of generalized solutions of convex extreme problems proposed below.

Theorem 8.6. *Let X be a convex, bounded, and closed subset of a Banach space $(E, \| \times \|_E)$. Then there exists such a linear subspace $F \subset E^*$ such that:*

(1) $(F, \| \times \|_{E^*})$ is a separable linear normed space.
(2) F is a subspace of characteristic one, i.e.

$$\forall x \in E : \quad \|x\|_E = \sup_{y \in F \cap S_1(E^*)} |\langle y, x \rangle_{E^*, E}|.$$

(3) an arbitrary point $x \in X \setminus E$ is strictly Hausdorff from the set X by an element of the subspace F, i.e.

$$\forall x \in E \setminus X \ \exists y \in F : \ \langle y, x \rangle_{E^*, E} > \sup_{\tilde{x} \in X} \langle y, \tilde{x} \rangle_{E^*, E}.$$

Proof. Let us consider a set $M = \{x_n : n \in \mathbb{N}\}$ which is countable and everywhere dense in E. Let us construct a system of functionals $\{y_n\} \subset E^*$ such that

$$\forall n \in \mathbb{N} : \ \|y_n\|_{E^*} = 1, \ \langle y_n, x_n \rangle_{E^*, E} = \|x_n\|_E.$$

Denote by F_0 a linear span of the set $\{y_n : n \in \mathbb{N}\}$. Let $x \in E$. For any $\varepsilon > 0$ consider such an element $x_n \in M$ that $\|x - x_n\|_E < \varepsilon$. Then,

$$\langle y_n, x \rangle_{E^*, E} = \langle y_n, x_n + x - x_n \rangle_{E^*, E} = \|x_n\|_E +$$
$$+ \langle y_n, x - x_n \rangle_{E^*, E} > \|x_n\|_E - \varepsilon \geq \|x\|_E - 2\varepsilon.$$

Therefore, F_0 is a subset E^* of characteristics one.

Without loss of generality, we can consider that $\theta \in X$ and $X \subset S_r(E)$, where $r > 0$. Let us consider a set $M \setminus X = \{z_n\}$. Note that

$$\delta_n = \rho(z_n, X) = \inf_{x \in X} \|z_n - x\|_E > 0.$$

Now, let us construct a set for every $n \in \mathbb{N}$:

$$X_n = \bigcup_{x \in X} \left(x + S_{\delta_n/2}(E) \right).$$

Let μ_n be a Minkowski functional of the set X_n. Let us define the following functional on the linear subspace $L_n = \{\lambda z_n : \lambda \in \mathbb{R}\}$:

$$y_n'(\lambda z_n) = \lambda \qquad \forall \lambda \in \mathbb{R}.$$

From the fact that $y_n'(z_n) = 1$ and $z_n \in E \setminus X_n$ it follows that $y_n'(z_n) < \mu_n(z_n)$. By the Hahn–Banach Theorem there exists a functional $y_n'' \in E^*$ such that:

$$\langle y_n'', x \rangle_{E^*, E} = y_n'(x) \qquad \forall x \in L_n,$$
$$\langle y_n'', x \rangle_{E^*, E} \leq \mu_n(x) \qquad \forall x \in E,$$

Let us show that $\|y_n''\|_{E^*} \leq 2/\delta_n$. To do that let us majorize values μ_n on the ball $S_1(E)$. Since $S_{\delta_n/2}(E) \subset X_n$, then

$$\mu_n(x) \leq \mu_{S_{\delta_n/2}(E)}(x) \qquad \forall x \in E,$$

where $\mu_{S_{\delta_n/2}(E)}$ is a Minkowski functional of the set $S_{\delta_n/2}(E)$. By the definition of a Minkowski functional the inequality $\mu_{S_{\delta_n/2}(E)}(x) \leq \delta_n/2$ holds for $x \in S_1(E)$. It implies the estimation $\|y_n''\|_{E^*} \leq 2/\delta_n$.

Let us show that functionals from the set $\{y_n'' : n \in \mathbb{N}\}$ strictly separate points of the set $E \setminus X$ from the set X. Indeed, let $x \in E \setminus X$ and $\alpha = \rho(x,X) = \inf_{x' \in X} \|x - x'\|_E > 0$. Let us consider a point $z_n \in M \setminus X$ such that $\|z_n - x\|_E < \varepsilon$ for $\varepsilon \in (0,\alpha)$. Then for the functional $y_n'' \in E^*$ the following inequality holds

$$\langle y_n'', x \rangle_{E^*,E} = \langle y_n'', z_n \rangle_{E^*,E} + \langle y_n'', x - z_n \rangle_{E^*,E} \geq 1 - \|y_n''\|_{E^*} \|x - z_n\|_E > 1 - \frac{2}{\delta_n}\varepsilon.$$

Since

$$-\varepsilon < \alpha - \delta_n < \varepsilon, \tag{8.6}$$

we get the estimation

$$\langle y_n'', x \rangle_{E^*,E} > 1 - \frac{2\varepsilon}{\alpha - \varepsilon}. \tag{8.7}$$

Let us prove the following inequality:

$$\sup_{x' \in X} \langle y_n'', x' \rangle_{E^*,E} \leq 1 - C_n, \tag{8.8}$$

where $C_n = C(r, \delta_n)$ is a number from the interval $(0,1)$. Let us select a number $C_n \in (0,1)$ satisfying the condition

$$(1 - C_n)\left(1 + \frac{\delta_n}{2r}\right) > 1,$$

i.e.

$$C_n \in \left(0, \frac{\delta_n}{2r + \delta_n}\right). \tag{8.9}$$

Let us show that (8.9) implies (8.8). Indeed, if the inequality $\langle y_n'', x' \rangle_{E^*,E} > 1 - C_n$ holds for some point $x' \in X$, then $x'' = x' + \frac{\delta_n}{2r}x' \in X_n$. Therefore, $\langle y_n'', x'' \rangle_{E^*,E} \leq 1$. From the other hand, we have

$$\langle y_n'', x'' \rangle_{E^*,E} = \langle y_n'', x' \rangle_{E^*,E}\left(1 + \frac{\delta_n}{2r}\right) > (1 - C_n)\left(1 + \frac{\delta_n}{2r}\right) > 1.$$

So, we have a contradiction.

Taking into account (8.6), (8.8), and (8.9), we obtain the inequality

$$\sup_{x' \in X} \langle y_n'', x' \rangle_{E^*,E} \leq 1 - \frac{\alpha - \varepsilon}{2r + \alpha + \varepsilon}. \tag{8.10}$$

The inequalities (8.7) and (8.10) allow to conclude that selecting $\varepsilon > 0$ small enough for some y_n'' we can get the inequality

$$\langle y_n'', x \rangle_{E^*,E} > \sup_{x' \in X} \langle y, x' \rangle_{E^*,E},$$

which required to be proved.

Let us denote by F a linear span of the set $F_0 \cup \{y_n'' : n \in \mathbb{N}\}$. It is clear that the linear normed space $(F, \| \times \|_{E^*})$ is separable and other statements of the theorem are also true. $\square$

Remark 8.50. If the Banach space $(E, \| \times \|_E)$ is reflexive, then the closure of the linear subset $F \subset E^*$ constructed in Theorem 8.6 coincides with E^*, as a Banach space topologically conjugate to a reflexive Banach space does not contain any proper, closed and total linear subspaces.

Theorem 8.7. *Let X be a non-empty subset of a separable Banach space $(E, \| \cdot \|_E)$, $f : E \to \mathbb{R}$ be a continuous convex functional. Then there exist a linear subspace $F \subset E^*$ such that:*

(1) $(F, \| \cdot \|_{E^*})$ is a separable linear normed space.
(2) If for any $y \in F$ $\langle y, x_n \rangle_{E^*, E} \to \langle y, x \rangle_{E^*, E}$ as $n \to \infty$ $(x \in X, x_n \in X)$, then

$$f(x) \le \varliminf_{n \to \infty} f(x_n).$$

Proof. It is known that a continuous convex functional is locally Lipschitz; in addition, it has a subdifferential in the sense of convex analysis [15].

Let $\{x_n' : n \in \mathbb{N}\}$ be a countable and dense subset of X. For any $n \in \mathbb{N}$ let us select an arbitrary functional $y_n' \in \partial f(x_n)$ and consider a set F, which is a linear span of the set $\{y_n' : n \in \mathbb{N}\}$.

Let us show that the theorem is true for F. Of course, it is necessary to prove the second statement only. Let $x \in X$ and the sequence $x_n \in X$ be such that

$$\forall y \in F : \langle y, x_n \rangle_{E^*, E} \to \langle y, x \rangle_{E^*, E}.$$

Let us take x_k' and consider the difference

$$
\begin{aligned}
f(x_n) - f(x) &= f(x_n) - f(x_k') + f(x_k') - f(x) \\
&\ge \langle y_k', x_n - x_k' \rangle_{E^*, E} + f(x_k') - f(x) \\
&= \langle y_k', x_n - x \rangle_{E^*, E} + \langle y_k', x - x_k' \rangle_{E^*, E} + f(x_k') - f(x).
\end{aligned}
$$

Passing to the lower limit as $n \to \infty$, we obtain

$$\varliminf_{n \to \infty} f(x_n) - f(x) \ge \langle y_k', x - x_k' \rangle_{E^*, E} + f(x_k') - f(x). \tag{8.11}$$

Let us show that selecting x_k' properly we can make the right-hand side of (8.11) arbitrarily small, and the theorem will be proved. Let the functional f on the ball $x + S_\delta(E)$ satisfy the Lipschitz condition with constant $L = L(x, \delta) \ge 0$. Select $x_n' \in x + S_\delta(E)$ in a such way that $\|x_k' - x\|_E \to 0$. Then we get

$$\varliminf_{n \to \infty} f(x_n) - f(x) \ge -(\|y_k'\|_{E^*} + L)\|x_k' - x\|_E. \tag{8.12}$$

The fact that the functional f is Lipschitz continuous on $x + S_\delta(E)$ implies that the set $\{\|y'_k\|_{E^*}\}$ is bounded, and therefore, the right-hand side of (8.12) tends to zero. $\qquad\square$

The following theorem is an direct corollary of Theorems 8.6 and 8.7.

Theorem 8.8. *Let X be a non-empty, convex, bounded, and closed subset of a separable Banach space $(E, \|\cdot\|_E)$, $f_k : E \to \mathbb{R}$ be continuous convex functionals ($k \in \mathbb{N}$). Then there exists a linear subspace $F \subset E^*$ such that:*

(1) $(F, \|\cdot\|_{E^*})$ is a separable Banach space.
(2) F is a subspace E^* of characteristics one.
(3) an arbitrary point $x \in X \setminus E$ is strictly Hausdorff from the set X by an element of the subspace F.
(4) if for all $y \in F$ $\langle y, x_n \rangle_{E^*, E} \to \langle y, x \rangle_{E^*, E}$ as $n \to \infty$ ($x \in X$, $x_n \in X$), then

$$\forall k \in \mathbb{N} \; f(x) \leq \varliminf_{n \to \infty} f(x_n).$$

Proof. The theorem holds for a set F, which is a closed linear span of the set $\bigcup\limits_{k=0}^{\infty} F_k$, where F_0 is a linear subspace E^* satisfying assumptions of Theorem 8.6, F_k is a linear subspace, which is constructed for the kth functional and satisfies conditions of Theorem 8.7. $\qquad\square$

Let us consider generalized extreme elements of convex functionals. Let $(E, \|\cdot\|_E)$ be a separable Banach space. Let us consider the minimization problem

$$f(x) \to \inf_{x \in X}, \tag{8.13}$$

where $X \neq \varnothing$ is a convex, bounded, and closed subset of the space E, functional f is continuous and convex on the set X. Denote by $\inf_X f$ the infimum of f on X, and denote by $\arg\inf_X f = \{x \in X : f(x) = \inf_X f\}$ the set of classical solutions of Problem (8.13).

If there are no additional propositions, Problem (8.13) can have no solutions. Our aim is to construct a generalized statement of the problem (8.13) which has a solution.

Let us construct a Banach space $(F, \|\cdot\|_{E^*})$ satisfying Theorem 8.8 for the set X and functional f. Denote by $\|\cdot\|_F$ the narrowing of the norm $\|\cdot\|_{E^*}$ onto F. The linear subspace $F \subset E^*$ is total. Therefore, we can consider in E a Hausdorff and locally convex topology $\sigma(E, F)$. The fundamental system of neighbors of this topology is the following collection of sets:

$$W(y_1, \ldots, y_n; \varepsilon) = \{x \in E : |\langle y_k, x \rangle_{E^*, E}| < \varepsilon, \; y_k \in F, \; 1 \leq k \leq n\}.$$

Theorem 8.8 implies that the functional f is lower continuous on X with respect to the topology $\sigma(E, F)$, and the set X is closed in the topology $\sigma(E, F)$.

Let us consider a Banach space $(F^*, \|\cdot\|_{F^*})$ conjugate to the space $(F, \|\cdot\|_F)$. This space can be used as an extension of the space $(E, \|\cdot\|_E)$. If the Banach space $(E, \|\cdot\|_E)$ is reflexive (in this case Problem (8.13) is solvable), then (see Remark 8.50) the space $(F^*, \|\cdot\|_{F^*})$ coincides with $(E, \|\cdot\|_E)$.

The following simple statement describes the relation between the spaces $(E, \|\cdot\|_E)$ and $(F^*, \|\cdot\|_{F^*})$.

Statement 8.4. *The space $(E, \|\cdot\|_E)$ is linearly and isometrically embedded into the space $(F^*, \|\cdot\|_{F^*})$.*

Proof. A point $x \in E$ induces a linear functional jx on F:

$$(jx)(y) = \langle y, x \rangle_{E^*, E} \quad \forall y \in F.$$

It is clear that $jx \in F^*$. The operator $j : E \to F^*$ is linear. The fact that it is isometrical follows from the following equalities:

$$\|jx\|_{F^*} = \sup_{\|y\|_F=1} |(jx)(y)| = \sup_{y \in F \cap S_1(E^*)} |\langle y, x \rangle_{E^*, E}| = \|x\|_E \quad \forall x \in E.$$

Here we have taken into account that F is a subspace in E^* of characteristics one.　　　□

Remark 8.51. It is easy to see that for all $x \in E$ $jx = \pi x|_F$, where $\pi : E \to E^{**}$ is a canonical embedding of E into the second conjugate space E^{**}.

Statement 8.5. *The set $j(S_1(E))$ is sequentially dense in $S_1(F^*)$ with respect to the topology $\sigma(F^*, F)$.*

Proof. Denote by M^s a sequential closure of the set $M \subset F^*$ in topology $\sigma(F^*, F)$. Let us show that $j(S_1(E))^s = S_1(F^*)$. The closure and sequential closure of a convex bounded subset of a conjugate space to a separable Banach space coincide in the topology $\sigma(F^*, F)$ [86]. Therefore, it is sufficient to prove that $S = S_1(F^*)$, where S is a $\sigma(F^*, F)$-closure of the set $j(S_1(E))$.

The fact that the ball $S_1(F^*)$ is $\sigma(F^*, F)$-closed implies that $S \subset S_1(F^*)$. The set S is convex. Let us show that $S \supset S_1(F^*)$. Suppose that there exists such a point that $\widetilde{x}_0 \in S_1(F^*) \setminus S$. Then there exist a $\sigma(F^*, F)$-continuous linear functional $\widetilde{y}$ and positive numbers c and ε such that

$$c + \varepsilon \leq \widetilde{y}(\widetilde{x}_0), \quad \sup_{\widetilde{x} \in S} \widetilde{y}(\widetilde{x}) \leq c.$$

By the Banach theorem on weakly continuous linear functional [7] for $\widetilde{y}$ there exist a unique $y \in F$ such that $\widetilde{y}(\widetilde{x}) = \langle \widetilde{x}, y \rangle_{F^*, F}$ for all $\widetilde{x} \in F^*$. Since $j(S_1(E)) \subset S_1(F^*)$, then $\langle y, x \rangle_{E^*, E} \leq c$ for all $x \in S_1(E)$. The central symmetry of the ball $S_1(E)$ implies that $|\langle y, x \rangle_{E^*, E}| \leq c$ for any $x \in S_1(E)$, i.e. $\|y\|_F \leq c$. Then, we have $|\widetilde{y}(\widetilde{x}_0)| = |\langle \widetilde{x}_0, y \rangle_{F^*, F}| \leq \|\widetilde{x}_0\|_{F^*} \|y\|_F \leq c$ and that contradicts the inequality $\widetilde{y}(\widetilde{x}_0) \geq c + \varepsilon$. Therefore, $S \supset S_1(F^*)$ and hence $S = S_1(F^*)$　　　　□

Statement 8.6. *The set $j(E)$ is sequentially dense in F^* with respect to the topology $\sigma(F^*, F)$.*

Proof. Since $j(E)^s$ is a linear subspace in F^*, there exists the embedding $j(S_1(E))^s \subset j(E)^s$ and, by Statement 8.5, $j(S_1(E))^s = S_1(F^*)$, then $j(E)^s$ coincides with the entire space F^*. $\qquad\square$

Statement 8.7. *The bounded set $M \subset F^*$ is relatively sequentially compact in the topology $\sigma(F^*, F)$.*

Proof. It follows from the Banach–Alaoglu Theorem and the fact that the space F^* is conjugate to a separable Banach space $(F, \|\cdot\|_F)$. $\qquad\square$

Let us define the generalization of problem (8.13).

Let us consider the set $\widetilde{X}$, which is a sequential closure of the set $j(X)$ in the topology $\sigma(F^*, F)$. The set $\widetilde{X} \subset F^*$ is convex and $\sigma(F^*, F)$-compact.

Note that the elements of E, which does not belong to X, cannot be elements of the set $\widetilde{X}$. More precisely, if $x \in E \setminus X$, then $jx \in F^* \setminus \widetilde{X}$. Let us suppose the contrary. Then there exists a sequence $x_n \in X$ such that $jx_n \to jx$ in the topology $\sigma(F^*, F)$. But the fact that $x \in E \setminus X$ implies that there exists a functional $y \in F$ such that $\langle y, x \rangle_{E^*, E} > \sup_{x' \in X} \langle y, x' \rangle_{E^*, E}$, in particular, $\langle jx, y \rangle_{F^*, F} > \sup_{n \in \mathbb{N}} \langle jx_n, y \rangle_{F^*, F}$.

Let us construct the continuation of the functional f onto the set $\widetilde{X}$. For all $\tilde{x} \in \widetilde{X}$ put

$$\widetilde{f}(\tilde{x}) = \inf \left\{ \lim_{n \to \infty} f(x_n) : x_n \in X, \; jx_n \to \tilde{x} \text{ in the topology } \sigma(F^*, F) \right\}. \quad (8.14)$$

Lemma 8.2. *The following statements hold:*

(1) *The functional $\widetilde{f}$ is convex on $\widetilde{X}$.*
(2) $\widetilde{f}|_{j(X)} = f.$
(3) $\inf_X f = \inf_{\widetilde{X}} \widetilde{f}.$
(4) *The functional $\widetilde{f}$ is lower $\sigma(F^*, F)$-semicontinuous on $\widetilde{X}$.*

Proof. Statement 1) immediately follows from the convexity of f and the definition of $\widetilde{f}$. Indeed, let us consider points $\tilde{x}' \in \widetilde{X}$, $\tilde{x}'' \in \widetilde{X}$. Let us fix an arbitrary $\varepsilon > 0$. It follows from (8.14) (the definition of $\widetilde{f}$) that there exist such sequences $x_n' \in X$ $x_n'' \in X$, that

$$jx_n' \to \tilde{x}' \text{ and } jx_n'' \to \tilde{x}'' \text{ in the topology } \sigma(F^*, F),$$

$$f(x_n') < \widetilde{f}(\tilde{x}') + \varepsilon \text{ and } f(x_n'') < \widetilde{f}(\tilde{x}'') + \varepsilon.$$

Let us take $\lambda \in [0, 1]$ and pass to the lower limit in the inequality

$$f(\lambda x_n' + (1 - \lambda) x_n'') \leq \lambda f(x_n') + (1 - \lambda) f(x_n'') < \lambda \widetilde{f}(\tilde{x}') + (1 - \lambda) \widetilde{f}(\tilde{x}'') + \varepsilon$$

$$\widetilde{f}(\lambda \tilde{x}' + (1 - \lambda) \tilde{x}'') \leq \lim_{n \to \infty} f(\lambda x_n' + (1 - \lambda) x_n'') \leq \lambda \widetilde{f}(\tilde{x}') + (1 - \lambda) \widetilde{f}(\tilde{x}'') + \varepsilon.$$

Since $\varepsilon > 0$ is arbitrary, we have:

$$\widetilde{f}(\lambda \widetilde{x}' + (1 - \lambda)\widetilde{x}'') \leq \lambda \widetilde{f}(\widetilde{x}') + (1 - \lambda)\widetilde{f}(\widetilde{x}'').$$

Let us prove Statement 2). Let $x \in X$. Consider a sequence $x_n \in X$ such that $jx_n \to jx$ in the topology $\sigma(F^*, F)$. The fact that f is lower $\sigma(E, F)$-semicontinuous on X implies that $f(x) \leq \varliminf_{n \to \infty} f(x_n)$. Passing to infimum in the inequality, we have $f(x) \leq \widetilde{f}(jx)$. Taking a stationary sequence $x_n = x$, we get the opposite inequality $f(x) \geq \widetilde{f}(jx)$. Therefore, $\widetilde{f}|_{j(X)} = f$.

Statement 2) implies the inequality $\inf_X f \geq \inf_{\widetilde{X}} \widetilde{f}$. Suppose that $\inf_X f > \inf_{\widetilde{X}} \widetilde{f}$. Then for some $\varepsilon > 0$ there exists a point $\widetilde{x} \in \widetilde{X}$ such that $\inf_X f > \widetilde{f}(\widetilde{x}) + \varepsilon$. The definition $\widetilde{f}$ implies that there exists a point $x \in X$ such that $\widetilde{f}(\widetilde{x}) + \varepsilon > f(x)$. Therefore, $\inf_X f > f(x)$. This contradiction proves Statement 3).

Let us prove Statement 4). Let $\widetilde{x} \in \widetilde{X}$. Consider a sequence $\widetilde{x}_n \in \widetilde{X}$ such that $\widetilde{x}_n \to \widetilde{x}$ in the topology $\sigma(F^*, F)$. Let us show that

$$\widetilde{f}(\widetilde{x}) \leq \varliminf_{n \to \infty} \widetilde{f}(\widetilde{x}_n). \tag{8.15}$$

If $\widetilde{x}_n \in j(X)$ for all $n \in \mathbb{N}$, then (8.15) immediately follows from (8.14). Suppose that $\widetilde{x}_n \in \widetilde{X} \setminus j(X)$ for all $n \in \mathbb{N}$. It is known that the topology $\sigma(F^*, F)$ is metrized by some metric d on the set $\widetilde{X}$ [7]. Let us take an arbitrary $\varepsilon > 0$. The fact that $j(X)$ is $\sigma(F^*, F)$-dense in $\widetilde{X}$ and (8.14) implies that there exists a sequence of points $x_n \in X$ such that

$$d(jx_n, \widetilde{x}_n) \leq d(\widetilde{x}_n, \widetilde{x}), \tag{8.16}$$

$$f(x_n) < \widetilde{f}(\widetilde{x}_n) + \varepsilon. \tag{8.17}$$

Inequality (8.16) implies that jx_n converges to $\widetilde{x} \in \widetilde{X}$ in the topology $\sigma(F^*, F)$. Taking into account (8.14), let us pass to the lower limit in (8.17):

$$\widetilde{f}(\widetilde{x}) \leq \varliminf_{n \to \infty} f(x_n) \leq \varliminf_{n \to \infty} \widetilde{f}(\widetilde{x}_n) + \varepsilon.$$

Thus, inequality (8.15) holds.   □

Let us set up a correspondence between Problem (8.13) and the following minimization problem

$$\widetilde{f}(\widetilde{x}) \to \inf_{\widetilde{x} \in \widetilde{X}}, \tag{8.18}$$

We will call it a generalized definition of Problem (8.13) or F^*-extension of Problem (8.13).

Remark 8.52. It should be stressed that if the space $(E, \| \cdot \|_E)$ is reflexive, then F^*-extension of Problem (8.13), i.e. (8.18), coincides with the original problem (8.13).

Remark 8.53. Using Theorem 8.8, we can construct an extension of the problem definition with the help of a general space F^* for an arbitrary countable family of convex extreme problems

$$f_k(x) \to \inf_{x \in X}, \ k \in \mathbb{N}.$$

Definition 8.40. An element $\tilde{x} \in \tilde{X}$ is called *a generalized solution* of Problem (8.13), if $\tilde{f}(\tilde{x}) = \inf_{\tilde{X}} \tilde{f}$. Denote by $\arg\inf_{\tilde{X}} \tilde{f}$ the set of all generalized solution of Problem (8.13).

The following theorem holds.

Theorem 8.9. *The set* $\arg\inf_{\tilde{X}} \tilde{f}$ *of generalized solution of Problem* (8.13) *is nonempty, convex and* $\sigma(F^*, F)$-*compact.*

Proof. The convexity of the set $\arg\inf_{\tilde{X}} \tilde{f}$ follows from the convexity of $\tilde{X}$ and $\tilde{f}$.

Let us show that $\arg\inf_{\tilde{X}} \tilde{f} \neq \varnothing$. Consider an arbitrary sequence of points $\tilde{x}_n \in \tilde{X}$ minimizing the functional $\tilde{f}$ for F^*-extension of Problem (8.13), i.e.

$$\tilde{f}(\tilde{x}_n) \to \inf_{\tilde{X}} \tilde{f}.$$

The set $\tilde{X}$ is compact in the topology $\sigma(F^*, F)$, therefore, there exist such a subsequence $\{\tilde{x}_{n_k}\}$ and a point $\tilde{x} \in \tilde{X}$ that

$$\tilde{x}_{n_k} \to \tilde{x} \text{ in the topology } \sigma(F^*, F).$$

The fact that the functional $\tilde{f}$ is lower $\sigma(F^*, F)$-semicontinuous implies that

$$\tilde{f}(\tilde{x}) \leq \varliminf_{k \to \infty} \tilde{f}(\tilde{x}_{n_k}) = \inf_{\tilde{X}} \tilde{f};$$

hence, $\tilde{x} \in \arg\inf_{\tilde{X}} \tilde{f}$.

The compactness in the topology $\sigma(F^*, F)$ can be proved in a similar way. $\qquad\square$

Let us ascertain the relation between generalized solutions, classical solutions and minimizing sequences of Problem (8.13).

Theorem 8.10. *The following statements hold:*

(1) $j(\arg\inf_X f) = j(X) \cap \arg\inf_{\tilde{X}} \tilde{f}$.
(2) If $\{x_n\}$ *is a minimizing sequence of Problem* (8.13), *then there exist such an element* $\tilde{x} \in \arg\inf_{\tilde{X}} \tilde{f}$ *and subsequence* $\{x_{n_k}\}$ *that* $jx_{n_k} \to \tilde{x}$ *in the topology* $\sigma(F^*, F)$.
(3) If $\tilde{x} \in \arg\inf_{\tilde{X}} \tilde{f}$, *then there exists such a minimizing sequence of Problem* (8.13) $\{x_n\}$ *that* $jx_n \to \tilde{x}$ *in the topology* $\sigma(F^*, F)$.

Proof. The first statement follows from Statements 2) and 3) of Lemma 8.2. Let us prove Statement 2). Let $x_n \in X$ and $f(x_n) \to \inf_X f$. The fact that the set $\tilde{X}$ is $\sigma(F^*, F)$-compact implies that there exists a subsequence $\{x_{n_k}\}$ such that $jx_{n_k} \to \tilde{x}$

in the topology $\sigma(F^*,F)$, $\tilde{x} \in \tilde{X}$. It follows from the fact that the functional $\tilde{f}$ is lower $\sigma(F^*,F)$-semicontinuous that

$$\tilde{f}(\tilde{x}) \leq \varliminf_{k \to \infty} \tilde{f}(jx_{n_k}) = \varliminf_{k \to \infty} f(x_{n_k}) = \inf_X f = \inf_{\tilde{X}} \tilde{f},$$

i.e. $\tilde{x} \in \arg\inf_{\tilde{X}} \tilde{f}$.

Finally, let us prove Statement 3). Let $\tilde{x} \in \arg\inf_{\tilde{X}} \tilde{f}$. The fact that $j(X)$ in $\tilde{X} \subset F^*$ is $\sigma(F^*,F)$-dense and the functional $\tilde{f}$ is lower $\sigma(F^*,F)$-semicontinuous implies that there exists a sequence of points $x_n \in X$ such that $jx_n \to \tilde{x}$ in the topology $\sigma(F^*,F)$ and

$$\inf_{\tilde{X}} \tilde{f} = \tilde{f}(\tilde{x}) \leq \varliminf_{n \to \infty} \tilde{f}(jx_n) = \varliminf_{n \to \infty} f(x_n).$$

Passing to a subsequence (when it is necessary) and conserving the old denotations, we get:

$$jx_n \to \tilde{x} \text{ in the topology } \sigma(F^*,F),$$
$$\lim_{n \to \infty} f(x_n) = \inf_X f,$$

which required to be proved.   □

Let us formulate a sequential analogue of the classical condition of extreme for Problem (8.13). Suppose that the functional f is differentiable by Gateau on the set X.

Theorem 8.11. *Let $\tilde{x} \in \arg\inf_{\tilde{X}} \tilde{f}$. Then there exists a sequence $\{x_n\}$, $x_n \in X$ such that:*

$$jx_n \to \tilde{x} \text{ in the topology } \sigma(F^*,F), \tag{8.19}$$
$$f(x_n) \to \inf_X f = \tilde{f}(\tilde{x}), \tag{8.20}$$
$$\lim_{n \to \infty} \langle f'(x_n), x - x_n \rangle_{E^*,E} \geq 0 \quad \forall x \in X. \tag{8.21}$$

Proof. Theorem 8.10 implies that there exists a sequence $x_n' \in X$ such that

$$jx_n' \to \tilde{x} \text{ in the topology } \sigma(F^*,F), \tag{8.22}$$
$$0 \leq f(x_n') - \inf_X f < 1/n. \tag{8.23}$$

According to the Ekeland's variational principle [15] for an arbitrary $n \in \mathbb{N}$ there exists such a point $x_n \in X$, that

$$f(x_n) \leq f(x_n'), \tag{8.24}$$
$$\|x_n - x_n'\|_E \leq \sqrt{1/n}, \tag{8.25}$$
$$f(x_n) < f(x) + \sqrt{1/n}\|x - x_n\|_E \quad \forall x \in X \setminus \{x_n\}. \tag{8.26}$$

From (8.22) and (8.25) we get (8.19), and from (8.23) and (8.24) we get (8.20). Let us substitute the point x with the point $x_n + \tau(x - x_n) \in X$ $(x \in X, \ \tau \in (0,1))$ in (8.26):

$$f(x_n + \tau(x - x_n)) - f(x_n) > -\tau\sqrt{1/n}\|x - x_n\|_E.$$

Dividing the latter inequality by τ and passing to the limit as $\tau \to 0$, we have:

$$\langle f'(x_n), x - x_n \rangle_{E^*,E} \geq -\sqrt{1/n}\|x - x_n\|_E \quad \forall x \in X.$$

Tending n to infinity and taking into account the fact that the set X is bounded, we get inequality (8.21). □

Remark 8.54. If we do not demand the smoothness of f, then the inequality (8.21) in Theorem 8.11 should be replaced either by

$$\varliminf_{n\to\infty} f'(x_n; x - x_n) \geq 0 \quad \forall x \in X,$$

where $f'(x; p)$ is a directional derivative of f in direction p at the point x, either by the inequality

$$\varliminf_{n\to\infty} \left\{ \sup_{y \in \partial f(x_n)} \langle y, x - x_n \rangle_{E^*,E} \right\} \geq 0 \quad \forall x \in X.$$

Remark 8.55. Because of the reasonings above it becomes clear that we could embed the space E into the second conjugate space E^{**} and, repeating the corresponding closuring and extensions we could get the generalized solutions of Problem (8.13) from the space E^{**}. However, as a rule the elements of the space E^{**} have non-constructive description only. That is why we have selected that way that leads to embedding E into F^*.

8.5 Some Remarks

At first, let us discuss the possibility to apply the scheme described above to the convex maximization problem.

Let us consider an extreme problem

$$f(x) \to \sup_{x\in X}, \tag{8.27}$$

where $X \neq \varnothing$ is a convex, bounded and closed subset of a separable Banach space E, the functional $f : E \to \mathbb{R}$ is convex and Lipschitz on bounded subsets of E.

Let us note that the Lipschitz property of a convex functional on bounded sets is equivalent to the boundedness of this functional on bounded sets. Indeed, let $M > 0$ and $K = \sup_{x \in S_{2M}(E)} |f(x)|$, then $2K/M$ is a Lipschitz constant for f on the ball $S_M(E)$ [15].

If the functional f is not sequentially continuous in the topology $\sigma(E, E^*)$, then there exists a non-empty convex and bounded set $X \subset E$ where the functional f does not attain its supremum [65].

Theorem 8.12. *Let $(E, \| \cdot \|_E)$ be a linear normed space, $f : E \to \mathbb{R}$ be a continuous convex functional. If the functional f is not sequentially continuous in the topology $\sigma(E, E^*)$, then there exists a bounded, closed and convex set $X \subset E$, where the functional f does not attain its supremum.*

Proof. Let $x^* \in E$ be a point, at which the functional f is not sequentially continuous in the topology $\sigma(E, E^*)$, i.e. there exist such a number $\varepsilon > 0$ and sequence $x_n \in E$, that $x_n \to x^*$ weakly in E and $|f(x_n) - f(x^*)| \geq \varepsilon$ for every $n \in \mathbb{N}$. Without loss of generality, we can consider that $f(x^*) = 0$.

There exists such a number $n^* \in \mathbb{N}$ that the inequality $f(x_n) \geq \varepsilon$ holds for all $n \geq n^*$. Indeed, otherwise there exists such a subsequence $\{x_{n_k}\}$, that $f(x_{n_k}) < -\varepsilon$, and, respectively, $\lim_{n \to \infty} f(x_n) \leq -\varepsilon$ and that contradicts to the fact that the functional f is lower semicontinuity.

For an arbitrary number $n \geq n^*$ there exists such a point

$$y_n \in [x_n, x^*] = \{y \in X : y = \lambda x_n + (1 - \lambda) x^*, \ \lambda \in [0, 1]\},$$

that $f(y_n) = \alpha_n \varepsilon$, where $\{\alpha_n\}$ is a sequence of real numbers from the interval $(0, 1)$, which steadily converges to 1. It is clear that $y_n \to x^*$ weakly in E. Let us consider a closed, convex and bounded set $X = \overline{\text{conv}}\{y_n : n \geq n^*\}$. The fact that f is convex and continuous implies that $X \subset \{x \in E : f(x) \leq \varepsilon\}$.

Let us show that the functional f does not attain its supremum on the set X. Suppose that there exists $\bar{x} \in X$: $f(\bar{x}) = \sup_{x \in X} f(x)$. In particular, $f(\bar{x}) \geq f(y_n) = \alpha_n \varepsilon$ for all $n \geq n^*$, whence,

$$f(\bar{x}) \geq \sup_{n \geq n^*} \alpha_n \varepsilon = \varepsilon. \tag{8.28}$$

Let us consider $\{z_p\}$ such that $z_p \in \text{conv}\{y_n : n \geq n^*\}$, $z_p \to \bar{x}$ strongly in E. The fact that the functional f is continuous and (8.28) imply

$$\varepsilon \leq f(\bar{x}) = \lim_{p \to \infty} f(z_p).$$

If we show that $z_p \xrightarrow[p \to \infty]{} x^*$ weakly in E, then we get $\bar{x} = x^*$ and an absurd inequality

$$f(x^*) = 0 < \varepsilon \leq f(\bar{x}) = f(x^*) = 0.$$

For an arbitrary $p \in \mathbb{N}$ the vector z_p has the form $z_p = \sum_{n=n^*}^{\infty} \lambda_{p,n} y_n$, where $\lambda_{p,n} \in [0, 1]$, $\sum_{n=n^*}^{\infty} \lambda_{p,n} = 1$ and $\lambda_{p,n} = 0$ for a fixed p starting from some n.

Let us prove that $\lambda_{p,n} \to 0$ as $p \to \infty$ for an arbitrary $n \geq n^*$. Taking into account the convexity of the functional f for every $m \geq n^*$, we can write the inequality

$$f(z_p) = f\left(\sum_{n=n^*}^{\infty} \lambda_{p,n} y_n\right) \leq \sum_{n=n^*}^{\infty} \lambda_{p,n} f(y_n) = \sum_{n=n^*}^{\infty} \lambda_{p,n} \alpha_n \varepsilon$$

$$\leq \lambda_{p,m} \alpha_m \varepsilon + \sum_{n=n^*, n \neq m}^{\infty} \lambda_{p,n} \varepsilon$$

$$= \left(\sum_{n=n^*}^{\infty} \lambda_{p,n} + (\alpha_m - 1) \lambda_{p,m}\right) \varepsilon$$

$$= (1 + (\alpha_m - 1) \lambda_{p,m}) \varepsilon.$$

Passing to the lower limit as p and fixing the number $m \geq n^*$, we have:

$$\varliminf_{p \to \infty} f(z_p) \leq \varliminf_{p \to \infty} (1 + (\alpha_m - 1) \lambda_{p,m}) \varepsilon = \left(1 + (\alpha_m - 1) \varlimsup_{p \to \infty} \lambda_{p,m}\right) \varepsilon.$$

Hence, $\varlimsup_{p \to \infty} \lambda_{p,m} \leq 0$ and, taking into account the fact that $\lambda_{p,m}$ is non-negative, we get

$$\lim_{p \to \infty} \lambda_{p,m} = 0. \tag{8.29}$$

Let us take a functional $l \in E^*$ and introduce a denotation $C = \sup_{x \in X} |\langle l, x \rangle_{E^*, E}|$. Then for all $n \geq n^*$:

$$|\langle l, x^* \rangle_{E^*, E}| \leq C \qquad |\langle l, y_n \rangle_{E^*, E}| \leq C.$$

The weak convergency of the sequence $\{y_n\}$ to x^* implies that for an arbitrary $\varepsilon' > 0$ there exists $n' \geq n^*$, such that

$$|\langle l, y_n - x^* \rangle_{E^*, E}| < \varepsilon'/2 \quad \forall n \geq n'.$$

For all $p \in \mathbb{N}$

$$\left|\sum_{n=n'}^{\infty} \lambda_{p,n} \langle l, y_n - x^* \rangle_{E^*, E}\right| \leq \sum_{n=n'}^{\infty} \lambda_{p,n} |\langle l, y_n - x^* \rangle_{E^*, E}| < \left(\sum_{n=n'}^{\infty} \lambda_{p,n}\right) \frac{\varepsilon'}{2} \leq \frac{\varepsilon'}{2}.$$

Taking into account (8.29), we conclude that $\exists p' \in \mathbb{N}: 0 \leq \lambda_{p,n} < \frac{\varepsilon'}{4Cn'}$ for all $p \geq p'$, $n^* \leq n < n'$. Therefore, for all $p \geq p'$:

$$|\langle l, z_p - x^* \rangle_{E^*, E}| = \left|\sum_{n=n^*}^{\infty} \lambda_{p,n} \langle l, y_n - x^* \rangle_{E^*, E}\right|$$

$$\leq \sum_{n=n^*}^{n'-1} \lambda_{p,n} |\langle l, y_n - x^* \rangle_{E^*, E}| + \sum_{n=n'}^{\infty} \lambda_{p,n} |\langle l, y_n - x^* \rangle_{E^*, E}|$$

$$< \sum_{n=n^*}^{n'-1} \frac{\varepsilon'}{4Cn'} 2C + \frac{\varepsilon'}{2} < \varepsilon'.$$

Thus, we proved that $z_p \to x^*$ in the topology $\sigma(E, E^*)$. $\qquad\square$

The analysis of typical problem definitions in optimal control theory and estimation theory shows that the condition of weak sequential continuity of the functional f, as a rule, does not hold and Problem (8.27) frequently has not solutions.

Let us use the approach described above to get some "regularization" of Problem 8.27.

Let us construct spaces $(F, \|\cdot\|_F)$ and $(F^*, \|\cdot\|_{F^*})$. Then, we will enclose the set $j(X)$ in the space F^* up to the set $\widetilde{X}$ and we will extend the functional f onto F^* in the following way:

$$\widetilde{f}(\widetilde{x}) = \inf\left\{\lim_{n\to\infty} f(x_n) : x_n \in X, \ jx_n \to \widetilde{x}\ \text{in}\ \sigma(F^*,F),\ \{x_n\}\ \text{is bounded}\right\}.$$

The extended functional $\widetilde{f}$ is convex and lower semi-continuous with respect to the topology $\sigma(F^*,F)$. The following inequality holds:

$$\widetilde{f}(\widetilde{x}) \le \sup\left\{f(x) : x \in S_{\|\widetilde{x}\|_{F^*}}(E)\right\} \qquad \forall \widetilde{x} \in F^*.$$

Thus, the functional $\widetilde{f}$ is bounded on bounded subsets of the space F^*, therefore, it has Lipschitz property on the set $\widetilde{X}$.

Let us consider the problem

$$\widetilde{f}(\widetilde{x}) \to \sup_{\widetilde{x}\in\widetilde{X}}.$$

This problem can have not solutions. However, it is posed in a conjugate Banach space and the feasible set $\widetilde{X}$ is compact in the topology $\sigma(F^*,F)$.

The results of the work [103] imply that for any $\varepsilon > 0$ there exists such a functional $y \in F$ that $\|y\|_F < \varepsilon$ and the problem

$$\widetilde{f}(\widetilde{x}) + \langle \widetilde{x}, y\rangle_{F^*,F} \to \sup_{\widetilde{x}\in\widetilde{X}}.$$

has non-empty set of solutions.

Let F be a subspace of the space E^* of characteristics one. Then, the functional $E \ni x \mapsto \|x\|_E$ is lower semi-continuous with respect to the topology $\sigma(E,F)$. Indeed, let $x \in E$. For any $\varepsilon > 0$ there exists $y \in F \cap S_1(E^*)$ such that

$$\|x\|_E - \varepsilon \le \langle y,x\rangle_{E^*,E}.$$

If $x_n \to x$ in the topology $\sigma(E,F)$, then

$$\lim_{n\to\infty} \|x_n\|_E - \|x\|_E = \lim_{n\to\infty}\left(\|x_n\|_E - \|x\|_E - \langle y, x_n - x\rangle_{E^*,E}\right)$$

$$= \lim_{n\to\infty}\left(\|x_n\|_E - \langle y,x_n\rangle_{E^*,E} - \|x\|_E + \langle y,x\rangle_{E^*,E}\right) \ge -\varepsilon,$$

as $\|x_n\|_E - \langle y, x_n \rangle_{E^*,E} \geq 0$. The fact that $\varepsilon > 0$ is arbitrary implies that

$$\|x\|_E \leq \lim_{n \to \infty} \|x_n\|_E.$$

Consider the best approximation problem

$$\|x - y\|_E \to \inf_{x \in X},$$

where $X \neq \varnothing$ is convex, bounded and closed subset of the space E, $y \in E \setminus X$. This means that the construction of the generalized definitions of the best approximation problems depend only on the properties of the set X.

Let us consider in detail the example of the application of generalized extreme elements. Let us formulate an extreme problem in the space $L_1(-1,1)$:

$$f(x) = \int_{-1}^{1} |x(t)| dt - 2 \int_{-1}^{1} (1 - t^2) x(t) dt \to \inf, \tag{8.30}$$

$$x \in S_1(L_1(-1,1)) = \left\{ x \in L_1(-1,1) : \|x\|_{L_1(-1,1)} = \int_{-1}^{1} |x(t)| dt \leq 1 \right\}. \tag{8.31}$$

Let us show that problem (8.30), (8.31) has no solutions in the space $L_1(-1,1)$. Indeed, from the one hand, it is clear that

$$\forall x \in S_1(L_1(-1,1)) : f(x) \geq \|x\|_{L_1(-1,1)} - 2\|x\|_{L_1(-1,1)} = -\|x\|_{L_1(-1,1)} \geq -1$$

and the values of the functional f on the elements of the sequence $x_n = n\chi_{[0,1/n]} \in S_1(L_1(-1,1))$ satisfy the relations

$$f(x_n) = -1 + \frac{2}{3n^2} \to -1 \quad \text{as} \quad n \to \infty.$$

Thus,

$$\inf_{S_1(L_1(-1,1))} f = -1.$$

From the other hand, the functional f does not attain the value -1 on the ball $S_1(L_1(-1,1))$. Indeed, if we suppose that $f(x) = -1$ for some $x \in S_1(L_1(-1,1))$, then

$$\int_{-1}^{1} |x(t)| dt = \int_{-1}^{1} (1 - t^2) x(t) dt = 1. \tag{8.32}$$

However, there exists $\delta > 0$ such that

$$\int_{-1}^{-\delta} |x(t)| dt > 0 \quad \text{or} \quad \int_{\delta}^{1} |x(t)| dt > 0,$$

otherwise $x = 0$ almost everywhere on $[-1, 1]$. We get the chain of the inequalities, that contradicts to (8.32)

$$\int_{-1}^{1}(1-t^2)x(t)\mathrm{d}t = \int_{-1}^{-\delta}(1-t^2)x(t)\mathrm{d}t + \int_{-\delta}^{\delta}(1-t^2)x(t)\mathrm{d}t + \int_{\delta}^{1}(1-t^2)x(t)\mathrm{d}t$$

$$\leq \int_{-1}^{-\delta}(1-\delta^2)|x(t)|\mathrm{d}t + \int_{-\delta}^{\delta}|x(t)|\mathrm{d}t + \int_{\delta}^{1}(1-\delta^2)|x(t)|\mathrm{d}t$$

$$= \int_{-1}^{1}|x(t)|\mathrm{d}t - \delta^2\int_{-1}^{-\delta}|x(t)|\mathrm{d}t - \delta^2\int_{\delta}^{1}|x(t)|\mathrm{d}t < \|x\|_{L_1(-1,1)}.$$

Let us consider problems (8.30) and (8.31) in a generalized definition. As a subspace $F \subset (L_1(-1,1))^*$, which satisfies Theorems 8.6 and 8.7, we can choose the set of the functionals induced by the elements of $C([-1,1])$, i.e.

$$y \in F \iff \exists \bar{y} \in C([-1,1]) : \langle y, x\rangle_{L_\infty, L_1} = \int_{-1}^{1}\bar{y}(t)x(t)\mathrm{d}t \qquad \forall x \in L_1(-1,1).$$

The mapping $F \ni y \mapsto \bar{y} \in C([-1,1])$ is a linear isometrical isomorphism between the spaces $(F, \|\cdot\|_{L_\infty})$ and $\big(C([-1,1]), \|\cdot\|_{C([-1,1])}\big)$. By the Riesz theorem [31] the conjugate space $(F^*, \|\cdot\|_{F^*})$ is isometrically isomorphous to the space $M(-1,1)$ of all Borel measures of finite variate defined on the segment $[-1,1]$. The set $S_1(L_1(-1,1))$ can be closed in the Banach space $M(-1,1)$ up to the set

$$S_1(M(-1,1)) = \{\mu \in M(-1,1) : \|\mu\|_{M(-1,1)} = var(\mu) \leq 1\}.$$

The functional f can be extended onto the entire space $M(-1,1)$.

So, we get the generalized problem

$$\tilde{f}(\mu) = \int_{-1}^{1}\mathrm{d}|\mu|(t) - 2\int_{-1}^{1}(1-t^2)\mathrm{d}\mu(t) \to \inf,$$

$$\mu \in M(-1,1), \ var(\mu) \leq 1.$$

for which Theorems 8.9 and 8.10 hold. By the way, the generalized solution of this problem is a Dirac δ-measure, lumped at the point 0.

The requirement that the Banach space $(E, \|\cdot\|_E)$ must be separable is not a principal restriction. The construction of generalized solutions of convex extreme problems can be used without this condition. Of course, we can use more universal topological mathematical tools – nets and filters. The extension of the functional f onto the set $\tilde{X}$ (now, it is a closure of $j(X)$ in the topology $\sigma(F^*, F)$) can be constructed in the following way:

$$\tilde{f}(\tilde{x}) = \inf\left\{\lim_{\alpha} f(x_\alpha) : x_\alpha \in X, \ jx_\alpha \to \tilde{x} \text{ in the topology } \sigma(F^*, F)\right\},$$

or

$$\widetilde{f}(\widetilde{x}) = \sup_{V \in V(\widetilde{x})} \left(\inf_{x \in j^{-1}(V \cap \widetilde{X})} f(x) \right),$$

where $V(\widetilde{x})$ is a fundamental system of neighborhoods of the point $\widetilde{x} \in F^*$ in the topology $\sigma(F^*, F)$. The functional $\widetilde{f}$ is called lower semi-continuous regularization of the functional f [53].

Let us consider the possibility of using the scheme of F^*-extension in the problems of minimization of non-convex functionals. The extreme problem has the following form:

$$f(x) \to \inf_{x \in X}, \tag{8.33}$$

where, in contrast to (8.13), the convexity of the functional f is not supposed. Let us consider the subspace $F \subset E^*$ satisfying Theorem 8.6. If the functional f is lower semi-continuous on X in the topology $\sigma(E, F)$, then it is possible to construct a generalized definition of Problem (8.33), which is correct in the space F^*. Therefore, the key problem consists in the investigation of the conditions of lower $\sigma(E, F)$-semicontinuity of the non-convex functional f on the set X for the given total subspace $F \subset E^*$.

A.Ioffe and V.Tikhomirov [28] considered the problem of minimization of real-valued functional f, defined on a Hausdorff topological space X, and formalized the concept of the variational problem extension. Namely, the problem $f \to \inf_X$ matches to a pair (X, f), called by the authors a variational [28]. A variational pair (Y, g) is called an extension of (X, f), if there exists a continuous mapping $i : X \to Y$ such that:

(1) The set $i(X)$ is dense in Y.
(2) $f(x) \geq g(i(x))$ $\qquad \forall x \in X$.
(3) $\forall y \in Y \, \forall V \in V(y)$ $\quad \inf_{x \in i^{-1}(V)} f(x) \leq g(y)$, where $V(y)$ is a fundamental system of neighborhoods of the point $y \in Y$.

If the functional g is lower semi-continuous and the space Y is compact, then the extension (Y, g) is called a regular extension of the variational pair (X, f).

The extensions of classic variational problems were studied in [28]. However, the authors systematically used an approach, which differs from the approach described above. The set of feasible solutions did not change, but the functional was conversed into a convex one in a special way

The definition of extension implies that $\inf_X f = \inf_Y g$. It is easy to show that if $y = i(x)$ is a locally optimal point of a pair (Y, g), then x is a locally optimal point of a pair (X, f). From the other hand, in general we can say nothing about the existence of $\min_X f$, if there exists $\min_Y g$, but frequently it is possible to formulate sufficient conditions of the existence of $\min_X f$, based on the existence of $\min_Y g$. For example, the theorems on existence of classical solutions of elliptic boundary value problems were proved. Their proofs were based on increasing smoothness of weak solutions from Sobolev spaces.

We constructed a regular extension of Problem (8.13) in the sense of Ioffe-Tikhomirov. The resulting problem $(\widetilde{P})$ is a relaxation of problem (8.13) in the sense of [15], as their infimums coincide, and all solutions $(\widetilde{P})$ are limits of minimizing sequences of Problem (8.13).

Also, it is interesting to study the scheme of generalization based on the F^*-extension on the game search saddle point problems, problems of Nash and Pareto equilibria, and problems of hierarchical optimization (the Stackelberg problem).

References

1. *Acosta, M.D., Paya, R.* Norm attaining and numerical radius attaining operators. Revista Mathematica de la Universidad Complutense de Madrid 2, 19-25 (1989).
2. *Akhmetov, D.R.*: On an isomorphism induced by the heat conductivity equation. Sibirian Math. J. *39*, 243-259 (1998). (In Russian)
3. *Alexandrov, P.S. (ed.)*: Hilbert problems. Nauka, Moscow (1969). (In Russian)
4. *An, L.H., Du, P.X., Duc, D.M., Tuoc, P.V.*: Lagrange multipliers for functions derivable along directions in a linear subspace. Proc. Amer. Math. Soc. 133 (2005), 595604.
5. *Berezansky, Yu.M.*: Expansion of Self-adjoint Operators on Eigenfunctions. Naukova Dumka, Kiev (1965). (In Russian)
6. *Bondarenko, P.S.*: On an issue of uniqueness of infinite system of linear equations. Matem Sb. *29*, 403-418 (1951). (In Russian)
7. *Burbaki, N.*: Topological Vector Spaces. IIL, Moscow (1959). (In Russian)
8. *Burbaki, N.*: General Topology. Main Structures. Nauka, Moscow (1968). (In Russian)
9. *Chikrii, A.A.*: Conflict-controlled Processes, Kluwer Academic Publisher, Boston-London-Dordrecht (1997).
10. *Chouila-Houri, A.*: Sur la Generalisation de la Notion de Commande d'un Systeme Guidable. Rev. Francaise Informat. Recherche Operationnelle. *4*, 7-32 (1967).
11. *Deineka, V.S., Sergienko, I.V., Skopetsky, V.V.*: Models and Methods for Solving problems with Conjugation Conditions. Naukova Dumka, Kiev (1998).
12. *Edwards, R.E.*: Functional Analysis: Theory and Applications. Holt, Rinehart and Winston, Inc., New York (1965).
13. *Efremovich, V.A.*: Infinitesimal spaces. Uspekhi math. nauk. *VI*, 4(44), 203-205 (1951). (In Russian)
14. *Efremovich, V.A.*: Geometry of proximity. 1. Math. Sb. *31 (73)*, 189-200 (1952). (In Russian)
15. *Ekeland, I., Temam, R.*: Convex Analysis and Variational Problems. North-Holland Publ. Company, Amsterdam-Oxford (1976).
16. *Faddeev, D.K., Faddeeva, V.N.*: Numerical Methods of Linear Algebra. IFML, Moscow (1963). (In Russian)
17. *Filippov, A.F.*: On some aspects of optimal regulation theory. Bulletin of Moscow University. *2*, 25-32 (1959).
18. *Gabov, S.A., Malysheva, G.Yu., and Sveshnikov, A.G.*: On some equation of dynamic of viscous stratified liquid. Diff Eq. *20*, 1156-1165 (1984). (In Russian)
19. *Gabov, S.A., Orazov, B.B.*: On equation $\frac{\partial^2}{\partial t^2}[u_{xx} - u] + u_{xx} = 0$ and some related problems. J. of Math. Phys. and Comp. Math. *26*, 92-102 (1986). (In Russian)
20. *Gabov, S.A., Sveshnikov, A.G.*: Linear Problems of Non-stationary Inner Waves. Nauka, Moscow (1990). (In Russian)
21. *Gabov, S.A.*: New Problems of Mathematical Wave Theory. Nauka, Moscow (1998). (In Russian)

D.A. Klyushin et al., *Generalized Solutions of Operator Equations and Extreme Elements*, 195
Springer Optimization and Its Applications 55, DOI 10.1007/978-1-4614-0619-8,
© Springer Science+Business Media, LLC 2012

22. *Gamkrelidze, R.V.*: On sliding optimal regimes. Dokl. AN SSSR. *143*, 1243-1246 (1962). (In Russian)
23. *Grenander, U.*: Random Processes and Statistical Inferences. IL, Moscow (1961). (In Russian)
24. *Hayden, T.*: Representation theorems in reflexive Banach spaces. Math. Z. *104*, 405-406 (1968).
25. *Hilbert, D.*: Mathematical problems. Bulletin of Amer. Math. Soc. *37*, 437-479 (2000).
26. *Hotelling, H.*: Analysis of a complex of statistical variables into principal components. Mathematical problems. J. Educ. Psych. *24*, 417-441 and 417-441 (1933).
27. *Ikezi, H.*: Ion acoustic soliton experiments in a plasma. In: Solitons in action. Academic Press, New York (1979).
28. *Ioffe, A.D., Tikhomirov, V.M.*: Extension of variational problems. Trudy MMO, *18*, 187-246 (1968). (In Russian)
29. *Kantorovuch, L.V., Akilov G.P.*: Functional Analysis. Nauka, Moscow (1984). (In Russian)
30. *Kantorovich, L.V., Krylov, V.I.*: Approximate Methods of Hogher Analysis. GIFML, Moscow, 1962. (In Russian)
31. *Khelemsky, A.Ya.*: Lectures on Functional Analysis. MCCME, Moscow (2004). (In Russian)
32. *Kislov, N.V.*: Projection theorem and its appliction to non-homogeneous boundary value problems. Dokl. AN SSSR *265*, 31-34 (1982). (In Russian)
33. *Klyushin, D.A., Kuschan, A.A., Lyashko, S.I., Nomirovskii, D.A., and Petunin, Yu.I.*: Generalized solution of some operator equations in Banach spaces. J. of Comp. and App. Math. 1, 29-50 (2001). (In Russian)
34. *Klyushin, D.A., Petunin, Yu.I.*: A concept of generalized solution of non-linear operator equations in metric spaces. J. of Comp. and App. Math. 1, 11-23 (2002). (In Russian)
35. *Klyushin, D.A., Kuschan, A.A., Lyashko, S.I., Nomirovskii, D.A., and Petunin, Yu.I.*: Generalized solving for some operator equations in Banach spaces. Bulletin of Kiev University. Cybernetics. 3, 47-49 (2002). (In Ukrainian)
36. *Kook, Yu.V., Petunin, Yu.I.*: Observable linear estimations of mathematical expectation of a random process. Dokl. AN SSSR. *209*, 37-39 (1973). (In Russian)
37. *Krasnoselsky, M.A.*: Topological Methods in Theory of Nonlinear Integral Equations. ITTL, Moscow (1956). (In Russian)
38. *Krasnoselsky, M.A.*: On solving equations with self-adjoint operator by the method of successive approximations *15*, 161-165 (1960).
39. *Krein, S.G.*: Linear Equations in Banach Space. Nauka, Moscow (1971). (In Russian)
40. *Krein, S.G. (ed.)*: Hanbook on Functional Analysis. Nauka, Moscow (1972). (In Russian)
41. *Krein, S.G., Petunin, Yu.I., Semenov, E.M.*: Interpolation of Linear Operators. Nauka, Moscow (1978). (In Russian)
42. *Kuntsevich, V.M., Lychak, M.M.*: Guaranteed estimates, adaptation and robustness in control systems. Springer-Verlag, Berlin (1992).
43. *Krein, S.G., Petunin, Yu.I.*: Scale of Banach spaces. Uspekhi Mat. Nauk. 2, 89-169 (1966)
44. *Ladyzhenskaya, O.A.*: Mathematical Aspects of Dynamics of Viscous Incompressible Liquid. Nauka, Moscow (1970). (In Russian)
45. *Ladyzhenskaya, O.A.*: Boundary Value Problems in Mathematical Physics. Nauka, Moscow (1973). (In Russian)
46. *Landesman, E.M.* A Generalized Lax-Milgram Theorem. Proceedings of the American Mathematical Society. *19*, 339-344 (1868).
47. *Lavrentjev, M.M., Romanov, V.G., Shishatsky, S.P.*: Non-correct Problems in Mathematical Physics and Analysis. Nauka, Moscow (1980).
48. *Lax P.D., Milgram N.* Parabolic equations. Contributions to the theory of partial differential equations. Ann. Math. Studies. *33*, 167-190 (1954).
49. *Lindenstrauss J.* On operators which attain their norm. Israel J. Math. *3*, 139-148 (1963).
50. *Lions J.-L.*: Sur les problemes mixtes pour certains systemes paraboliques dans les ouverts non cylindriques. Annales de l'institut Fourier. *7*, 143-182 (1957).
51. *Lions, J.-L.*: Optimal Control of Systems Governed by Partial Differential Equations. Springer, Berlin-New York (1971).

52. *Lions, J.-L.*: Control of distributed singular systems. Trans-Inter-Scientia, Tonbridge (1985).
53. *Loran, P.-J.*: Approximation and Optimization. Hermann, Paris (1972).
54. *Lyashko, I.I., Didenko, V.P., and Tsitritsky, O.E.*: Noise filtration. Naukova Dumka, Kiev (1979). (In Russian)
55. *Lyashko, I.I., Demchenko, V.F.*: Generalized statement of problems of heat and mass transport in layared media. Preprint 87-14. Glushkov Institute of Cybernmetics, Kiev (1987). (In Russian)
56. *Lyashko, I.I., Demchenko, V.F., and Demchenko, L.I.*: Numerical Simulation of Heat and Mass Transport processes. UMK VS, Kiev (1988). (In Russian).
57. *Lyashko, I.I., Demchenko, L.I., and Mistetsky, G.E.*: Numerical solving for problems of heat and mass transport inporous media. Naukova Dumka, Kiev (1991). (In Russian)
58. *Lyashko, S.I*: On solvability of pseudo-parabolical equations. Izvestiya Vyshikh Uchebnykh Zavedeniy. Mathematics. *9*, 71-72 (1985). (In Russian)
59. *Lyashko, S.I, Vityuk, N.Ya.*: Pulse-point control of some distributed systems. Dokl. AN USSR. Ser. A. Physics, Mathematics and Engineering. *8*, 61-63 (1985). (In Russian)
60. *Lyashko, S.I, Red'ko, S.E.*: Optimal pulse-point control of dynamics of viscous stratified liquid. Diff. Eq. *23*, 1890-1897 (1987). (In Russian)
61. *Lyashko, S.I, Red'ko, S.E.*: Approximate solution of viscous liquid dynamics problem. J. Math. Phys. and Comp. Math. *5*, 720-729 (1987). (In Russian)
62. *Lyashko, S.I*: Generalized Control of Linear Systems. Naukova Dumka, Kiev (1998). (In Russian)
63. *Lyashko, S.I.*: Generalized Optimal Control of Linear Systems with Distributed Parameters. Kluwer Academic Publishers. Boston- Dordrecht (2002).
64. *Lyashko, S.I, Nomirovskii,; D.A.*: Generalized solution and optimal control in systems describing the dynamicof a viscous liquid fluid. Diff. Eq. *39*, 90-98 (2003).
65. *Lyashko, S.I, Semenov, V.V., and Katsev, M.V.*: Some Remarks Concerning Supremunm Attainment. J. Automat. Inf. Scien. *38*, 1-7 (2006).
66. *Lyashko, S.I., Semenov, V.V.* On one theorem of M.A. Krasnoselski. Cybernetics and System Analysis, *5*, 180-183 (2010).
67. *Lyashko, S.I., Klyuhin, D.A., Nomirovskii, D.A., Semenov, V.V.* Identification og age-structured contamination sources in ground water. In: Optimal Control of Age-Structured Populations in Economy, demography, adn the Environment (ed. R.Boucekkine, N.Hritonenko, and Yu.Yatsenko), Routledge, Lonon and New York (2010).
68. *Mair, B.A., Ruymgaart, F.H., and Urrabazo, T.*: Some Comments on Wicksell's Problem. Journal of Statistical Planning and Inference. *87*, 31-42 (2000).
69. *Malitski, Yu.V, Semenov, V.V.* On theory of generalized solutions of operator equations. In: Proceedings of III Int. Conf. "Computational and Applied Mathematics" (in memory of academician I.I.Lyashko).– P. 52. – Kiev (2009).
70. *McShane, E.J.* Generalized curves, Duke Math. J., 6 (1940), 513-536.
71. *McShane, E.J.* Relaxed Controls and Variational Problems. SIAM Journal of Control. *5*, 438-485 (1967).
72. *Mordukhovich, B.S.*: Variational Analysis and Generalized Differentiation. I: Basic Theory, Grundlehren Series (Fundamental Principles of Mathematical Sciences), Vol. 330, 584 pp., Springer-Verlag, Berlin (2006).
73. *Nomirovskii, D.A.*: Numerical methods of optimization and modelling in pseudohyperbolic systems. Thesis of Ph.D. Kiev (1999). (In Ukrainian)
74. *Nomirovskii, D.A., Petunin, Yu.I., and Savkina, M.Yu.*: Generaized extreme elements in Banach space J. Comp. and App. Math *89*, 71-79 (2003). (In Russian)
75. *Nomirovskii, D.A.*: On generalized solvability of linear systems. Dopovidi NANU. *10*, 26-23 (2004). (In Russian)
76. *Nomirovskii, D.A.*: On homeomorphisms realized by certain partial differential operators. Ukr. Math. J. *56*, 2017-2027 (2004).
77. *Nomirovskii, D.A.*: Generalized solvability of parabolic systems with nonhomogeneous transmission conditions of nonideal contact type. Diff. Eq. *40*, 1467-1477 (2004).

78. *Nomirovskii, D.A.*: On uniquness of generalized solutions of operator equations. Bulletin of Kiev Univ. Ser. Physics and Math. *4*, 223-227 (2004). (In Russian)

79. *Nomirovskii, D.A.*: Unique solvability of pseudohyperbolic equations with singular right-hand sides. Math. Notes. *80*, 550-562 (2006).

80. *Nomirovskii, D.A.*: On the unique solvability of wave systems in various classes of distributions Diff. Eq. *42*, 882-895 (2006).

81. *Nomirovskii, D.A.*: Approximate method for solving the boundary value problem for a parabolic equation with inhomogeneous transmission conditions of nonideal contact type. Comp. Math. and Math. Phys. *46*, 995-1006 (2006).

82. *Nomirovskii, D.A.*: Convergence of Galerkin method for a wave equation with singilar right-hand side. Ukr. Math. J. *58*, 876-886 (2006).

83. *Nomirovskii, D.A.* Generalized solvability and optimization of a parabolic system with a discontinuous solution Journal of differential equations. *233*, 1-21 (2007).

84. *Obi, G.M.M.* An algebraic extension of the Lax-Milgram theorem. Pacific Journal of Mathematics. *86*, 543-552 (1980).

85. *Palienko, L.I.*: Simulation and generalized optimization of pseudo-hyperbolic systems. Thesis of Ph. D. Kiev (1999). (In Ukrainian)

86. *Petunin, Yu.I., Plichko, A.N.*: Theory of characteristics of subspaces and its applications. Vyscha Shkola, Kiev (1980). (In Russian)

87. *Petunin, Yu.I., Semeyko, N.G.*: Generalized Vixell problem. In: Proceedings of IV In. Vilnius Conf on Prob. Theor. and Math. Stat. *III*, 28-29 (1985). (In Russian)

88. *Petunin Yu.I., Semeiko N.G.* Random cap process and generalized Wikcell problem on the surface of a sphere. SERDICA. *17*, 81-91 (1991).

89. *Petunin, Yu.I.*: On a concept of generalized solution of operator equations in Banach space. Ukr. Math. J. *48*, 1286-1290 (1996). (In Russian)

90. *Petunin, Yu.I., Ryabichev, O.M., Akchurin, O.M.*: One concept of conditioned of system of linear algebraical equations. Bulletin of Kiev University. Ser. Physics and Math. *4*, 180-187 (1998). (In Russian)

91. *Petunin, Yu.I., Savkina, M. Yu.*: New concept of generalized solution of operator equation in Banach space. J. Comp. and Appl. Math. *81*, 93-99 (1997). (In Russian)

92. *Petunin, Yu.I., Semeyko, N.G., Yatsenko, V.P.*: Investigation of morphometrical characteristics of cpmplexes of neclear shell of sensor neuron by methods of spherical stochastics geometry. Cybernetics and System Analysis, *6*, 175-182 (2006). (In Russian)

93. *Pletner, Yu.D.*: On oscillation of flat two-sided disk on interface between two stratified liquids. J. Comp. Math. and Math. Phys. *30*, 736-750 (1990). (In Russian)

94. *Ramaswamy, S.*: The Lax-Milgram Theorem for Banach Spaces. I // Proc. Japan Acad., 56, Ser. A (1980), 462-464.

95. *Ramaswamy, S.*: The Lax-Milgram Theorem for Banach Spaces. II // Proc. Japan Acad., 7, Ser. A (1981), 29-33.

96. *Ramm. A.G.*: Random fields estimation theory. Longman Scientific and Technical, Wiley, Harlow-New York, 1990.

97. *Robertson, A.P., Robertson, W.*: Topological vector spaces. Cambridge Tracts in Math. and Math. Phys., no. 53, Cambridge Univ. Press, New York (1964).

98. *Schachermayer, W.*: Norm attaining operators and renorming of Banach spaces. Israel J. Math. *44*, 210-212 (1983).

99. *Schultz, C.*: Iterative Berechnung der reziproken matrix. Zeitschrift fuer Angewandte Mathematik und Mechanik. *13*, 57-59 (1933).

100. *Semenov, V.V.*: Models and methods of generalized optimization of linear systems with distributed parameters. Thesis of Ph.D. Kiev (2002). (In Ukrainian)

101. *Semenov, V.V.*: On density of set of linear operators and bilinear forms attaining norm. Bulletin of Kiev University. Ser. Physics and Math. *3*, 255-259 (2003). (In Ukrainian)

102. *Semenov, V.V.*: Generalized extrenme elements of convex functional. Bulletin of Kiev University. Ser. Physics and Math. *3*, 189-193 (2007). (In Ukrainian)

103. *Semenov, V.V., Katsev, M.V.*: Linear variational principle in convex minimization. Dopovidi NANU. *3*, 51-58 (2007). (In Ukrainian)

104. *Semenov, V.V.*: Projection theorem for Banach and locally convex spaces. Cybern. Syst. Anal. 44, 722-728 (2008).

105. *Semenov, V.V.*: Convex maximization and Schachermayer property α. Cybern. Syst. Anal. 49, 309-313 (2009).

106. *Semenov, V.V.*: Variational problems and generalized optimization of linear systems. Thesis of Doctor in Physics and Mathematics, Kiev (2010).

107. *Skopetsky, V.V., Stoyan, V.A., Krivonos, Yu.G.*: Mathematical Simulation of Direct and Inverse Problems of Distrubuted Systems Dynamics. Naukova Dumka, Kiev (2002). (In Ukrainian)

108. *Smirnov, Yu.M*: On proximity spaces. Math. Sb. *31 (73)*, 543-574 (1952). (In Russian)

109. *Sobol, I.M.*: Monte-Carlo Numerical Methods. Nauka, Moscow (1973). (In Russian)

110. *Stegall, C.*: Optimization of functions on certain subsets of Banach spaces. Math. Ann. *236*, 171-176 (1978).

111. *Tikiliainen, A.A.*: On hyroscopic waves in media with stream and time-dependent rotation. J. Comp. Math. and Math. Phys. *30*, 270-277 (1990). (In Russian)

112. *Tikhonov, A.N., Arsenin, V.Ya.*: Method for Solving Non-Correct Problems. Nauka, Moscow (1974). (In Russian).

113. *Vishik, M.I.*: On strongly elliptic systems of differential equations. Math. Sb. *29*, 615-676 (1951). (In Russian)

114. *Vulikh, B.Z.*: Introduction to Functional Analysis. IFML, Moscow (1958). (In Russian)

115. *Warga J.*: Relaxed Variational Problems. J. Math. Anal. Appl. *4*, 111-128 (1962).

116. *Warga, J.*: Optimal Control of Differential and Functional Equations. Academic Press, New York (1972).

117. *Young L.C.*: Generalized curves and the existence of an attained absolute minimum in the calculus of variations, C. R. Sci. et des Letters de Varsovie, classe III, 30 (1937), 212-234.

118. *Young, L.C.*: Young, L. C. Lectures on Calculus of Variations and Optimal Control Theory. W. B. Saunders, Philadelphia (1969)

119. *Yosida, K.*: Functional Analysis, Springer-Verlag, New York (1980).

120. *Zgurovsky, M.Z., Mel'nik, V.S.*: Nonlinear Analysis and Control of Physical Processes and Fields. Springer, Berlin (2004).

121. *Zizler, V.*: On some extremal problems in Banach spaces. Math. Scand. *32*, 214-224 (1973).

Index

D.A. Klyushin et al., *Generalized Solutions of Operator Equations and Extreme Elements*,
Springer Optimization and Its Applications 55, DOI 10.1007/978-1-4614-0619-8,
© Springer Science+Business Media, LLC 2012